Munich Studies on Innovation and Competition

Volume 16

The Munich Studies on Innovation and Competition present fundamental research on legal systems that have been created with the objective of promoting and safeguarding innovation and competition as the most important factors for economic growth and prosperity. Accordingly, this series will include monographic works in English in the different fields of intellectual property and competition law such as patent, trademark and copyright law, as well as unfair competition and antitrust law. Rather than describing what the law is, the series strives to contribute to the scholarship on how these legal systems should develop so as to promote innovation and competition. Therefore, its outlook is both international, by not focussing on any specific national legal system, and interdisciplinary. In particular, studies are encouraged that also incorporate the methods and findings of other disciplines such as economics, sociology and anthropology.

More information about this series at http://www.springer.com/series/13275

Daria Kim

Access to Non-Summary Clinical Trial Data for Research Purposes Under EU Law

Springer

Daria Kim
Max Planck Institute for Innovation and Competition
Munich, Germany

ISSN 2199-7462 ISSN 2199-7470 (electronic)
Munich Studies on Innovation and Competition
ISBN 978-3-030-86777-5 ISBN 978-3-030-86778-2 (eBook)
https://doi.org/10.1007/978-3-030-86778-2

This Springer imprint is published by the registered company Springer Nature Switzerland AG.
The registered company address is: Gewerbestrasse 11, 6330 Cham, Switzerland

I dedicate this work to all individuals volunteering as clinical trial subjects.

Preface

Access to clinical trial data has been subject to a long-standing policy, legislative and general public debate. Notwithstanding potential benefits for medical research and drug development, many jurisdictions have struggled to implement and enforce legal rules governing clinical trial data accessibility even at the summary results level. Policy initiatives have been strongly opposed by research-based drug companies arguing that mandatory data disclosure impedes their innovation incentives.

Conventionally, policymakers approached access to clinical trial data from the perspective of transparency and research ethics. This study offers a complementary view and considers access to individual patient-level trial data for secondary analysis as a matter of research and innovation policy. Such an approach appears to be particularly relevant against the backdrop of a data-driven economy where digital data is acknowledged as a valuable economic resource.

Overall, the study seeks to define how the rules of access to de-identified individual patient-level data should be designed to reconcile the policy objectives of leveraging research potential of data through secondary analysis, on the one hand, and protecting economic incentives of research-based drug companies, on the other hand.

Munich, Germany Daria Kim

Acknowledgements

This book is based on a dissertation project carried out in pursuit of Dr. iur. degree at the University of Augsburg. I would like to thank Professor Dr. Ulrich M. Gassner for his invariably kind support and trust as the *Doktorvater* ('dissertation father', the doctoral supervisor). In addition to the *Doktorvater*, I would like to thank the 'dissertation godfather'—Professor Dr. Josef Drexl—who continues to provide invaluable guidance to date. Thanks also go to Dr. Filipe Fischmann, Dr. Gintarė Surblytė-Namavičienė and Dr. Axel Waltz, who acted as encouraging and caring academic advisors at various points in time.

The Max Planck Institute for Innovation and Competition provided the best possible research conditions and has been home for several years now. I would like to thank all of the colleagues and researchers who have shared their physical premises and intellectual space. Working with Professor Dr. Josef Drexl and Professor Dr. Reto M. Hilty has been foundational, expanding and rewarding. Professor Dr. Joseph Straus, Professor Dr. Hans Ullrich, Dr. Heiko Richter and Dr. Roberto Romandini will always remain the 'tuning fork' for the research quality. Charles Heard worked his editorial magic on numerous occasions. There is no better description of Dr. Eva Bastian's role than as the 'guardian angel'.

Throughout the study, I reached out to many medical researchers whose enthusiastic replies and a keen interest in the project made it tangible, relevant and needed. I deeply appreciate the exchanges with Professor Dr. med. Joerg Hasford, Professor Philippe Guerin, Professor Franz König, Professor Joel Lexchin, Dr. Sarah Nevitt and Dr. med. Arnoud Templeton, who provided insightful answers to my (often basic) questions.

This book might not have come about but for my fateful encounters with Professor Gawdat Bahgat and Dr. Tshimanga Kongolo, and but for the enriching and inspiring experience of working with Professor Bryan C. Mercurio.

I am thankful to Wang Xiangyu, Dr. Owais H. Shaikh, Raymond Kwok and Ivan Stepanov for helpful discussions about (clinical trial) data, IP and beyond. For the unwavering support 'behind the scenes' during the study, I would like to extend my sincere thanks to Dr. Barbara Schmid-Neuhaus, Dr. Angelika Bayer, the

Guggenheimer family, Dr. Natalia Łukaszewicz, Preston Richard, Dr. Alina Wernick and Angela Zacher.

My Family—Elena, Alexey, Kirill and Anya Kim—is the greatest source of inspiration.

Contents

Abbreviations

art/arts	Article/articles
ATC	Anatomical Therapeutic Chemical
CCI	Commercially Confidential Information
CFR	The Charter of Fundamental Rights
CHMP	Committee for Medicinal Products for Human Use
CIOMS	The Council for International Organizations of Medical Sciences
CJEU	The Court of Justice of the European Union
COM	Communication
CONSORT	Consolidated Standards of Reporting Trials
CSDR	Clinical Study Data Request
CSR	Clinical Study Report
Dir	Directive
e.g.	*exempli gratia*/for example
EC	The European Commission
EFPIA	The European Federation of Pharmaceutical Industries and Associations
EMA	The European Medicines Agency
EPO	The European Patent Office
et al.	*et alii* (*et aliae*)/and others
EU	The European Union
EudraCT	European Union Drug Regulating Authorities Clinical Trials
ff	*folio*/and the following
FOIA	Freedom of Information Act
FTA	Free Trade Agreement
GCP	Good Clinical Practice
GDPR	The General Data Protection Regulation
GRADE	Grades of Recommendation, Assessment, Development, and Evaluation
GSK	GlaxoSmithKline
i.e.	*id est*/that is to say
ibid	*ibidem*/in the same place

ICH	The International Council for Harmonisation of Technical Requirements for Pharmaceuticals for Human Use
ICMJE	The International Committee of Medical Journal Editors
IFPAM	The International Federation of Pharmaceutical Manufacturers and Associations
IP	Intellectual Property
IPD	Individual Patient Data
n/nn	Footnote/footnotes
NBER	The National Bureau of Economic Research
NCE	New Chemical Entity
NDA	New Drug Application
OECD	The Organisation for Economic Co-operation and Development
p/pp	Page/pages
para/paras	Paragraph/paragraphs
PhRMA	The Pharmaceutical Research and Manufacturers of America
PUMA	Paediatric Use Marketing Authorisation
R&D	Research and Development
RCT	Randomised Clinical Trial
REACH	The EU Regulation on Registration, Evaluation, Authorisation and Restriction of Chemicals
Reg	Regulation
RTLA	Reach Through License Agreement
SPC	Supplementary Protection Certificate
SUSAR	Suspected Unexpected Serious Adverse Reactions
SWD	Staff Working Document
TBA	Technical Board of Appeal
TFEU	The Treaty on the Functioning of the European Union
TRIPS	The Agreement on Trade-Related Aspects of Intellectual Property Rights
UNDP	The United Nations Development Programme
US	The United States of America
USFDA	The Food and Drug Administration of the United States of America
vol	Volume
WHO	The World Health Organization
WIPO	The World Intellectual Property Organization
WTO	The World Trade Organization

Chapter 1
Introduction

Abstract This chapter introduces the policy dilemma concerning access to non-summary clinical trial data examined by the study. It outlines the general regulatory context, motivation and the overall structure of the study.

> When data are stored, institutions must have a governance system [. . .] for future use of these data in research.[1]

> Creative discovery comes from unlikely journeys through the information space; No go zones restrict the right to roam.[2]

A clinical trial is designed and conducted to test a hypothesis regarding the effects of medical intervention on the state of health. A sound research hypothesis relevant to state-of-art scientific knowledge and clinical practice can emerge from searching the literature and examining data from earlier studies.[3] The more evidence from past trials is available, the more precisely new research questions can be formulated, the more efficiently—and with greater certainty—the subsequent trials can be designed.[4]

Thousands of clinical trials are conducted each year worldwide, producing volumes of research data.[5] Hundreds of thousands of participants are being

[1]CIOMS (2016), p. 47.

[2]Cameron (2001), p. 32.

[3]Cummings et al. (2007), p. 18.

[4]The International Council for Harmonisation of Technical Requirements for Pharmaceuticals for Human Use (ICH) (1998) ICH harmonised tripartite guideline. Statistical principles for clinical trials. E9, p. 5.

[5]As of 26 March 2021, 372,291 studies conducted in 219 countries since 2008 have been listed in the ClinicalTrials.gov database. See U.S. National Library of Medicine. Trends, charts, and maps. https://clinicaltrials.gov/ct2/resources/trends. Accessed 26 Mar 2021. Approximately 4000 trials are authorised in the European Economic Area each year. See European Medicines Agency. Clinical trials in human medicines. http://www.ema.europa.eu/ema/index.jsp?curl=pages/special_topics/general/general_content_000489.jsp&mid=WC0b01ac058060676f. Accessed 26 Mar 2021.

D. Kim, *Access to Non-Summary Clinical Trial Data for Research Purposes Under EU Law*, Munich Studies on Innovation and Competition 16,
https://doi.org/10.1007/978-3-030-86778-2_1

enrolled.[6] Data collected in clinical trials presents one of the major sources of medical knowledge. Its social value subsists in informing clinical practice and guiding subsequent medical research and drug development. *Individual patient-level* data ('source data') constitutes 'a powerful body of evidence'[7] that has 'the potential to be an unprecedented resource'[8] for research. Such potential can be realised only through secondary data analysis. The results can extend far beyond the benefit-risk balance of the investigational product, for which the analysed data was initially generated.

Access to clinical trial data has been subject to a long-standing public, policy and legislative debate. Benefits of broad access stem from enhanced transparency, improved research reproducibility, better informed clinical practice, eliminated duplicative research and the potential to generate new research hypotheses. Notwithstanding such advantages, many jurisdictions have struggled to implement and enforce legal rules governing access to clinical trial data, even at the level of summary results.

In 2015, the European Medicines Agency (EMA) announced an unparalleled publication policy.[9] It became the first[10]—and, to date, remains the only—drug regulator globally committed to the proactive release of the non-summary clinical trial data. Such data comprises clinical study reports (CSRs) and de-identified individual patient data (IPD).[11] However, the EMA's initiative left several important questions open-ended. Is *erga omnes* disclosure of IPD a proportionate and effective measure vis-à-vis the policy objectives of ensuring research reproducibility and promoting data-driven drug discovery and innovation? What is the scope of trial sponsors' 'legitimate economic interests' that should be protected? Is transparency regulation a suitable regulatory framework for leveraging the research potential of trial data for drug innovation through exploratory data analysis?

Given these uncertainties as a starting point, the study explores legal issues arising when access to IPD from past trials is sought for research purposes, particularly for secondary exploratory data analysis. While the primary analysis of IPD is directed at the benefit-risk assessment of a medical intervention investigated in a trial, secondary analysis can pursue confirmatory and exploratory objectives. A

[6] European Commission (17 Jul 2012) Impact assessment report on the revision of the 'Clinical Trials Directive' 2001/20/EC accompanying the document Proposal for a Regulation of the European Parliament and of the Council on clinical trials on medicinal products for human use, and repealing Directive 2001/20/EC, SWD(2012) 200 final [hereinafter SWD(2012) 200 final], vol. II, pp. 13–14.

[7] Stewart and Tierney (2002), p. 89.

[8] Krumholz et al. (2014), p. 502.

[9] EMA (21 Mar 2019) European Medicines Agency policy on publication of clinical data for medicinal products for human use. Policy/0070 [hereinafter EMA publication policy 0070]. The policy was revised in March 2019 to account for the relocation of the EMA to Amsterdam.

[10] European Medicines Agency (20 Oct 2016) Opening up clinical data on new medicines. EMA provides public access to clinical reports. EMA/650519/2016, p. 1.

[11] Doshi and Jefferson (2016).

secondary confirmatory analysis is directed at validating the trial outcome and original conclusions and, thus, can be instrumental in promoting transparency in clinical research and regulatory decision-making. Secondary exploratory data analysis can address questions beyond the benefit-risk assessment of the original investigational product and facilitate both academic research and industrial R&D.

Among various types of trial-related data and information, IPD comprises the most comprehensive records and, thus, constitutes a highly valuable resource for medical research and drug R&D. Such data remains, for the most part, 'below the waterline'[12] and inaccessible for secondary research purposes.

Conventionally, access to clinical trial data has been approached as a matter of transparency and research ethics, which importance is indisputable. However, issues concerning data accessibility stretch beyond what can be accommodated under those frameworks. While concerns regarding research reproducibility and transparency have dominated the public and policy discourse around clinical trial data, access for exploratory analysis has been gaining prominence in the context of emerging data economy[13] and data-driven innovation.[14]

This study offers a complementary perspective considering access to IPD for research purposes as a matter of innovation policy in the pharmaceutical sector. Such approach appears especially pertinent against the EU policy agenda for building a European data economy. Therefore, the analysis focuses mainly on the regulatory framework at EU level. The objective of leveraging the economic potential of digital data to enhance social benefits, including in the area of public health, has been promoted by the European Commission through several instruments and initiatives, such as 'Towards a thriving data-driven economy',[15] 'A Digital Single Market Strategy for Europe',[16] 'Building a European Data Economy'[17] and 'Towards a

[12] Doshi et al. (2013).

[13] The term 'data economy' is defined as 'an ecosystem of different types of market players – such as manufacturers, researchers and infrastructure providers – collaborating to ensure that data is accessible and usable'. European Commission (10 Jan 2017) Communication from the Commission to the European Parliament, the Council, the European Economic and Social Committee and the Committee of the Regions 'Building a European Data Economy', COM(2017) 9 [hereinafter COM (2017) 9], p. 2.

[14] Data-driven innovation is characterised by 'the capacity of businesses and public sector bodies to make use of information from improved data analytics to develop improved services and goods that facilitate everyday life of individuals and of organisations, including SMEs'. European Commission (2 Jul 2014) Communication from the Commission to the European Parliament, the Council, the European Economic and Social Committee and the Committee of the Regions 'Towards a thriving data-driven economy', COM(2014) 442 final [hereinafter COM(2014) 442 final], p. 5.

[15] ibid pp. 3, 6, 8.

[16] European Commission (6 May 2015) Communication from the Commission to the European Parliament, the Council, the European Economic and Social Committee and the Committee of the Regions 'A Digital Single Market Strategy for Europe', COM(2015) 192 final [hereinafter COM (2015) 192 final], p. 14 ff.

[17] European Commission, COM(2017) 9 final, p. 2.

common European data space'.[18] The goal of enabling 'free flow' and 'free movement' of data is acknowledged with regard to both personal[19] and non-personal data.[20]

While free flow of data can be, theoretically, achieved without policy intervention, contract-based solutions might fail where the negotiation positions of market participants are sharply unequal.[21] The risk of data being 'locked-in' should be examined in the context of a particular data ecosystem and taking into account sector specificities and the need to balance distinct public and private interests at stake.

The polarised debate regarding access to and 'ownership' of digital data[22] highlights the uncertainty as to how rights in data, as a digital resource for innovation, should be designed and allocated. In the case of clinical trial data, at issue is whether secondary IPD analysis should remain subject to the authorisation of data holders, who in most cases are the sponsors of trials capable of exercising control over third-party access to non-summary ('source') data.[23] In economic terms, this translates into the question of whether de-identified IPD should be treated as an 'excludable' or a 'non-excludable' resource for medical research and drug R&D.

Proponents of access argued that secondary IPD analysis could 'speed up scientific discoveries by relieving investigators of the burden of collecting new data [and] lead to promising new treatments or a better understanding of disease through various "big data" projects'.[24] Research-based drug companies claimed that non-summary clinical trial data constitutes their (intellectual) property and that mandatory disclosure of data impedes their innovation incentives.[25] In view of these conflicting propositions, this study seeks to define how legal rules of access to IPD should be designed to maximise the research potential of data through secondary analysis on the one hand while protecting innovation incentives of drug originator companies on the other hand.

[18]European Commission (25 Apr 2018) Communication from the Commission to the European Parliament, the Council, the European Economic and Social Committee and the Committee of the Regions 'Towards a common European data space' COM(2018) 232 final [hereinafter COM(2018) 232 final], p. 3 ff.

[19]Regulation (EU) No 2016/679 of the European Parliament and of the Council of 27 April 2016 on the protection of natural persons with regard to the processing of personal data and on the free movement of such data, and repealing Directive 95/46/EC (General Data Protection Regulation) (4 May 2016) OJ L 119, pp. 1–88 [hereinafter GDPR], Recitals 6, 53, 123, 170.

[20]Regulation (EU) No 2018/1807 of the European Parliament and of the Council of 14 November 2018 on a framework for the free flow of non-personal data in the European Union (28 Nov 2018) OJ L 303/59, rec 10, 13, 39.

[21]European Commission, COM(2017) 9 final, p. 10 ff; European Commission, COM(2018) 232 final, p. 9.

[22]For an overview, see Chap. 2 at Sect. 2.2.6.2.

[23]On the legal determinants of control over clinical trial data, see Chap. 4 at Sect. 4.2.

[24]Multi-Regional Clinical Trials Center at Harvard University (2014) Overview of data disclosure initiatives: current and ongoing data transparency activities in the pharmaceutical industry. https://www.regulations.gov/comment/FDA-2013-N-0271-0031. Accessed 26 Mar 2021.

[25]For an overview of arguments, see Chap. 5 at Sect. 5.1.

The policy dilemma as to whether to intervene by access measures or not is thorny in the case of data gathered in industry-sponsored trials. Such trials represent a situation where *inherently* public goods—medical research and drug innovation—are supplied by the private sector. Private provisioning of public goods often results in tension between private and public interests and can result in policy trade-offs. The conditions under which access to non-summary trial data can posit a trade-off with innovation incentives of commercial drug sponsors merits an in-depth analysis.

This study explores the above-outlined questions according to the following structure. Part I sets the scene by outlining the context and recent developments pertaining to the problem at issue (Chap. 2) and sketching a primer on the role of secondary IPD analysis in scientific research and drug R&D (Chap. 3). Part II presents an analysis *de lege lata*. Chapter 4 examines the applicable legal and regulatory framework at EU level and identifies legal determinants of control over and access to clinical trial data. Furthermore, the implications of third-party access to IPD for innovation instruments are clarified, given the research-based drug companies' claim that mandatory disclosure of non-summary test data impedes their innovation incentives (Chap. 5). Part III provides an analysis *de lege ferenda* and addresses the question of whether policy intervention by access measures can be justified on the grounds of promoting drug innovation. Chapter 6 dissects specific concerns due to the restricted access to IPD and defines the 'intervention logic' of access measures. Chapter 7 outlines a theoretical framework relevant for the analysis of the access-innovation policy dilemma. Chapter 8 integrates theoretical insights into a normative analysis of how the rules should be designed to enable access to IPD as input for drug research and innovation. Chapter 9 evaluates the legislative options for designing access to IPD relative to the innovation-related policy objectives. Chapter 10 synthesises research findings.

In the context of this study, the term 'clinical trial data'[26] refers to the quantitative measurements and qualitative characteristics gathered in a clinical trial that can be analysed by applying methods of biomedical statistics. It does not include biological materials. The notion of 'research' is not limited to academic research but encompasses R&D undertaken by the private sector.[27] The problem at the centre of the study is distinct from access to affordable medicine, much debated due to pharmaceutical patents and test data exclusivity.[28] The presented analysis reflects the state of the law as it stands on 1 May 2021.

[26] The term 'clinical trial data' is used throughout the study as a mass noun and, therefore, in the singular form.

[27] Clinical trials constitute part of translational research. McGartland Rubio et al. (2010).

[28] Implications of access to IPD for research purposes for patent protection and test data exclusivity are considered in Chap. 5 at Sects. 5.3 and 5.4.1, respectively.

References

Cameron G (2001) Scientific data, the electronic era, intellectual property. In: Workshop report on IPR (intellectual property rights) aspects of internet collaborations. European Commission, Luxemburg, pp 31–33

CIOMS (2016) International ethical guidelines for health-related research involving humans. CIOMS, Geneva

Cummings SR, Browner SB, Hulley SB (2007) Conceiving the research question. In: Hulley SB (ed) Designing clinical research, 3rd edn. Lippincott Williams & Wilkins, Philadelphia, pp 17–26

Doshi P, Jefferson T (2016) Open data 5 years on: a case series of 12 freedom of information requests for regulatory data to the European Medicines Agency. Trials 17:78. https://doi.org/10.1186/s13063-016-1194-7

Doshi P et al (2013) Restoring invisible and abandoned trials: a call for people to publish the findings. BMJ 346:f2865. https://doi.org/10.1136/bmj.f2865

Krumholz HM et al (2014) Sea change in open science and data sharing. Leadership by industry Industry. Circ Cardiovasc Qual Outcomes 7:499–504. https://doi.org/10.1161/CIRCOUTCOMES.114.001166

McGartland Rubio D et al (2010) Defining translational research: implications for training. Acad Med 85(3):470–475. https://doi.org/10.1097/ACM.0b013e3181ccd618

Stewart LA, Tierney JF (2002) To IPD or not to IPD? Advantages and disadvantages of systematic reviews using individual patient data. Eval Health Prof 25(1):76–97. https://doi.org/10.1177/0163278702025001006

Part I
Setting the Scene

Chapter 2
The Context and the Problem

Abstract This chapter presents in detail the what, why and how of the study. It starts with the basic definitions and general aspects of clinical trials necessary for understanding the problem at issue. Next, the policy discourse and recent developments concerning access to non-summary trial data are outlined. In the EU, the controversy was intensified by the transparency policies of the European Medicines Agency (EMA), the evolving case law of the Court of Justice of the European Union (CJEU) on disclosure of clinical study reports by the EMA, the investigations of the European Ombudsman and the policy initiatives aimed to promote a data-driven economy. The Study's objectives, questions and analytical approach are outlined against the backdrop of the legal scholarship on the subject of access to non-summary clinical trial data.

2.1 Clinical Trials: General Aspects

2.1.1 Basic Definitions

A *clinical trial* is an experimental, health-related study that evaluates a treatment effect in humans, whereby participants are prospectively assigned to one or more health interventions.[1] A *treatment effect* is an effect on the state of health, which is attributed to medical intervention.[2] Such intervention can involve drugs, biological materials (e.g. cells), medical devices, preventive care, behavioural treatments, surgical procedures, etc.[3] In contrast to non-interventional studies, in which

[1]Council of Europe (2012), p. 53.

[2]ICH (5 Feb 1998) ICH harmonised tripartite guideline. Statistical principles for clinical trials. E9, p. 35.

[3]World Health Organization. Clinical trials. https://www.who.int/topics/clinical_trials/en/. Accessed 26 Mar 2021.

D. Kim, *Access to Non-Summary Clinical Trial Data for Research Purposes Under EU Law*, Munich Studies on Innovation and Competition 16,
https://doi.org/10.1007/978-3-030-86778-2_2

investigators observe the natural course of events, clinical trials are initiated with a pre-specified allocation of the type of intervention to trial participants.[4]

The definition of a *clinical study* under the EU Clinical Trials Regulation[5] reads as follows. A clinical study means any investigation involving human participants and intending:

(a) to discover or verify the clinical, pharmacological or other pharmacodynamic effects of one or more medicinal products;
(b) to identify any adverse reactions to one or more medicinal products; or
(c) to study the absorption, distribution, metabolism and excretion of one or more medicinal products;

with the objective of ascertaining the safety and/or efficacy of those medicinal products.[6]

A *clinical trial* is a clinical study that, apart from the prerequisites mentioned above, fulfils one of the following conditions:

(a) the assignment of the subject to a particular therapeutic strategy is decided in advance and does not fall within normal clinical practice of the EU member state concerned;
(b) the decision to prescribe the investigational medicinal products is taken together with the decision to include the subject in the clinical study; or
(c) diagnostic or monitoring procedures in addition to normal clinical practice are applied to the subjects.[7]

Clinical trials are carried out according to a *trial protocol* that states the research hypotheses and questions and describes the study objectives, design, methodology, and overall implementation.[8] *Trial subjects* are individuals participating in a clinical trial, either as the recipients of an investigational medicinal product or within a control group.[9] An *investigational medicinal product* refers to 'a medicinal product which is being tested or used as a reference, including as a placebo, in a clinical trial'.[10]

---

[4] Sprague and Bhandari (2009), p. 279.

[5] Regulation (EU) 536/2014 of the European Parliament and of the Council of 16 April 2014 on clinical trials on medicinal products for human use, and repealing Directive 2001/20/EC (27 May 2014) OJ L 158 [hereinafter the EU Clinical Trials Regulation].

[6] Reg (EU) No 536/2014, art (2)(1). This definition corresponds to the one adopted by the International Council for Harmonisation of Technical Requirements for Pharmaceuticals for Human Use. See ICH (10 June 1996) ICH Harmonised tripartite guideline. Guideline for good clinical practice. E6, p. 3.

[7] Reg (EU) No 536/2014, art (2)(2).

[8] Reg (EU) No 536/2014, art (2)(22). Annex I(D) details the particularities of a trial protocol.

[9] Reg (EU) No 536/2014, art 2(2)(17).

[10] Reg (EU) No 536/2014, art (2)(5).

A trial *sponsor* can be an individual or an organisation responsible for initiating, managing and financing a clinical trial,[11] while trial investigators are individuals responsible for conducting a trial at a clinical trial site.[12] Two categories of sponsors are distinguished: commercial sponsors (mostly represented by large multinational pharmaceutical companies) and non-commercial sponsors (such as universities, foundations, academic and research institutions). The main difference is that trials by non-commercial sponsors are usually initiated for purposes other than the development, authorisation and commercialisation of medicinal products.[13] Four trial phases are conventionally distinguished.

- Phase I trials primarily address safety aspects, assess tolerance, define pharmacokinetics and pharmacodynamics, estimate activity, explore drug metabolism and drug interactions. Such information can allow investigators to make initial estimations of the appropriate dosage range and administration schedule.[14]
- Phase II trials explore the mechanism of action and therapeutic efficacy of an investigational product for the target indication, estimate the optimal dose strength and provide the basis for the confirmatory study design.[15]
- Phase III trials are usually designed and conducted as randomised controlled trials to confirm efficacy, establish therapeutic benefits (superiority) by comparing an investigational product against the standard treatment or placebo, safety profile, the dose-response relationship, and to provide the adequate basis for assessing the benefit-risk balance to support drug marketing authorisation.[16] Upon the successful completion of this phase, a drug can be approved for marketing.
- Phase IV studies monitor the therapeutic use of the approved medicinal products to evaluate long-term safety and effectiveness, refine the understanding of the benefit-risk relationship in general or special populations or environments, specify dosing recommendations and identify less common adverse reactions.[17]

The majority of trials are confirmatory by design as they test a predefined hypothesis. Usually, they are preceded by exploratory studies,[18] in which a hypothesis can be generated that a certain medical intervention can affect the outcome of

---

[11] Reg (EU) No 536/2014, art (2)(14). See also European Commission, SWD(2012) 200 final, vol. I, p. 11 (explaining that the terms 'sponsor' or 'sponsored' refer to the responsibility for conducting trials and should not be confused with 'funder' or 'funded' as a clinical trial can be funded by another entity than the sponsor).

[12] Reg (EU) No 536/2014, art (2)(15).

[13] European Commission, SWD(2012) 200 final, vol. I, p. 14.

[14] ICH (17 July 1997) ICH Harmonised tripartite guideline. General considerations for clinical trials. E8, p. 3; CIOMS (2005), pp. 232–233.

[15] ICH (17 July 1997) ICH Harmonised tripartite guideline. General considerations for clinical trials. E8, p. 3.

[16] Ibid.

[17] Ibid.

[18] ICH (5 Feb 1998) ICH Harmonised tripartite guideline. Statistical principles for clinical trials. E9, p. 4.

interest.[19] Even though the results of exploratory studies are inconclusive and should be interpreted cautiously,[20] they make an important contribution to the total body of evidence in a particular research area.[21] A single trial can investigate multiple endpoints and have both confirmatory and exploratory aspects.[22]

In general terms, clinical trial data can cover any data collected during an interventional study. More specifically, it refers to quantitative measurements or qualitative characteristics obtained according to the pre-specified variables that can be aggregated and statistically analysed 'to show general trends or values'.[23]

2.1.2 The Social Value of Clinical Trials

Medical research *in vivo* exposes study participants to substantial health risks and can only be justified if it brings value to society. Even though clinical trials are often conducted to obtain drug marketing authorisation, this is not their sole purpose. Trials lie at the heart of the concept of 'evidence-based medicine'; they are carried out to advance medical knowledge, inform clinical practice and follow-on research[24] and promote public health.[25] Their primary objective is to generate reliable and robust data that bears 'direct relevance for understanding [...] a significant health problem'[26] and estimate the true effect of medical intervention.[27]

The social value of clinical trials subsists in *reliable* and *valid knowledge* that can inform multiple stakeholders' decision-making, including patients, medical practitioners, researchers, policy-makers, and funders.[28] Therefore, ensuring that studies are designed and conducted according to high scientific standards is 'essential for maintaining the integrity of the research enterprise and its ability to fulfil its social

[19] Stoney and Lee Johnson (2018), p. 251.

[20] Results of exploratory data analysis cannot form the basis for the proof of efficacy. ICH (5 Feb 1998) ICH harmonised tripartite guideline. Statistical principles for clinical trials. E9, pp. 4, 27.

[21] Ibid p. 4.

[22] ICH (5 Feb 1998) ICH Harmonised tripartite guideline. Statistical principles for clinical trials. E9, p. 4. The types of clinical trial data, endpoints and data analysis are explained in more detail in Chap. 3.

[23] EMA (15 Oct 2018) External guidance on the implementation of the European Medicines Agency policy on the publication of clinical data for medicinal products for human use. EMA/90915/2016, version 1.4, p. 8. https://www.ema.europa.eu/en/documents/regulatory-procedural-guideline/external-guidance-implementation-european-medicines-agency-policy-publication-clinical-data_en-3.pdf. Accessed 26 Mar 2021. Notably, neither the EU Clinical Trials Directive nor the EU Clinical Trials Regulation defines the term 'clinical trial data'.

[24] European Commission, SWD(2012) 200 final, vol. I, p. 11.

[25] CIOMS (2016), p. 1.

[26] CIOMS (2016), p. 1.

[27] On reliability and robustness of clinical trial data, see Chap. 4 at Sect. 4.1.2.1.

[28] CIOMS (2016), p. 1.

function'.[29] Thus, the scientific value of a study forms a component of its overall social value. However, scientific quality in and of itself does not render a trial socially valuable. For instance, a trial can be well-designed yet lack social value if its endpoints are irrelevant for clinical practice[30] or if the research question can be answered satisfactorily based on the evidence gathered in prior studies.[31]

2.1.3 Clinical Trials in the Regulatory Context

2.1.3.1 Clinical Trial Approval

Before a clinical trial can be conducted, an application for the authorisation undergoes an ethical and scientific review[32] regarding the trial admissibility given the anticipated benefits for public health and risks for trial subjects.[33] In Europe, such assessment is performed by national competent authorities and ethics committees.[34] The review considers aspects such as the relevance of a study vis-à-vis the target patient population, the state of scientific research in the relevant area, and the ability to generate reliable and robust data.[35] The EU Clinical Trials Regulation details the requirements for the trial authorisation that apply to all clinical trials conducted in the EU.[36]

2.1.3.2 Drug Marketing Authorisation

Where trials are conducted to develop and commercialise medicinal products, data on safety and efficacy is submitted to drug authorities to support an application for drug marketing authorisation.[37] In the EU, a drug can be authorised via three regulatory pathways: the centralised procedure, the mutual recognition procedure

---

[29] Ibid p. 2.

[30] Ibid p. 1.

[31] Chalmers and Glasziou (2009), p. 87.

[32] Reg (EU) No 536/2014, art 4.

[33] Reg (EU) No 536/2014, rec 13, art 6.

[34] While the EU Clinical Trials Regulation does not harmonise how the assessment tasks and responsibilities should be divided between national competent authorities and ethics committees (see Reg (EU) No 536/2014, rec 18; art 4), the situation differs across the EU Member States in this regard. See generally Doppelfeld (2009).

[35] Reg (EU) No 536/2014, art 6(1)(b)(i).

[36] Reg (EU) No 536/2014, art 1. General aspects of the EU Clinical Trials Regulation are outlined in Chap. 4 at Sect. 4.1.2.1.

[37] Dir 2001/83/EC, art 8(3)(i).

and the decentralised procedure.[38] The grant of a marketing authorisation is contingent on assessing the beneficial therapeutic effects of a medical intervention vis-à-vis its potential risks for health (also known as the 'benefit-risk balance').[39]

Specific aspects of drug marketing authorisation are discussed in more detail in Chap. 4 in relation to the regulatory provisions applicable to clinical trial data, Chap. 5 in the context of regulatory mechanisms of incentivising innovation, and Chap. 6 with regard to the issue of transparency in research and regulatory decision making.

2.1.4 Clinical Trials as a Part of Industry R&D

Clinical trials form a part of translational research that transforms the findings from basic science into a clinical setting.[40] While studies can be initiated for solely scientific research purposes, the major share of trials—especially phase III trials—is sponsored by drug companies and conducted to develop and commercialise new drugs.[41] Even though drug R&D is characterised by increasing industry-academia collaboration,[42] randomised clinical trials (RCTs) are predominantly carried out as part of industrial R&D. Such state of affairs is a matter of historical development.[43] Throughout its evolution, the pharmaceutical industry has strengthened its organisational capabilities in managing large-scale clinical trials, regulatory affairs and marketing.[44]

Clinical trials constitute the most cost- and time-intensive part of drug R&D.[45] Any figures are likely to be outdated by the time this study might reach the reader. A 2016 study estimated the total out-of-pocket and capitalised R&D cost per new drug to amount to 1395 million and 2,558 million in 2013 US dollars, respectively.[46] These figures include the costs of the compounds abandoned during testing. The

---

[38] For a detailed explanation of drug marketing authorisation procedures, see European Commission (2018) Notice to applicants, vol. 2A, rev 9.

[39] Dir 2001/83/EC, art 1(28), (28a); Reg 726/2004/EC, art 2.

[40] Two stages of translational research are distinguished: first, knowledge from basic research is transferred to clinical research; second, findings from clinical studies are applied in clinical practice. Rubio McGartland et al. (2010), pp. 471–472.

[41] For statistics, see European Commission, SWD(2012) 200 final, vol. II, p. 8.

[42] Cockburn (2009), p. 165.

[43] Dosi and Mazzucato (2006), p. 5.

[44] Ibid p. 2 (with further references).

[45] Ibid p. 9. See also European Commission (8 July 2009) Pharmaceutical sector inquiry report. Final report, p. 55 ff. https://ec.europa.eu/competition/sectors/pharmaceuticals/inquiry/staff_working_paper_part1.pdf. Accessed 26 Mar 2021.

[46] DiMasi et al. (2016), p. 31. The study was based on a survey conducted among ten leading pharmaceutical companies; it assessed the costs of drug R&D projects that, for the most part, resulted in regulatory approvals between the 2000s and early 2010s. Notably, the pre-approval out-

European Federation of Pharmaceutical Industries claims that 16.4, 9.7, 10.6, 28.9, 3.3, and 11.6 per cent of R&D investments of research-based pharmaceutical companies are allocated to pre-clinical studies, clinical phase I, phase II, phase III, marketing approval and pharmacovigilance, respectively.[47] While estimates can diverge due to the varying definitions of what counts as 'drug R&D costs',[48] drug innovation is generally known as a highly risky and costly enterprise.[49]

2.2 The Debate Over Access to Clinical Trial Data

The issue of access to clinical trial data has been subject to a long-standing debate in many jurisdictions. This section surveys recent developments and outlines the most contested issues.

2.2.1 Concerns Related to Restricted Access to Clinical Trial Data

The following passage from the 2015 report of the Health and Medicine Division of the US National Academy of Science articulates the main concerns associated with the ability of trial sponsors to control access to primary research data.

> Vast amounts of data are generated over the course of a clinical trial; however, a large portion of these data is *never published* in peer-reviewed journals. Today [...] researchers other than the trialists have limited access to clinical trial data that could be used to *reproduce* published results, carry out *secondary analyses*, or combine data from different trials in systematic reviews. Public well-being would be enhanced by the additional knowledge that could be gained from these analyses. [...] Sharing clinical trial data might accelerate the *drug discovery* and development process, reducing *redundancies* and facilitating the identification and validation of *new drug targets* or surrogate endpoints. In short, there are today many *missed opportunities* to gain scientific knowledge from clinical trial data that could strengthen the evidence base for the treatment decisions of physicians and patients. In economic terms, these missed opportunities result in a *suboptimal return* on the

of-pocket cost was found to be a 166% increase, and the capitalised cost was a 145% increase in real dollars over the estimates published by the same authors earlier (DiMasi et al. 2003).

[47] European Federation of Pharmaceutical Industries and Associations (2020) The pharmaceutical industry in figures. Key data, p. 8. https://efpia.eu/media/554521/efpia_pharmafigures_2020_web.pdf. Accessed 26 Mar 2021.

[48] The pharmaceutical industry has been criticised, among other things, for overstating R&D costs and counting promotional expenses as 'R&D investment'. See e.g. Light and Warburton (2005).

[49] Among the driving factors of high costs of drug R&D are the complexity of scientific problems, the high rate of failure and the increasing stringency of the regulatory oversight. See DiMasi et al. (2016), p. 26 ff; Ahn (2014), p. 83 ff; Pammolli et al. (2011).

altruism and contributions of clinical trial participants, the efforts of clinical trialists and research staff, and the financial resources invested by study funders and sponsors.[50]

Various interrelated yet distinct concerns can be discerned within this statement. Some relate to transparency and reproducibility of research; others point towards the 'underutilised potential' of non-summary data. Let us consider each in turn.

2.2.2 *Transparency Issues*

In broad terms, transparency in clinical trials can be understood as the extent to which primary evidence is available to scientists, clinicians, regulatory bodies and the general public.[51] By transparency concerns, one usually refers to misrepresented, unreported or unpublished trial results.[52] Such malpractices adversely affect clinical care, undermine public trust in medical research[53] and generate waste in medical research.[54]

Transparency concerns often arise due to commercial sponsorship of trials due to the potential conflict of interest—the risk that commercial sponsors' financial interests 'may unduly influence professional judgments [and] threaten the integrity of scientific investigations'.[55] In this regard, the accessibility of primary ('source') research data for independent confirmatory analysis has been viewed as an ethical and scientific imperative.[56]

2.2.3 *Levels of Transparency in Clinical Trials*

In the context of clinical trials, four levels of transparency are distinguished, namely, prospective trial registration, reporting and publication of summary-level trial results, accessibility of clinical study reports and individual-patient level data.[57]

---

[50] Institute of Medicine of the National Academies (2015), p. 18 (emphasis added) (with further references).

[51] House of Commons (2013), p. 30.

[52] For a detailed discussion, see Chap. 6 at Sect. 6.4.2.

[53] House of Commons (2013), pp. 30–31. See also Friedman et al. (2015), p. 479 ff.

[54] Chalmers and Glasziou (2009), p. 88.

[55] Lo and Field (2009), p. 2. Empirical studies examining this proposition are reviewed in Chap. 6 at Sect. 6.4.2.3.

[56] Bauchner et al. (2016). On the role of secondary data analysis in improving research quality, see Chap. 6 at Sect. 6.4.2.5.

[57] See e.g. All trials registered, all results reported, p. 1. http://www.alltrials.net//wp-content/uploads/2013/09/What-does-all-trials-registered-and-reported-mean.pdf. Accessed 26 Mar 2021. For a similar approach, see House of Commons (2013), p. 34 ff.

These levels correspond to different types of clinical trial data and degrees of data 'granularity'.[58]

2.2.3.1 Trial Registration

Trial registration refers to the publication of the trial identification information and summary information about trial design, conduct and administration that allow to track changes, trial completion and results reporting. As proclaimed under the Declaration of Helsinki: 'Every research study involving human subjects must be registered in a *publicly accessible* database before recruitment of the first subject'.[59]

The idea of the prospective registration of clinical trials has been widely supported by the medical research community[60] and several international initiatives.[61] Public health benefits stem from the increased awareness of planned or ongoing trials that allows avoiding unnecessary duplication of research addressing similar or identical questions,[62] facilitating the recruitment of trial participants[63] and identifying 'knowledge gaps' within the 'research landscape'.[64] Notwithstanding such advantages, evidence shows that the level of registered trials remains quite low.[65]

2.2.3.2 Reporting and Publication of Trial Results

Summary-level study results refer to the conclusions derived through the primary data analysis usually published in academic journals and trial registers. Despite numerous institutional calls and initiatives, the problem of unreported trials

---

[58] Zarin and Tse (2016).

[59] Declaration of Helsinki – Ethical Principles for Medical Research Involving Human Subjects (1964), para 35 (emphasis added).

[60] In 2004, the International Committee of Medical Journal Editors adopted a decision to require, as a pre-condition for considering a manuscript for publication, that all clinical trials are registered in a public trials registry before the first patient enrolment. See DeAngelis et al. (2004).

[61] Among such initiatives are: the International Clinical Trials Registry Platform (ICTRP) launched and administered by the WHO (http://www.who.int/ictrp/en/) and the International Standard Randomised Controlled Trial Number (ISRCTN) Registry (http://www.isrctn.com/). Major clinical trial registration platforms include the Clinical Trials Register (https://www.clinicaltrialsregister.eu) and the ClinicalTrials.gov (https://clinicaltrials.gov/). Accessed 26 Mar 2021.

[62] WHO. Trial registration. Why is trial registration important? http://www.who.int/ictrp/trial_reg/en/. Accessed 26 Mar 2021.

[63] Ibid.

[64] Zarin and Tse (2016).

[65] See e.g. Killeen et al. (2014).

subsists.[66] As acknowledged by the WHO, 'a substantial number of clinical trials remain unreported several years after study completion, even in the case of large randomized clinical trials'.[67]

Non-reporting hinders the *totality of evidence* on a clinical question[68] and, consequently, the strength of the clinical recommendations developed based on systematic reviews and meta-analyses.[69] Hence, timely dissemination of study outcomes is critical for clinical practice, apart from ensuring research quality[70] and informing the design of new trials.[71]

2.2.3.3 Accessibility of Non-Summary Data

The notion of non-summary clinical trial data comprises clinical study reports (CSRs) and individual patient data (IPD). Neither CSRs nor IPD are usually publicly accessible.[72] CSRs are prepared for regulatory purposes and submitted for drug marketing authorisation. It contains a detailed protocol that sets out the study objectives, design, methodology, statistical considerations for efficacy and safety analyses, etc. Commentators argued that the accessibility of these documents could

---

[66] See e.g. Ben Goldacre et al. (2018) (finding that only 49.5% of the results in the sample of 7274 trials were reported on the due date). Notably, the reporting rate was substantially higher among commercial than non-commercial sponsors. The study also found '[e]xtensive evidence [...] of errors, omissions, and contradictory entries' in the EU Clinical Trials Register.

[67] WHO. WHO Statement on Public Disclosure of Clinical Trial Results, p. 1 http://www.who.int/ictrp/results/WHO_Statement_results_reporting_clinical_trials.pdf?ua=1. Accessed 26 Mar 2021. On the issue of non-publication of trial results, see e.g. Jones et al. (2011); Macleod et al. (2014).

[68] In meta-analysis, the term 'totality of evidence' refers to the assembly of all studies that have compared the treatment of interest in the target population, which are analysed 'to determine an internally consistent set of estimated relative treatment effects between all treatments'. Dias et al. (2018), p. 3. See also Council of Europe (2012), para 6.C.20.2 (stating that '[i]f some of this relevant evidence remains unpublished the totality of evidence is biased and therefore unreliable'; consequently, patients 'may then continue to receive treatments that are actually harmful, or conversely not receive treatments that would benefit them').

[69] On these types of secondary data analysis, see Chap. 3 at Sect. 3.2.2.7.

[70] Opinions and evidence diverge as to the effectiveness of trial registration in this regard. See DeAngelis et al. (2004), p. 1250 (reporting that 'the International Committee of Medical Journal Editors (ICMJE) proposes comprehensive trials registration as a solution to the problem of selective awareness and announces that all eleven ICMJE member journals will adopt a trials-registration policy to promote this goal'). But see Odutayo et al. (2017) (finding 'little evidence of a difference in positive study findings between registered and non-registered clinical trials').

[71] See Council of Europe (2012), para 6.C.20.2 (emphasising that 'systematic accumulation and analysis of research results is essential for developing medical treatments').

[72] Doshi et al. (2013) (noting that the reported or published trial findings represent 'the tip of the iceberg', while CSRs, case report forms, trial protocol, an investigator's brochure and IPD remain 'below the waterline'). On the transparency requirements under the EU Clinical Trials Regulation, see Chap. 4 at Sect. 4.3.1.1.

make an important contribution to transparency in trials, even where IPD is not shared.[73]

The main advantage of IPD over summary information is that its analysis—apart from verifying the original conclusions—can support a broad range of research activities, such as generating and testing secondary hypotheses, developing and evaluating novel statistical methods[74] and optimising the design of the subsequent trials.[75]

2.2.4 International Norm-Setting Initiatives Promoting Transparency in Clinical Research

Transparency in research that involve human subjects has been promoted as an ethical and scientific norm by multiple instruments of soft international law. For instance, the 1996 Guideline for Good Clinical Practice developed by the International Council for Harmonisation of Technical Requirements for Pharmaceuticals for Human Use (ICH) states that a trial protocol should contain publication policy unless such policy is addressed in a separate agreement.[76] The 1997 Oviedo Convention for the Protection of Human Rights[77] requires with regard to trial results that

- a report or *summary* shall be submitted to the ethics committee or the competent body upon completion of the research;
- research *conclusions* shall be made available to participants in a reasonable time, on request;
- the research *results* shall be publicised in a reasonable time.[78]

The 2012 Guide for research ethics committee members of the Council of Europe recommends that research ethics committees

> must [. . .] be assured that the researchers have formulated a *publication policy* and that they have negotiated the policy with any external *research sponsors* so that they are *not contractually inhibited* from disseminating their results.[79]

[73] Hoffmann et al. (2017).

[74] Chapter 3 surveys different types of secondary analysis of clinical trial data.

[75] Institute of Medicine of the National Academies (2015), p. 99; Zarin and Tse (2016); Jones et al. (2013).

[76] ICH (10 June 1996) ICH Harmonised tripartite guideline. Guideline for good clinical practice. E6, para 6.15.

[77] Convention for the protection of human rights and dignity of the human being with regard to the application of biology and medicine: convention on human rights and biomedicine (4 Apr 1997, Oviedo).

[78] Additional protocol to the Oviedo Convention concerning biomedical research (25 Jan 2005) CETS No. 195, art 28.

[79] Council of Europe (2012), para 6.C.20.2 (emphasis added).

The Guide particularly emphasises the importance of making available research results, irrespective of whether they are 'positive, negative or inconclusive'.[80] In particular, it states:

> Suppression of results not only *distorts the research endeavour* if other research groups are unaware of them but also can *directly affect patients*, who may be recruited needlessly to take part in *unnecessarily repetitive* research. In addition, *systematic accumulation* and analysis of research results is essential for developing medical treatments [. . .] Rather, progress depends on new research being carried out and interpreted in the context of systematic reviews of all other relevant and reliable evidence. If some of this relevant evidence remains unpublished the *totality of evidence* is biased and therefore unreliable. Patients may then continue to receive treatments that are actually harmful, or conversely not receive treatments that would benefit them.[81]

Along similar lines, the 2013 Declaration of Helsinki—Ethical Principles for Medical Research Involving Human Subjects states:

> Researchers, authors, sponsors, editors and publishers all have ethical obligations with regard to the publication and dissemination of the results of research. Researchers have a duty to make publicly available the results of their research on human subjects and are *accountable* for the completeness and accuracy of their reports. All parties should adhere to accepted guidelines for ethical reporting. *Negative* and *inconclusive* as well as positive results must be published or otherwise made publicly available.[82]

According to the 2016 International Ethical Guidelines for Health-Related Research Involving Humans adopted by the Council for International Organizations of Medical Sciences (CIOMS), whenever research in humans produces scientific knowledge, there 'must be assurance [. . .] that the scientific knowledge gained will be distributed and available for the benefit of the population'.[83] The CIOMS further endorses that researchers, sponsors, research ethics committees, funders, editors and publishers have to ensure public accountability of clinical research to

> maximize benefits accruing from health research, reduce risks to future volunteers from *undisclosed* harms identified in previous clinical studies, *reduce biases* in evidence-based decision-making, improve *efficiency* of resource allocation for both research and development and financing of health interventions and promote societal trust in health-related research.[84]

Notably, while these provisions call for transparency in clinical trials, they mainly refer to *summary*-level data.

---

[80] Ibid.

[81] Ibid (emphasis added).

[82] Declaration of Helsinki—ethical principles for medical research involving human subjects (Helsinki, June 1964), para 36 (emphasis added).

[83] CIOMS (2016), p. 4.

[84] Ibid p. 91 (emphasis added).

2.2.5 *Institutional Developments*

2.2.5.1 Editorial Campaign

The problem of biased publication and non-reporting of trial results has been, in part, attributed to the publishers' reluctance to accept negative results for publication.[85] This issue has been addressed by the International Committee of Medical Journals Editors (ICMJE)[86] in its 'Recommendations for the Conduct, Reporting, Editing, and Publication of Scholarly Work in Medical Journals' developed to set 'the best practice and ethical standards in the conduct and reporting of research [. . .] and to help authors, editors, and others involved in peer review and biomedical publishing create and distribute accurate, clear, reproducible, unbiased medical journal articles'.[87] In addition, the ICMJE called on editors to require authors to submit research protocols and plans for statistical analysis and encourage them 'to make such documents publicly available at the time of or after publication'.[88] As far as primary research data is concerned, the ICMJE issued 'Data Sharing Statements for Clinical Trials' intending to foster an environment where sharing de-identified IPD becomes a norm.[89]

2.2.5.2 Funding Institutions

Several institutional funders, including the UK Medical Research Council, the US National Institutes of Health, the Bill and Melinda Gates Foundation and the Wellcome Trust Foundation, adopted policies that either support or require clinical trial data sharing.[90] Such initiatives are less relevant for the present study as the recipients of the grants are typically public research institutions. Nevertheless, these institutional policies are worth mentioning as they indicate that data accessibility is topical not only in industry-sponsored trials.[91]

---

[85] CIOMS (2005), p. 47.

[86] The ICMJE is a working group of editors of general medical journal including Annals of Internal Medicine, British Medical Journal, Bulletin of the World Health Organization, Deutsches Ärzteblatt, The Lancet, etc.

[87] ICMJE (Dec 2019) Recommendations for the conduct, reporting, editing, and publication of scholarly work in medical journals, p. 1. http://www.icmje.org/icmje-recommendations.pdf. Accessed 26 Mar 2021.

[88] Ibid.

[89] Taichman et al. (2017).

[90] For an overview, see Naci et al. (2015).

[91] See e.g. Rathi et al. (2012).

2.2.6 *Access to Data as a Digital Resource in the Context of Data-Driven Innovation*

2.2.6.1 The Promises of 'Big Data'

The importance of access to data as a digital resource in the context of data-driven innovation and the new economic era hailed as 'data revolution'[92] can hardly be underestimated. Digital data has been viewed as 'an important source of value creation',[93] 'the new oil',[94] 'digital gold',[95] and the source of growth that transforms 'the paradigm [...] in knowledge creation and decision making'.[96] The 2015 OECD report envisages that digital data will enhance economic efficiency and productivity, competitiveness and social well-being across economic sectors.[97] Echoing this vision, the European Commission aspires that

> thriving data-driven economy will contribute to the well-being of citizens as well as to socio-economic progress through new business opportunities and through more innovative public services.[98]

In the healthcare sector, the 2017 'Conclusions on Health in the Digital Society'[99] issued by the Council of the European Union point out that 'progress in implementing the data-driven digital solutions in the health sector remains limited'.[100] In this regard, the Council notes that limited access to and use of large databases for research and innovation constitute a barrier to 'scaling up the potential in digital health and connected care'[101] and emphasises the importance of developing 'common data structures, coding systems and terminologies, as well as common standards for measuring clinical and patient reported outcomes, [to] improve *semantic interoperability*, *quality* and *comparability* of data'.[102] Further, the Commission and the EU Member States are invited to 'work together with the aim of improving

[92]The UN Secretary-General's Independent Expert Advisory Group on the Data Revolution for Sustainable Development (2014) A World that counts. Mobilising the data revolution for sustainable development. https://www.undatarevolution.org/wp-content/uploads/2014/12/A-World-That-Counts2.pdf. Accessed 26 Mar 2021.

[93]OECD (2015), p. 178.

[94]Ibid, p. 180.

[95]de Franceschi and Lehmann (2015), p. 51.

[96]OECD (2015), p. 132.

[97]Ibid p. 17.

[98]European Commission, COM(2017) 9 final, p. 2.

[99]The Council of the European Union (21 Dec 2017) Council conclusions on health in the digital society—making progress in data-driven innovation in the field of health. OJ 2017/C 440/05.

[100]Ibid para 12.

[101]Ibid.

[102]Ibid para 42 (emphasis added).

access to larger European datasets [. . .] for health research and innovation purposes, while ensuring a high level of data protection'.[103]

2.2.6.2 Legal and Policy Debate Concerning 'Ownership' of Sensor-Generated Data

The question of who 'owns' sensor-generated data has been at the centre of policy, economic and legal debate in the EU.[104] In 2017, the European Commission conducted a public consultation on Building a European Data Economy.[105] The initiative was implemented within the framework of the Digital Single Market and formed a component of the objective to maximise the growth potential of the digital economy[106] and enhance social benefits in various fields, including public health.[107]

At the outset, it should be clarified that the relevance of the debate regarding ownership and access rights in industrial data is limited for this study. The main reason is that IPD does not fall within the definition of 'machine-generated data'—i.e. data 'created *without the direct intervention* of a human by computer processes, applications or services, or by sensors processing information received from equipment, software or machinery, whether virtual or real'.[108] Even though IPD can often be obtained with the help of sensors of medical devices, it does not occur in an automated way comparable, for instance, with sensors embedded in a car that collect data on weather or road conditions while a car is driving.

At the same time, the policy and legal discourse on the regulation of a data economy provides a pertinent context for this study and accentuates its topicality, especially given the overall policy objective of leveraging the potential of digital data to enhance social benefits, including in public health.[109] Given that such benefits can be achieved only through data analysis, it will be pertinent to consider to what extent the applicable regulatory framework is conducive to realising the potential of non-summary clinical trial data.

---

[103] Ibid para 46.

[104] Literature on the regulatory framework for digital data is expansive. For an in-depth analysis, see Drexl (2017a, b, 2018); Richter and Hilty (2018); Weber and Thouvenin (2018); For an overview of the European Commission's initiatives on the 'emerging issue of data ownership', see Kim (2017a), p. 155 ff.

[105] European Commission (10 Jan 2017) Public consultation on building the European data economy. https://ec.europa.eu/digital-single-market/en/news/public-consultation-building-european-data-economy. Accessed 26 Mar 2021.

[106] European Commission, COM(2015) 192 final, p. 14.

[107] European Commission, COM (2014) 442 final, p. 12.

[108] European Commission, COM(2017) 9 final, p. 9 (emphasis added).

[109] European Commission, COM(2014) 442 final, pp. 2, 3, 6, 8.

2.2.6.3 'Big Data' Analysis in Public Healthcare and Drug R&D

Several studies highlight the promise of 'big data' to generate unprecedented benefits for public healthcare. The 2011 report by the McKinsey Global Institute explores the potential of 'big data' to transform the healthcare sector, particularly by promoting comparative effectiveness research, development of clinical decision support systems and services of remote patient monitoring and supporting new business models.[110] Computational approaches based on data analysis play a prominent role in drug R&D, including the discovery of activity-response correlations, personalisation of treatments and design of new trials.[111] The 2015 report by the Organisation for Economic Co-operation and Development (OECD) envisages that advanced methods of data analysis applied to large volumes of different types of health-related data (e.g. clinical, genetic, behavioural, etc.) can yield 'extraordinary insights into the natural history of diseases and their diagnosis, prevention and treatment'.[112]

2.2.6.4 Data-Sharing Policies and Practices Adopted by the Pharmaceutical Industry

While industry initiatives for IPD sharing might still be in the 'infancy' stage,[113] they were hailed as a 'sea change'.[114] GlaxoSmithKline (GSK) was first among the drug companies to adopt a data sharing policy in 2012 and provide access to anonymised IPD through a secure portal to external researchers based on the review of the submitted research proposals.[115] Subsequently, data-sharing policies have been implemented by the leading pharmaceutical companies, including Janssen Pharmaceuticals, Pfizer, Novartis, Roche, Sanofi, Merck, Bayer, Bristol-Myers Squibb, AbbVie, Eli Lilly and AstraZeneca. In 2013, a consortium of pharmaceutical companies, including GSK, Astellas, Bayer, Boehringer Ingelheim, Eisai, Lilly, Novartis, Roche, Sanofi, Takeda, UCB and ViiV Healthcare, launched an online platform intending to assist researchers in finding and negotiating access to

---

[110] James Manyika et al. (2011) Big data: the next frontier for innovation, competition, and productivity. McKinsey & Company. https://www.mckinsey.com/business-functions/mckinsey-digital/our-insights/big-data-the-next-frontier-for-innovation. Accessed 26 Mar 2021.

[111] Ibid pp. 44–48.

[112] OECD (2015), p. 363. See generally Raghupathi and Raghupathi (2014). On the role of 'big data' analysis in drug R&D, see also Chap. 3 at Sect. 3.3.1.

[113] Sudlow et al. (2016), p. 34.

[114] Krumholz et al. (2014), pp. 499–504. For an overview of data-sharing policies of drug companies, see Institute of Medicine of the National Academies (2015), pp. 16–18.

[115] GlaxoSmithKline. Data transparency. https://www.gsk.com/en-gb/behind-the-science/innovation/data-transparency/. Accessed 26 Mar 2021.

de-identified IPD obtained in trials sponsored by these companies, if the review panel approves a research proposal.[116]

In 2014, the Pharmaceutical Research and Manufacturers of America (PhRMA) and the European Federation of Pharmaceutical Industries and Associations (EFPIA) issued 'Principles for Responsible Clinical Trial Data Sharing' that articulated several commitments, including the commitment to grant access to *de-identified* patient-level data and trial protocols related to drugs and indications, which have already been approved for marketing in the US and the EU, for 'legitimate research purposes' upon request from 'qualified' scientific and medical researchers.[117] The Principles also stipulate that it is 'appropriate' for a trial sponsor/data holder to refuse to share the requested data in situations involving competitive concerns and potential conflict of interests.[118] Although the commitments under the joint 'Principles for Responsible Clinical Trial Data Sharing' are not binding on the members of both associations,[119] they are reflected in many data-sharing policies adopted by individual drug companies.[120]

Furthermore, the Yale University Open Data Access Project (YODA) launched in 2011 should be mentioned as an outstanding example of the industry-academia collaboration. It started with the partnership between Medtronic and Yale University and, later on, was joined by Johnson & Johnson. The collaboration model developed throughout the project aims to promote responsible and sustainable IPD sharing.[121]

2.3 Diversity of Policy Approaches and Academic Views

2.3.1 The Controversy Over Disclosure of Non-Summary Clinical Trial Data in the EU

2.3.1.1 Investigations of the European Ombudsman

Within the EU institutional framework, broad access to clinical trial data has been strongly advocated by the European Ombudsman. According to Emily O'Reilly,

---

[116] Clinical Study Data Request. About us. https://www.clinicalstudydatarequest.com/Default.aspx. Accessed 26 Mar 2021.

[117] PhRMA and EFPIA (18 July 2013) Principles for responsible clinical trial data sharing. Our commitment to patients and researchers, pp. 1, 4. https://www.efpia.eu/media/25189/principles-for-responsible-clinical-trial-data-sharing.pdf. Accessed 26 Mar 2021.

[118] Ibid p. 4 (emphasis added).

[119] The PhRMA and EFPIA encouraged their members to adhere to the joint data-sharing principles. See Goldacre et al. (2017).

[120] On the data-sharing policies of drug companies, see Chap. 5 at Sect. 5.1.3 and Chap. 6 at 6.3.2.

[121] The YODA project, policies & procedures to guide external investigator access to clinical trial data. http://yoda.yale.edu/policies-procedures-guide-external-investigator-access-clinical-trial-data. Accessed 26 Mar 2021.

CSRs generally do not 'fall within the commercial interests exception',[122] and 'the entirety of a clinical study report should ultimately be disclosed'.[123] Several disputes were investigated by the European Ombudsman between 2005 and 2020[124] concerning the refusals of the EMA to grant access to clinical trial data submitted by the pharmaceutical companies, including GSK,[125] Biogen Idec,[126] and AbbVie.[127] When justifying its decisions to refuse access to CSRs, the EMA used to invoke the exception for the protection of commercial interests under the EU Transparency Regulation[128] and the obligation under Article 39(3) of the TRIPS Agreement.[129] Most investigations resulted in the Ombudsman's non-binding decisions requesting the EMA to reconsider the possibility to grant access to the requested data.[130] The Ombudsman's recommendations upon the inquiry initiated by the Nordic Cochrane Centre in 2007[131] triggered the landmark change in the EMA's approach to treating clinical trial data submitted for drug marketing authorisation.

[122] European Ombudsman (24 Nov 2010) Decision of the European Ombudsman closing his inquiry into complaint 2560/2007/BEH against the European Medicines Agency, para 80.

[123] Ibid paras 73–74.

[124] The search of the database of the Ombudsman's decisions (https://www.ombudsman.europa.eu) was conducted in November 2020. The majority of the identified inquiries were launched upon a request lodged by third parties.

[125] European Ombudsman (3 Apr 2008) Decision of the European Ombudsman on complaint 2371/2005/OV against the European Commission.

[126] European Ombudsman (28 Aug 2013) Decision of the European Ombudsman closing his inquiry into complaint 693/2011/(ELB)RA against the European Medicines Agency.

[127] European Ombudsman (8 Jun 2016) Inquiry OI/3/2014/FOR concerning the partial refusal of the European Medicines Agency to give public access to studies related to the approval of a medicinal product.

[128] Regulation (EC) No 1049/2001 of the European Parliament and of the Council of 30 May 2001 regarding public access to European Parliament, Council and Commission documents (31 May 2001) OJ L 145 [hereinafter the EU Transparency Regulation].

[129] European Ombudsman (24 Nov 2010) Decision of the European Ombudsman closing his inquiry into complaint 2560/2007/BEH against the European Medicines Agency, paras 14, 18.

[130] European Ombudsman (29 Apr 2010) Draft recommendation of the European Ombudsman in his inquiry into complaint 2493/2008/(BB)TS against the European Medicines Agency.

[131] European Ombudsman (19 May 2010) Draft recommendation of the European Ombudsman in his inquiry into complaint 2560/2007/BEH against the European Medicines Agency. The dispute arose after the EMA had denied Cochrane Centre access to CSRs and trial protocols related to two anti-obesity drugs, on the grounds of protecting the commercial interests of the original document holder—the pharmaceutical company AbbVie.

2.3.1.2 The EMA Transparency Policies

The EMA transparency policies jointly refer to the 2010 access policy 0043[132] providing for the request-based access and the 2015 publication policy 0043[133] providing for the proactive publication of non-summary clinical trial data submitted to the EMA under the centralised marketing authorisation procedure.[134] Acclaimed by the European Ombudsman as 'a paradigm shift',[135] the latter policy presents a remarkable departure from the presumptively confidential treatment of non-summary clinical trial data. According to the EMA Executive Director Guido Rasi, it set 'a new standard for transparency in public health and pharmaceutical research and development', and such 'unprecedented level of access to clinical reports will benefit patients, healthcare professionals, academia and industry'.[136] The promise is significant, given that commercial sponsors account for the major share of clinical trials in the EU,[137] and that application dossiers represent 'the mostly hidden and untapped source of detailed and exhaustive data',[138] which has been for the most part unavailable to the scientific community and the public at large.[139]

The EMA was the first regulatory authority worldwide to provide such broad access to data[140] and, at the time of writing, remains the only drug regulator intending to release proactively non-summary clinical trial data submitted by drug sponsors.[141] The policy objective is twofold: first, to enable public scrutiny by allowing external investigators to verify the original analysis and conclusions and, second, to allow 'the wider scientific community to make use of detailed and high quality clinical trial data to *develop new knowledge* in the interest of public

[132] EMA (4 Oct 2018) European Medicines Agency policy on access to documents. Policy/0043. EMA/729522/2016 [hereinafter EMA access policy 0043].

[133] EMA publication policy 0070 (see Chap. 1 and footnote 9).

[134] Ibid para 4.3 (specifying the effective dates of the policy implementation). EMA publication policy 0070 does not supersede EMA access policy 0043 and 'any natural or legal person may continue to submit a request for access to documents to the Agency independently of the proactive publication mechanisms established by this policy'. ibid p. 1.

[135] European Ombudsman (8 Jun 2016) Decision on own-initiative inquiry OI/3/2014/FOR concerning the partial refusal of the European Medicines Agency to give public access to studies related to the approval of a medicinal product, para 71.

[136] EMA (2 Oct 2014) Publication of clinical reports. EMA adopts landmark policy to take effect on 1 January 2015, EMA/601455/2014, p. 1. https://www.ema.europa.eu/en/documents/press-release/publication-clinical-reports_en.pdf. Accessed 26 Mar 2021.

[137] See European Commission, SWD(2012) 200 final, vol. II, p. 8.

[138] Doshi and Jefferson (2013).

[139] Institute of Medicine of the National Academies (2014), p. 2; R Ross et al. (2012). On the scope of data available under the data-sharing policies of drug companies, see Chap. 6 at Sect. 6.3.2.

[140] EMA (20 Oct 2016) Opening up clinical data on new medicines. EMA provides public access to clinical reports. EMA/650519/2016, p. 1; Doshi and Jefferson (2016).

[141] For an overview policy approaches in other jurisdictions, see below at Sect. 2.3.2 in this chapter.

health'.[142] By disclosing clinical data proactively, the EMA is pursuing the goal of making 'medicine development more *efficient* by establishing a *level* playing field that allows all medicine developers to learn from past successes and failures'.[143]

The 2015 policy is implemented in two steps: during the first phase, only CSRs are published; during the second stage, de-identified IPD is expected to be released, once the Agency 'find[s] the most appropriate way [of making data available] in compliance with privacy and data protection laws'.[144] Access can be granted under the condition that the released data is used for scientific, non-commercial research purposes and, explicitly, not for regulatory purposes.[145] Under the EMA access policy 0043, disclosure of non-summary clinical trial data has *erga omnes* effect.[146] Under the EMA publication policy 0070, access to data can be granted in downloadable and searchable formats[147] without the remuneration to be paid to initial trial sponsors. Not surprisingly, the policies were strongly opposed by the research-based sector of the biopharmaceutical industry. During the public consultation conducted by the EMA, industry representatives argued that disclosure of clinical data—even for non-commercial research purposes—impedes their innovation incentives.[148] They invoked inter alia trade secrets protection[149] and protection obligations under

[142] EMA publication policy 0070, p. 4 (emphasis added).

[143] Ibid (emphasis added).

[144] Ibid p. 7. See also EMA (2019) Questions and answers on the European Medicines Agency policy on publication of clinical data for medicinal products for human use. EMA/357536/2014, Rev. 2, p. 4. https://www.ema.europa.eu/en/documents/report/questions-answers-european-medicines-agency-policy-publication-clinical-data-medicinal-products_en.pdf. Accessed 26 Mar 2021. At the time of writing, no official information is available regarding when the second phase of EMA publication policy 0070 will commence.

[145] The model terms preclude using the accessed data 'to support an application to obtain a marketing authorisation and any extensions or variations thereof for a product anywhere in the world'. See EMA publication policy 0070, annex 1, para 3; annex 2, para 3. See also EMA. Clinical data available. https://clinicaldata.ema.europa.eu/web/cdp/background. Accessed 26 Mar 2021.

[146] Case C-513/16 *EMA v PTC Therapeutics International* [2017] ECLI:EU:C:2017:148, paras 118–120.

[147] Such option is available for the registered users under the terms of use for academic and other non-commercial research purposes. See EMA publication policy 0070, annex 2, para 3.

[148] See e.g. EMA (2 Oct 2014) Overview of comments received on 'Publication and access to clinical-trial data' (EMA/240810/2013). EMA/349245/2014, pp. 23, 25–26 (citing the submission by the German Association of Research-Based Pharmaceutical Companies); ibid pp. 70, 74 (citing the submission by the Danish Association of the Pharmaceutical Industry); EMA (2 Oct 2014) Overview of comments received on 'Publication and access to clinical-trial data' (EMA/240810/2013). EMA/342115/2014, pp. 22, 31 (citing the submission by the EFPIA); ibid p. 12 (citing the submission by the BioIndustry Association); EMA (2 Oct 2014) Overview of comments received on 'Publication and access to clinical-trial data' (EMA/240810/2013). EMA/344107/2014, p. 28 (citing the submission by the EuropaBio). For an overview of all submissions, see EMA (2 Oct 2014) Outcome of public consultation on 'Policy 0070 on publication and access to clinical-trial data' EMA/34238/2014. https://www.ema.europa.eu/en/documents/comments/comments-received-policy-0070-publication-access-clinical-trial-data_en.pdf. Accessed 26 Mar 2021.

[149] See e.g. EMA (2 Oct 2014) Overview of comments received on 'Publication and access to clinical-trial data' (EMA/240810/2013). EMA/349245/2014, pp. 10–13; EMA (2 Oct 2014)

the TRIPS Agreement regarding undisclosed information.[150] Notably, while pharmaceutical companies ardently opposed the EMA transparency policies,[151] they have been the most active users of the system.[152]

2.3.1.3 Evolving Case Law of the CJEU on Clinical Trial Data Disclosure

Since the adoption of the EMA access policy 0043, several pharmaceutical companies have objected to the EMA's decision to grant third-party access to CSRs submitted to the EMA for drug marketing authorisation. This case law was analysed elsewhere,[153] and the overall tendency can be summarised as follows. In earlier cases, the General Court ordered interim measures suspending the EMA's decisions 'to prevent serious and irreparable harm to the applicant's interests'.[154] Some decisions were later appealed,[155] but the proceedings were discontinued due to the change of circumstances.[156] In later disputes,[157] the CJEU ruled on merits and clarified several important substantive issues, including the applicability of trade secret protection to CSRs[158] and the existence of the general presumption of confidentiality for CSRs.[159] The overall implication of this case law, as it stands at the time of writing, is that the release of data held by the EMA is subject to the case-by-case and element-by-element assessment as to whether certain content might fall

Overview of comments received on 'Publication and access to clinical-trial data' (EMA/240810/2013). EMA/344107/2014, pp. 27–28 (citing arguments of the German Association of Research-Based Pharmaceutical Companies and the EuropaBio).

[150]In particular, such claim was raised by Pfizer. See EMA (2 Oct 2014) Overview of comments received on 'Publication and access to clinical-trial data' (EMA/240810/2013). EMA/344107/2014, p. 86.

[151]For an overview of arguments, see Chap. 5 at Sect. 5.1.

[152]See annex A to this study.

[153]Kim (2017b).

[154]Case T-235/15R *Pari Pharma v EMA* [2015] ECLI:EU:T:2015:587, para 58 (also emphasising 'that weighing up of interests, to be carried out by the Court adjudicating on the substance of the case, should not be confused with that carried out for the purpose of the present interim proceedings'). See also T-718/15 *PTC Therapeutics International v EMA* [2016] ECLI:EU:T:2016:425, para 122; Case T-44/13R *AbbVie v EMA* [2013] ECLI:EU:T:2013:221, para 48; Case T-73/13R *InterMune v EMA* [2013] ECLI:EU:T:2013:222, para 37.

[155]See e.g. Case C-406/16P(R) *EMA v Pari Pharma* [2016] ECLI:EU:C:2016:775.

[156]Pari Pharma and Novartis concluded an agreement allowing the latter to use the reports at issue for the purposes related to the court proceedings.

[157]Case T-718/15 *PTC Therapeutics International v EMA* [2018] ECLI:EU:T:2018:66; Case T-33/17 *Amicus Therapeutics v EMA* [2018] ECLI:EU:T:2018:595; Case T-235/15 *Pari Pharma v EMA* [2018] ECLI:EU:T:2018:65.

[158]See Chap. 4 at Sect. 4.2.2.1.

[159]See Chap. 4 at (ii).

within the exceptions provided under the EU Transparency Regulation, or meet the criteria of trade secrets protection.

Numerous disputes initiated by pharmaceutical companies challenging the EMA's transparency policies might cast doubt on its effectiveness and raise the question of whether the current transparency-based approach to non-summary clinical trial data accessibility provides a workable solution for the pursued policy objectives.[160] As argued by the EMA, 'the systematic suspension of any decision authorising the disclosure of documents to third parties [due to] interim relief [...] raises *a question of principle* as regards the effective application of Regulation No 1049/2001'.[161]

2.3.2 Policy Approaches in Other Jurisdictions

To date, the EMA remains 'the only regulator in the world that is routinely releasing original clinical trial data'.[162] In light of the CJEU case law related to the EMA's disclosure of clinical study reports, one could doubt whether the EU Member States would readily follow the EMA's lead. According to a 2015 study,[163] drug regulatory authorities in Germany, Italy, Spain, Sweden and the UK, while being generally committed to ensuring transparency, were neither actually releasing non-summary clinical trial data submitted to regulatory bodies nor intended to implement publication policies comparable to the one adopted by the EMA.[164]

Upon an inquiry conducted in 2012–2013, the UK Science and Technology Committee of the House of Commons was 'not in favour of placing anonymised individual patient-level data in the public domain in an *unrestricted* manner as [...] the risk to patient confidentiality is too great'.[165] It recommended that, besides patients' privacy protection, transparency measures should account for 'any intellectual property contained within clinical trial data and respect commercial sensitivities',[166] as well as 'mitigate the risk that clinical trial data would be re-analysed in an

---

[160] For the analysis, see Chap. 6 at Sect. 6.5.

[161] Case T-235/15R *Pari Pharma v EMA* ECLI:EU:T:2016:309, para 19 (emphasis added).

[162] Doshi and Jefferson (2016).

[163] Pugatch Consilium (2015), pp. 19–21. Another report concludes that the EMA's initiative 'is a stark contrast and break from preceding EMA practices' and 'in a broader context [...] also contrasts starkly with existing international practices'. Pugatch Consilium (2014) Heading in a different direction? The European Medicines Agency's policy on the public release of clinical trials data, pp. 15, 27. http://www.theglobalipcenter.com/wp-content/uploads/2014/05/EMA-Study-COMPLETE.pdf. Accessed 26 Mar 2021. See also Lemmens and Vacaflor (2018) (observing that, even though 'data sharing initiatives have been undertaken in several countries, the transparency of pharmaceutical clinical trials data has been hindered by its characterisation as commercially confidential information').

[164] Ibid p. 44.

[165] House of Commons (2013), p. 45 (emphasis added).

[166] Ibid p. 31.

inexpert or *irresponsible* way, potentially leading to regulatory decisions being undermined and *misleading* conclusions reaching the public domain'.[167] At the same time, the Committee did acknowledge that the scientific value of IPD is 'currently underutilised'[168] and that broader access can improve clinical decision-making, increase public trust in research and enhance scientific knowledge.[169]

In 2013, the US Food and Drug Administration (USFDA) issued a request for comments[170] proposing a system of proactive release of raw, de-identified and masked IPD.[171] As acknowledged by the agency, safety and effectiveness datasets submitted for drug marketing authorisation have 'tremendous potential to help address critical challenges and provide new opportunities for innovation in medical product development, including for human drugs, medical devices, and biological products'.[172] However, the pharmaceutical industry strongly opposed the proposal,[173] and the USFDA did not proceed further.[174]

In 2018, Health Canada launched a public consultation on the public release of clinical trial data.[175] Concerning disclosure of patient-level data, the authority stated that, due to high costs of data de-identification and the effects of de-identification on 'overall data utility, individual patient records [would] not be released proactively as part of th[e] initiative'.[176] Notably, the document explicitly stated that individual efficacy response data does not constitute confidential business information.[177]

In 2015, the World Health Organisation (WHO), one of the leading promoters of mandatory registration of trials and results reporting, updated its official statement on public disclosure of clinical trial results, based upon the public consultation

[167] Ibid (emphasis added).

[168] Ibid.

[169] Ibid.

[170] USFDA (4 June 2013) Availability of masked and de-identified non-summary safety and efficacy data—request for comments. Fed. Reg. 78(107). https://www.govinfo.gov/content/pkg/FR-2013-06-04/pdf/2013-13083.pdf. Accessed 26 Mar 2021.

[171] By 'masked data', the USFDA refers to 'data with information removed that could link it to a specific product or application'. ibid p. 33422.

[172] Ibid. Furthermore, the Agency contemplated the approaches to providing access to data, given its research value, to non-FDA experts and other interested parties 'in a way that would both safeguard the privacy interests of patients enrolled in clinical trials, and appropriately protect the commercial investments of sponsors'. ibid.

[173] See e.g. PhRMA (24 Jul 2013) Comments to Docket No. FDA-2013-N-0271: Availability of masked and de-identified non-summary safety and efficacy data. Request for comments. https://www.regulations.gov/document?D=FDA-2013-N-0271-0003. Accessed 26 Mar 2021.

[174] Westergren (2016), p. 909.

[175] Health Canada. Draft Guidance Document, Public Release of Clinical Information. 10 Apr 2018. https://www.canada.ca/content/dam/hc-sc/documents/programs/consultation-public-release-clinical-information-drug-submissions-medical-device-applications/draft-guide-public-release-clinical-information.pdf. Accessed 26 Mar 2021.

[176] Ibid p. 10.

[177] Ibid p. 25.

conducted in 2014-2015. Notably, the WHO took a somewhat cautious position concerning clinical trial data disclosure and did not call to share 'primary data',[178] even though it generally recognised that greater access to primary datasets could facilitate research and supported sharing 'health research datasets whenever appropriate'.[179]

2.3.3 *Academic Discourse*

2.3.3.1 General Medical Literature

Publications in general medical journals provide extensive arguments why accessibility of all clinical trial data is indispensable and should be a norm. Among the most comprehensive studies, two reports should be particularly mentioned: 'Maximizing Benefits, Minimizing Risks'[180] and 'Clinical Trial Data as the Basic Staple'[181] published by the US National Academy of Medicine[182] in 2015 and 2010, respectively. They explore the role of access to clinical trial data in healthcare and provide in-depth coverage of the related scientific, policy and regulatory issues.

The 2017 review by Hoffmann et al. identified 75 articles on clinical trial data sharing published between 2008 and 2017 in the leading general medical journals.[183] Notably, even though many of these publications mention various methods of promoting transparency in trials (e.g. trial registration, publication of summary results, disclosure of trial protocols, case report forms, methodology and other relevant background documentation), they focus mainly on IPD sharing.[184] Key healthcare benefits of accessibility of patient-level data running as a leitmotif through this literature include improved reproducibility of research, enhanced transparency in regulatory decision-making, the advancement of scientific knowledge and the elimination of duplicative research.[185]

More recently, the focus of the debate on access to clinical trial data has shifted from calls for IPD disclosure towards the concept of responsible[186] and

---

[178] Ibid.

[179] World Health Organization. WHO statement on public disclosure of clinical trial results. https://www.who.int/ictrp/results/WHO_Statement_results_reporting_clinical_trials.pdf?ua=1. Accessed 26 Mar 2021.

[180] Institute of Medicine of the National Academies (2015).

[181] Grossmann et al. (2010).

[182] Formerly, the Institute of Medicine of the National Academies.

[183] Hoffmann et al. (2017).

[184] Ibid.

[185] Chapter 3 provides examples that illustrate this proposition.

[186] Institute of Medicine of the National Academies (2014), pp. 8–16. See generally Berlin et al. (2014); Mello et al. (2013); Laine et al. (2007); Ross et al. (2012).

meaningful[187] data-sharing. The overall idea is that data-sharing policies and practices should accommodate concerns regarding the protection of personal data of trial participants and the protection of commercial interests of trial sponsors.[188] Besides, to enable secondary data analysis, recommendations were proposed concerning data standards and formats.[189]

Insights drawn from this body of literature further inform and illustrate the analysis of legal issues throughout this study.

2.3.3.2 Legal Discourse on Access to Clinical Trial Data

The review of legal scholarship on access to clinical trial data in the EU context identified several monographs and collections that inter alita provide commentary on the regulatory framework applicable to clinical trials as a subset of the Union *acquis* in the healthcare sector.[190] Overall, access to clinical trial data tends to be framed within research ethics, personal data protection and transparency.[191]

The applicable regulatory framework in the EU was criticised for its 'limited success in promoting transparency'.[192] The EU Clinical Trials Directive[193] did not address transparency concerns arising from pharmaceutical companies' factual control over clinical trial data allowing them to 'manage information flow' and restrain the independent assessment of trial data.[194] Furthermore, the EU Clinical Trials Regulation repealing the Clinical Trials Directive took only 'a modest step towards transparency',[195] lacking mandatory obligations to share primary research data and publish data from unsuccessful clinical trials.[196]

Access to clinical trial data appears to be more rigorously debated on the other side of the Atlantic. In 1980, McGarity and Shapiro raised the question regarding social welfare implications of confidential treatment of pharmaceutical test data.

---

[187] See e.g. Lusher et al. (2014), pp. 859, 860, 862.

[188] See e.g. Berlin et al. (2014); Piwowar et al. (2008).

[189] See e.g. Fletcher et al. (2013); Sudlow et al. (2016), pp. 32–33.

[190] de Ruijter (2019); Hervey et al. (2017); Stanton et al. (2016); Herring (2016); Jackson (2016); Hervey and McHale (2015); Flear (2015); Joly and Knoppers (2015); den Exter and Hervey (2012); den Exter and Földes (2014); van de Gronden et al. (2011); Hervey and McHale (2004).

[191] See e.g. Choy (2014), pp. 20–28; Hervey and McHale (2015), p. 319 ff.

[192] Hervey and McHale (2004), p. 430.

[193] Directive 2001/20/EC of the European Parliament and of the Council of 4 April 2001 on the approximation of the laws, regulations and administrative provisions of the Member States relating to the implementation of good clinical practice in the conduct of clinical trials on medicinal products for human use (4 Apr 2001) OJ L 121/34.

[194] Hervey and McHale (2015), p. 319.

[195] Ibid.

[196] Ibid p. 320.

They defined social costs of data disclosure as reduced incentives for drug R&D (due to originator companies' inability to recover the costs of generating test data) and the social cost of data confidentiality as a potential to 'hamper scientific progress, deny consumers the opportunity to make fully informed product use decisions, increase the risks that agency decisions based on faulty data or analysis will remain undiscovered, and encourage potentially hazardous duplicative human testing'.[197] Already at that time, the authors argued that innovation incentives of drug companies should be protected 'by means *other than nondisclosure* [of test data]'.[198] As a solution, the authors proposed 'to couple disclosure with generic "exclusive use periods" which guarantee a data submitter that no one else can use its data to register a product for a specific number of years'.[199]

The idea of test data exclusivity was supported by the US economists Casey, Marthinsen and Moss,[200] who viewed confidentiality of test data submitted for the drug marketing authorisation as 'an extremely important issue to the sponsor's return on investment in nonpatentable know-how' and argued that protection against disclosure is 'the secretive imperatives of the innovation process'.[201] Conversely, several scholars endorse the idea of treating clinical trial data as a public good,[202] emphasising that the non-excludable and non-rivalrous nature of trial information 'facilitates the proper functioning of a healthcare market'.[203] The legal discourse produced several legislative and policy proposals outlined below.

Removing Financial Ties with the Industry

Scholars often allude to the conflict of interest associated with the commercial sponsorship of trials that can compromise the integrity and reproducibility of research outcomes.[204] To address such concerns, Reichman suggested that

---

[197] McGarity and Shapiro (1980), p. 837.

[198] Ibid (emphasis added).

[199] Ibid p. 839. It should be noted that the proposal of McGarity and Shapiro dates back to 1980, i.e. before the Hatch-Waxman Act was adopted in 1984 and introduced the statutory protection for test data in the US.

[200] Casey et al. (1983).

[201] Ibid p. 201.

[202] See e.g. Reichman (2009), pp. 49–52; Lemmens and Telfer (2012), p. 89 ff. See generally Taubman (2008); Lewis et al. (2007).

[203] Reinhardt (2004).

[204] See e.g. Lemmens (2004), p. 652 (arguing that research commercialisation has affected 'all levels of the production, review and use of medical knowledge'); McGarity and Shapiro (1980), p. 840 ff; Reichman (2009), p. 50 (highlighting the issues of 'study design biases, and other questionable practices'); Shapiro (1978), p. 156 (pointing out that the system that protects data confidentiality 'contains an inherent bias that adversely affects the accuracy and acceptability of drug research'). Further, he observes that, notwithstanding 'the sponsor's financial interest in deriving precise information from clinical tests and the professionalism of researchers involved in

'[a] better alternative to calls for mandatory disclosure is to *remove the direct link* between the test sponsor (the drug company) and the drug testers'.[205] In particular, he proposed establishing an independent testing agency that would act as an intermediary between the industry and researchers and manage the funds collected from the pharmaceutical industry by awarding testing contracts to qualified scientists.[206] Accordingly, such 'separation of clinical trials from sponsorship could attenuate the conflict of interest problem, and it would better ensure objective processing with full disclosure of results [. . .]'.[207]

The idea of removing the linkage between the industry and clinical trials is not novel. In a 1978 article,[208] Shapiro suggested establishing a testing agency that would verify the results submitted by drug companies or 'replace industry testing altogether',[209] whereby the USFDA would have access to trial results provided by 'persons with no ties to the sponsor'.[210] He argued that, under such a system, trial sponsors would have an incentive 'to avoid deliberate distortions of data and [. . .] to minimize any distortions arising from possible bias'.[211]

Disclosure as a *quid pro quo* for Data Exclusivity Protection

While confidentiality of test data could be justified as a market barrier that protects innovation incentives of originator companies by delaying price competition,[212] its necessity in the presence of test data exclusivity protection has been questioned. In particular, Eisenberg argued that confidential treatment of test data 'has outlived its original justification'[213] and that regulatory exclusivity is 'a better way to protect

testing, the possibility remains that data submitted to the agency will be at least unconsciously biased and, at worse, fraudulent' ibid p. 181.

205 Reichman (2009), pp. 50–51 (emphasis added).

206 Ibid p. 51 (emphasis added).

207 Ibid.

208 Shapiro (1978).

209 Ibid pp. 175–176.

210 Ibid.

211 Ibid p. 176.

212 Casey et al. (1983), p. 201 (observing that 'investments [into R&D] will not be made unless sufficient entry *barriers* exist in the market to permit the capturing of returns on innovations', and that 'innovators will be unable to extract the necessary revenues from the sale of new products if the specialized knowledge and insight gained in the innovational process are disseminated or divulged immediately to competitors' (emphasis added)). See also Eisenberg (2011), p. 467 (explaining that prior to the adoption of the Hatch-Waxman Act, 'drug developers relied upon confidential treatment of the data they submitted to the Food and Drug Administration (FDA) in support of a New Drug Application (NDA) to keep their data out of the hands of generic competitors who might otherwise use it to get competing versions of the same products approved').

213 Eisenberg (2011), p. 491.

innovators from unfair competitive use of their data without the need for secrecy'.[214] Along the same lines, Ünlü argued that 'a simple and direct Congressional mandate for full disclosure is the next best remedy'[215] if the USFDA cannot change its confidentiality policy. He argued that a proper balance could be achieved if 'data exclusivity would prevent competitors and generic manufacturers from using data access as a means to gain regulatory approval while allowing research data to be mined for maximum benefit'.[216]

Amending the Freedom of Information Legislation

In the US, third-parties access to safety and efficacy information held by the USFDA can be refused based on Exemption 4 under the FOIA regarding 'commercial or financial information [which is] privileged or confidential', provided that the party objecting to disclosure 'meets the [. . .] burden of showing a likelihood of substantial competitive injury'.[217] Drawing on an incident where timely access to clinical trial data could save a human life, Boyce[218] proposed to reconsider the application of Exemption 4 of the FOIA to clinical trial data to restore the balance between public and private interests.

2.3.3.3 Comparative and International Law Perspectives

Westergren offers a comparative overview of regulatory developments and approaches to non-summary data sharing in the EU and the US[219] Overall, she concludes that, notwithstanding certain measures towards broader clinical trial data sharing in the US, 'the FDA's program is meager compared to the EMA's new proactive reporting regime'.[220]

From an international human rights perspective, the analysis by Lemmens explores the interface between the right to life, the right to information and the right to health under the international legal framework governing drug safety and efficacy information.[221] His work integrates historical, social, cultural, regulatory,

---

[214] Ibid.

[215] Ünlü (2010), p. 545.

[216] Ibid.

[217] *Government Accountability Project v US Dept of Health and Human Services*, 691 F Supp 2d 175–176 (DDC 2010). In an earlier case, the US Supreme Court recognised the proprietary nature of efficacy and safety data and qualified its disclosure by the USFDA as a regulatory taking (*Ruckelshaus v Monsanto*, 467 US 986 (1984)).

[218] Boyce (2005).

[219] Westergren (2016).

[220] Ibid p. 909.

[221] Lemmens (2013).

and economic aspects of trials within the international human rights framework. Highlighting deficiencies in the production, distribution and use of pharmaceutical knowledge (especially in industry-sponsored trials), he calls on the WHO to 'take up the torch and reinvigorate [...] the promotion of global knowledge governance through adequate transparency measures'.[222]

Lemmens and Telfer argued that access to trial information—in particular, concerning trial registration and results—should be recognised as a fundamental component of the right to health[223] and that such approach would provide a legal basis to counter obligations under international trade agreements and national rules that impose limitations on access to data.[224]

From an international law perspective, Kim analysed the compliance of the EMA policy on the proactive publication of non-summary clinical trial data with the obligations to protect test data under the TRIPS Agreement[225] and international investment law.[226]

Overall, the existence of barriers to access to clinical trial data appears to be well-acknowledged in the legal scholarship,[227] while access measures tend to be justified as a matter of transparency and public interest in access to safety and efficacy information. Some scholars alluded to the potential of data-sharing to promote drug R&D.[228] However, to the best of my knowledge, the question of how access and usage rights in IPD as a research resource should be optimally allocated and integrated within the sector-specific regulatory framework has not been explored in depth.

[222] Ibid p. 169.

[223] Lemmens and Telfer (2012).

[224] Ibid.

[225] Kim (2015). See also de Carvalho (2018), paras 39.137 ff.

[226] Kim (2016).

[227] See e.g. Lemmens (2013), p. 177 (noting that '[m]any countries [...] have failed to implement adequate transparency measures through enforceable and stringent regulation'); Scheineson and Lynn Sykes (2005), p. 525 (stating that the USFDA 'releases summaries and descriptions of [...] trials to the public, but not the complete underlying data' after the drug is approved for marketing); Brailer (2010), p. 53 (stating that researchers 'do not have the ability to actually get information in a raw, useful, assembled analyzable format').

[228] Eisenberg (2011), p. 487 (noting that the application of 'modern bioinformatics techniques to aggregations of these data could greatly increase what is learned from the data'); Ünlü (2010), p. 537 ff; McGarity and Shapiro (1980), p. 847 (observing that '[s]uppressing scientific data can also hamper innovation by preventing researchers from becoming fully apprised of scientific findings relevant to their work').

2.4 The Present Study Against the Background of Policy and Legal Discourse

Against the background sketched in this chapter, a distinct policy dilemma concerning access to non-summary clinical trial data for research and drug development purposes can be identified. While drug sponsors claim that mandatory disclosure would negatively affect innovation incentives, the research community argue that access to IPD could enhance the stock of medical knowledge that, in turn, would support drug discovery and development. Notably, drug companies also acknowledge that secondary analysis of IPD could make R&D more efficient as one would 'avoid some of the trial-and-error development'.[229] Given these seemingly conflicting propositions, the question of how to balance the competing interests at stake calls for further analysis. Accordingly, this study aims to contribute to the policy and legal discourse by

(i) providing a comprehensive overview of the legal determinants of the trial sponsors' control over and third-party access to IPD under the EU framework (Chap. 4);
(ii) analysing implications of IPD disclosure for the protection of innovation incentives in the pharmaceutical sector under the existing regulatory instruments (Chap. 5);
(iii) elaborating an innovation-based justification for policy intervention enabling access to IPD for research purposes (Chaps. 6–8);
(iv) proposing legal provisions on access to IPD under the EU framework that could reconcile the policy objective of maximising the research potential of IPD through secondary analysis, on the one hand, and protecting innovation incentives of drug companies, on the other hand (Chap. 9).

While focusing on the implications of access to IPD for innovation, this study does not intend to belittle transparency-related concerns. However, policy issues regarding access to IPD stretch beyond what can be accommodated under the transparency framework.[230] As long as the public health system relies on the private sector in drug innovation, it needs to account for the economic interests of drug sponsors, the protection of which should ultimately serve the public interest in innovative medicines. What appears unclear is how the scope of trial sponsors' legitimate economic interests in relation to IPD should be defined and how competing interests should be balanced, given the potential of secondary analysis of IPD from past trials to facilitate subsequent drug discovery and development.

In exploring these issues, the study draws on two methodological approaches: law in social context and law and economics of innovation. The former looks at legal

[229] Case T-718/15 R *PTC Therapeutics International v EMA* [2016] ECLI:EU:T:2016:425, para 92 (emphasis added).

[230] As discussed in Chap. 4 at Sect. 4.3.3.4.

rules through a lens of social norms, practices and institutions;[231] the latter examines and validates legal rules from an economic efficiency perspective.[232] Even though it is unsettled whether a law-and-economics analysis is a method within the law-in-social-context approach,[233] the two approaches are complementary:[234] the common denominator is the policy intention to design legal rules that would improve society's well-being.

The specific socio-economic context of this study is shaped by the implications of secondary IPD analysis for a broad circle of stakeholders, including trial subjects, patients, medical practitioners, drug developers, academic researchers and the society at large. Before the task of balancing such interests can be undertaken, the next chapter presents a primer on the specific applications of IPD analysis in medical research and drug R&D.

References

Ahn H (2014) Second generation patents in pharmaceutical innovation. Nomos, Baden-Baden

Bauchner H, Golub RM, Fontanarosa PB (2016) Data sharing. An ethical and scientific imperative Mar 22–29;315(12):1237–1239. https://doi.org/10.1001/jama.2016.2420

Berlin JA et al (2014) Bumps and bridges on the road to responsible sharing of clinical trial data. Clin Trials 11(1):7–12. https://doi.org/10.1177/1740774513514497

Boyce J (2005) Disclosure of clinical trial data: why exemption 4 of the Freedom of information act should be restored. Duke Law Technol Rev 4:1–14

Brailer D (2010) Clinical data as the basic staple of health learning. In: Grossmann C et al (eds) Clinical data as the basic staple of health learning: creating and protecting a public good. National Academy of Sciences, Washington DC, pp 50–55

Casey WL, Marthinsen JE, Moss LS (1983) Entrepreneurship, productivity, and the Freedom of Information Act: protecting circumstantially relevant business information. Lexington Books, Lexington

Chalmers I, Glasziou P (2009) Avoidable waste in the production and reporting of research evidence. Lancet 374(9683):86–89. https://doi.org/10.1016/S0140-6736(09)60329-9

Choy K (2014) Ethical considerations in clinical trials: a critique of the ICH-GCP Guideline. Dev World Bioeth 14:20–28

CIOMS (2005) Management of safety information from clinical trials. Report of CIOMS working group VI. CIOMS, Geneva

CIOMS (2016) International ethical guidelines for health-related research involving humans. CIOMS, Geneva

---

[231] Cryer et al. (2011), p. 86.

[232] Ibid pp. 38, 83–84.

[233] Ibid p. 84.

[234] Polinsky and Shavell (2008), p. 20 (stating that the economic analysis of law 'seeks to identify the effects of legal rules on the behaviour of relevant actors and to determine whether these effects are socially desirable', while the social desirability of such effects is assessed within the framework of welfare economics). See also Perry-Kessaris (2014), p. 177 ff; Perry-Kessaris (2015), p. 58 (employing the term 'econo-socio-legal' as a 'reminder of the interconnectedness that characterizes the intersections between economic and legal aspects of social life').

Cockburn IM (2009) Intellectual property rights and pharmaceuticals: challenges and opportunities for economic research. In: The economics of intellectual property. Suggestions for further research in developing countries and countries with economies in transition. WIPO, Geneva, pp 150–173
Council of Europe (2012) Guide for research ethics committee members. Council of Europe
Cryer R et al (2011) Research methodologies in EU and international law. Hart, Oxford
de Carvalho NP (2018) The TRIPS regime of patents and test data, 5th edn. Wolters Kluwer, Alphen aan den Rijn
de Franceschi A, Lehmann M (2015) Data as tradable commodity and new measures for their protection. Ital Law J 1(1):51–72
de Ruijter A (2019) EU health law and policy: The expansion of EU power in public health and health care. OUP, Oxford
DeAngelis C et al (2004) Clinical trial registration: a statement from the International Committee of Medical Journal Editors. N Engl J Med 351:1250–1251. https://doi.org/10.1056/NEJMe048225
den Exter A, Földes MÉ (2014) Casebook on European Union health law. Maklu, Apeldoorn, Antwerpen
den Exter A, Hervey TK (2012) European Union health law: treaties and legislation. Maklu Apeldoorn, Antwerpen
Dias S et al (2018) Network meta-analysis for decision-making. John Wiley & Sons, Hoboken
DiMasi JA, Hansen RW, Grabowski HG (2003) The price of innovation: new estimates of drug development costs. J Health Econ 22(2):151–185. https://doi.org/10.1016/S0167-6296(02)00126-1
DiMasi JA, Grabowski HG, Hansen RW (2016) Innovation in the pharmaceutical industry: new estimates of R&D costs. J Health Econ 47:20–33. https://doi.org/10.1016/j.jhealeco.2016.01.012
Doppelfeld E (2009) Aufgaben und Arbeitsweise medizinischer Ethik-Kommissionen. Bundesgesundheitsbl 52:387–393. https://doi.org/10.1007/s00103-009-0824-6
Doshi P, Jefferson T (2013) Clinical study reports of randomised controlled trials: an exploratory review of previously confidential industry reports. BMJ Open 2013;3:e002496. https://doi.org/10.1136/bmjopen-2012-002496
Doshi P, Jefferson T (2016) Open data 5 years on: a case series of 12 freedom of information requests for regulatory data to the European medicines agency. Trials 17:78. https://doi.org/10.1186/s13063-016-1194-7
Doshi P et al (2013) Restoring invisible and abandoned trials: a call for people to publish the findings. BMJ 346:f2865. https://doi.org/10.1136/bmj.f2865
Dosi G, Mazzucato M (2006) Introduction. In: Dosi G, Mazzucato M (eds) Knowledge accumulation and industry evolution. The case of pharma-biotech. CUP, Cambridge, pp 1–18
Drexl J (2017a) Designing competitive markets for industrial data – between propertisation and access. JIPITEC 8(4):257–292
Drexl J (2017b) On the future EU legal framework for the digital economy: a competition-based response to the 'ownership and access' debate. In: Lohsse S, Schulze R, Staudenmayer D (eds) Trading data in the digital economy: legal concepts and tools. Nomos and Hart, Baden-Baden and Oxford, pp 223–243
Drexl J (2018) Data access and control in the era of connected devices - study on behalf of the European Consumer Organisation BEUC. BEUC, Brussels
Eisenberg RS (2011) Secrecy in the age of regulatory exclusivity. In: Dreyfuss RC, Strandburg KJ (eds) The law and theory of trade secrecy. Edward Elgar Publishing, pp 467–491
Flear ML (2015) Governing public health: EU law, regulation and biopolitics. Hart, Oxford, Portland
Fletcher C et al (2013) European federation of statisticians in the pharmaceutical industry's position on access to clinical trial data. Pharm Statistics 12(6):333–336. https://doi.org/10.1002/pst.1603
Friedman LM et al (2015) Fundamentals of clinical trials, 4th edn. Springer, Heidelberg

Goldacre B et al (2017) Pharmaceutical companies' policies on access to trial data, results, and methods: audit study. BMJ 358:j3334. https://doi.org/10.1136/bmj.j3334
Goldacre B et al (2018) Compliance with requirement to report results on the EU clinical trials register: cohort study and web resource. BMJ 362:k3218. https://doi.org/10.1136/bmj.k3218
Grossmann C et al (2010) Clinical data as the basic staple of health learning: creating and protecting a public good. National Academy of Sciences, Washington DC
Herring J (2016) Medical law and ethics, 6th edn. OUP, Oxford
Hervey TK, McHale JV (2004) Health law and the European Union. CUP, Cambridge
Hervey TK, McHale JV (2015) European Union health law: themes and implications. CUP, Cambridge
Hervey TK, Young CA, Bishop LE (2017) Research handbook on EU health law and policy. Edward Elgar Publishing, Cheltenham
Hoffmann T et al (2017) Focus on sharing individual patient data distracts from other ways of improving trial transparency. BMJ 357:j2782. https://doi.org/10.1136/bmj.j2782
House of Commons, Science and Technology Committee (2013) Clinical trials – HC 104. Third report of session 2013-14, vol 1. The Stationary Office, London
Institute of Medicine of the National Academies (2014) Discussion framework for clinical trial data sharing. Guiding principles, elements, and activities. The National Academic Press, Washington DC
Institute of Medicine of the National Academies (2015) Sharing clinical trial data: maximizing benefits, minimizing risk. The National Academies Press, Washington DC
Jackson E (2016) Medical law: text, cases, and materials, 4th edn. OUP, Oxford
Joly Y, Knoppers BM (2015) Routledge handbook of medical law and ethics. Routledge, Abingdon, New York
Jones AP et al (2013) The use of systematic reviews in the planning, design and conduct of randomised trials: a retrospective cohort of NIHR HTA funded trials. BMC Med Res Methodol 13:50. https://doi.org/10.1186/1471-2288-13-50
Jones CW et al (2011) Non-publication of large randomized clinical trials: cross sectional analysis. BMJ 347:f6104. https://doi.org/10.1136/bmj.f6104
Killeen S et al (2014) Registration rates, adequacy of registration, and a comparison of registered and published primary outcomes in randomized controlled trials published in surgery journals. Ann Surg 259(1):193–196. https://doi.org/10.1097/SLA.0b013e318299d00b
Kim D (2015) Enabling access to clinical trial data: when is unfair use fair? Chicago-Kent J Intellect Prop Law 14(2):521–552
Kim D (2016) Protecting trade secrets under international investment law: what secrets investors should not tell states. J Marshall Rev Intell Prop Law 15(2):999–1027
Kim D (2017a) No one's ownership as the status quo and a possible way forward: a note on the public consultation on building a European Data Economy. GRUR Int 66(8/9):697–705
Kim D (2017b) Transparency policies of the European medicines agency: has the paradigm shifted? Med Law Rev 25(3):456–483
Krumholz HM et al (2014) Sea change in open science and data sharing. Leadership by industry Industry. Circ Cardiovasc Qual Outcomes 7:499–504. https://doi.org/10.1161/CIRCOUTCOMES.114.001166
Laine C et al (2007) Reproducible research: moving toward research the public can really trust. Ann Int Med 146(6):450–453. https://doi.org/10.7326/0003-4819-146-6-200703200-00154
Lemmens T (2004) Leopards in the temple: restoring scientific integrity to the commercialized research scene. J Law Med Ethics 32(4):641–657. https://doi.org/10.1111/j.1748-720X.2004.tb01969.x
Lemmens T (2013) Pharmaceutical knowledge governance: a human rights perspective. J Law Med Ethics 41(1):163–184
Lemmens T, Telfer C (2012) Access to information and the right to health: the human rights case for clinical trials transparency. Am J Law Med 38:63–112

Lemmens T, Vacaflor CH (2018) Clinical trial transparency in the Americas: the need to coordinate regulatory spheres. BMJ 362:k2493. https://doi.org/10.1136/bmj.k2493
Lewis TR, Reichman JH, So AD (2007) The case for public funding and public oversight of clinical trials. Econ Voice 4(1):1–4
Light DW, Warburton RN (2005) Extraordinary claims require extraordinary evidence. J Health Econ 24(5):1030–1033. https://doi.org/10.1016/j.jhealeco.2005.07.001
Lo B, Field MJ (2009) Conflict of interest in medical research, education, and practice. National Academies Press, Washington DC
Lusher SJ et al (2014) Data-driven medicinal chemistry in the era of big data. Drug Discov Today 19(7):859–868. https://doi.org/10.1016/j.drudis.2013.12.004
Macleod MR et al (2014) Biomedical research: increasing value, reducing waste. Lancet 383 (9912):101–104. https://doi.org/10.1016/S0140-6736(13)62329-6
McGarity TO, Shapiro SA (1980) The trade secret status of health and safety testing information: reforming agency disclosure policies. Harv Law Rev 93(5):837–888
Mello MM et al (2013) Preparing for responsible sharing of clinical trial data. N Engl J Med 369 (17):1651–1658. https://doi.org/10.1056/NEJMhle1309073
Naci H, Cooper J, Mossialos E (2015) Timely publication and sharing of trial data: opportunities and challenges for comparative effectiveness research in cardiovascular disease. Eur Heart J Quality Care Clin Outcomes 1(2):58–65. https://doi.org/10.1093/ehjqcco/qcv012
Odutayo A et al (2017) Association between trial registration and positive study findings: cross sectional study (epidemiological study of randomized trials – ESORT). BMJ 356:j917. https://doi.org/10.1136/bmj.j917
OECD (2015) Data-driven innovation: big data for growth and well-being. OECD Publishing, Paris. https://doi.org/10.1787/9789264229358-en
Pammolli F, Magazzini L, Riccaboni M (2011) The productivity crisis in pharmaceutical R&D. Nat Rev Drug Discov 10(6):428–438. https://doi.org/10.1038/nrd3405
Perry-Kessaris A (2014) The case for a visualized economic sociology of legal development. Curr Leg Probl 67(1):169–198. https://doi.org/10.1093/clp/cuu016
Perry-Kessaris A (2015) Approaching the econo-socio-legal. Ann Rev Law Soc Sci 11:57–74. https://doi.org/10.1146/annurev-lawsocsci-120814-121542
Piwowar HA et al (2008) Towards a data sharing culture: recommendations for leadership from academic health centers. PLoS Med 5(9):e183. https://doi.org/10.1371/journal.pmed.0050183
Polinsky MA, Shavell S (2008) Law, economic analysis of. In: Durlauf SN, Blume LE (eds) The new Palgrave dictionary of economics, vol 5, 3rd edn. Palgrave Macmillan, Basingstoke, pp 20–34
Pugatch Consilium (2015) Clinical trial data and disclosure policies. The European Union, Member States, and international best practices. US Chamber of Commerce, Washington DC
Raghupathi W, Raghupathi V (2014) Big data analytics in healthcare: promise and potential. Health Inf Sci Syst 2:3. https://doi.org/10.1186/2047-2501-2-3
Rathi V et al (2012) Sharing of clinical trial data among trialists: a cross sectional survey. BMJ 345: e7570. https://doi.org/10.1136/bmj.e7570
Reichman JH (2009) Rethinking the role of clinical trial data in international intellectual property law: the case for a public goods approach. Marq Intell Prop Law Rev 13(1):1–68
Reinhardt UE (2004) An information infrastructure for the pharmaceutical market. Health Affairs 23(1):107–112. https://doi.org/10.1377/hlthaff.23.1.107
Richter H, Hilty RM (2018) Die Hydra des Dateneigentums – eine methodische Betrachtung. In: Datenschutz S (ed) Dateneigentum und Datenhandel, Schriftenreihe Daten Debatten, vol 3. Erich Schmidt, Berlin, pp 241–260
Ross JS, Lehman R, Gross CP (2012) The importance of clinical trial data sharing: toward more open science. Circ Cardiovasc Qual Outcomes 5(2):238–240. https://doi.org/10.1161/CIRCOUTCOMES.112.965798
Rubio McGartland D et al (2010) Defining translational research: implications for training. Acad Med 85(3):470–475. https://doi.org/10.1097/ACM.0b013e3181ccd618

Scheineson MJ, Lynn Sykes M (2005) Major new initiatives require increased disclosure of clinical trial information. Food Drug Law J 60:525–546

Shapiro SA (1978) Divorcing profit motivation from new drug research: a consideration of proposals to provide the FDA with reliable test data. Duke Law J 155–183

Sprague S, Bhandari M (2009) Organisation and planning. In: Cox Gad S (ed) Clinical trials handbook. John Wiley & Sons, New Jersey, pp 161–184

Stanton C et al (2016) Health law: frameworks and context. CUP, Cambridge

Stoney CM, Johnson LL (2018) Design of clinical trials and studies. In: Gallin JI, Ognibene FP, Lee Johnson L (eds) Principles and practice of clinical research, 4th edn. Academic Press, London, pp 250–268

Sudlow R et al (2016) EFSPI/PSI working group on data sharing: accessing and working with pharmaceutical clinical trial patient level datasets – a primer for academic researchers. BMC Med Res Methodol 16(Suppl 1):73. https://doi.org/10.1186/s12874-016-0171-x

Taichman DB et al (2017) Data sharing statements for clinical trials: a requirement of the international committee of medical journal editors. Ann Int Med. https://doi.org/10.7326/M17-1028

Taubman A (2008) Unfair competition and the financing of public knowledge goods: the problem of test data protection. J Intell Prop Law Prac 3:591–606

Ünlü M (2010) It is time: why the FDA should start disclosing drug trial data. Mich Telecommun Technol Law Rev 16:511–545

van de Gronden JW et al (2011) Health care and EU law. T.M.C. Asser Press, Springer, The Hague

Weber RH, Thouvenin F (2018) Dateneigentum und Datenzugangsrechte – Bausteine der Informationsgesellschaft? ZSR 1:43–74

Westergren A (2016) The data liberation movement: regulation of clinical trial data sharing in the European Union and the United States. Houston J Int Law 38(3):887–912

Zarin DA, Tse T (2016) Sharing individual participant data (IPD) within the context of the trial reporting system (TRS). PLoS Med 13(1):e1001946. https://doi.org/10.1371/journal.pmed.1001946

Chapter 3
Secondary Analysis of Individual Patient-Level Clinical Trial Data: A Primer

Abstract Proponents of access to data argued that secondary analyses of clinical trial data—especially individual patient-level data—can generate knowledge far beyond original research hypotheses and the benefit-risk profile of investigational products. The chapter explores this proposition. Admittedly, such task goes beyond a legal inquiry. However, without a detailed understanding of the role of secondary IPD analysis in medical research and drug R&D, arguments—both *de lege lata* and *de lege ferenda*—regarding the applicable legal framework lack a substantive basis. The overview of potential implications of secondary IPD analysis presented here is by no means exhaustive. Instead, insights drawn from the general medical literature are systematised to inform and illustrate the subsequent legal analysis.

3.1 Clinical Trial Data

3.1.1 Definitions and General Aspects

Clinical trial data refers to measurements and characteristics gathered according to the pre-specified quantitative variables[1] that can allow trial investigators to objectively evaluate the research hypothesis regarding the effects of medical interventions

The research for this chapter is based on a literature review conducted within the databases of medical journals NEJM, JAMA, BMJ, PLoS Med, Ann Intern Med, Lancet and JAMA Intern Med. In March 2018, a workshop with researchers at the Medical University of Vienna was conducted. The author is especially thankful to Professor Franz Koenig at the Medical University of Vienna, Dr. Sarah Nevitt at the University of Liverpool and Dr. med. Arnoud Templeton at Claraspital Basel for valuable insights and clarifications provided in the course of research.

[1] ICH (5 Feb 1998) ICH Harmonised tripartite guideline. Statistical principles for clinical trials. E9, pp. 2–5. EMA (2 Mar 2016) External guidance on the implementation of the European Medicines Agency policy on the publication of clinical data for medicinal products for human use. EMA/90915/2016, p. 8.

D. Kim, *Access to Non-Summary Clinical Trial Data for Research Purposes Under EU Law*, Munich Studies on Innovation and Competition 16,
https://doi.org/10.1007/978-3-030-86778-2_3

on the state of health[2] (the *treatment effect*[3]). Such variables are also defined as *trial endpoints* (or trial outcome measures)[4] as they can indicate biological and pathogenic processes or pharmacologic responses to a therapeutic intervention. Three types of clinical endpoints are commonly distinguished:

- *primary* endpoints (primary or target variables)—measurements that can establish the *efficacy* of medical intervention;[5]
- *secondary* endpoints—measurements that are used to support the primary endpoint(s) or to define additional therapeutic effects of an intervention;[6]
- *exploratory* endpoints—measurements that can allow investigators to explore *new* research hypotheses.[7]

The main difference between primary and exploratory endpoints is that the former are *hypothesis testing* (confirmatory), while the latter are *hypothesis-generating* (exploratory).[8] Unlike primary endpoints, exploratory endpoints are optional and are usually not planned or defined prospectively.[9] However, data generated in a confirmatory trial can also be used for exploratory data analysis and lay the basis for new research hypotheses.[10]

[2]ICH (5 Feb 1998) ICH Harmonised tripartite guideline. Statistical principles for clinical trials. E9, pp. 2–5. See also CONSORT. Glossary. http://www.consort-statement.org/resources/glossary#H. Accessed 26 Mar 2021.

[3]The effect size refers to a quantitative measure that provides the statistical description of the study hypothesis and is derived from the outcome variable. Laake and Breien (2015), p. 114.

[4]See National Institutes of Health, Biomarkers Definition Working Group (2001), p. 91 (defining a clinical endpoint as 'a characteristic or variable that reflects how a patient feels, functions or survives', and a surrogate endpoint 'as a biomarker intended to act as a clinical endpoint'). For instance, endpoints used in cancer clinical trials often include survival, tumour response rate, progression-free survival and patient-reported endpoints such as quality of life. George et al. (2016), pp. 4–5.

[5]ICH (5 Feb 1998) ICH Harmonised tripartite guideline. Statistical principles for clinical trials. E9, p. 5. The main results of a clinical trial are based on the primary endpoints as the most relevant characteristics for the disease and intervention under examination. See Huque and Röhmel (2010), p. 3; Moyé (2003), p. 114.

[6]ICH (5 Feb 1998) ICH Harmonised tripartite guideline. Statistical principles for clinical trials. E9, p. 6. Secondary endpoints characterise additional benefits of the treatment under investigation. See Huque and Röhmel (2010), p. 3.

[7]Goffin (2009), pp. 13–14.

[8]Moyé (2003), p. 70. Hypotheses based on exploratory endpoints are often evaluated in the subsequent trials. ibid 156. See also Huque and Röhmel (2010), p. 3.

[9]Huque and Röhmel (2010), p. 3.

[10]ICH (5 Feb 1998) ICH Harmonised tripartite guideline. Statistical principles for clinical trials. E9, p. 4.

3.1.2 The Types of Clinical Trial Data

Legal and policy discourse regarding access to clinical trial data often refers to 'trial data' as a whole. However, it is useful to differentiate between various types of trial data—or, one can say, levels of data 'granularity',[11] as distinct policy and legal issues can arise concerning specific data types.

First of all, one should distinguish between *non-summary* and *summary*-level data. *Non-summary* or primary research data comprises individual patient-level data (IPD) and clinical study reports (CSRs).[12] IPD is recorded separately for each trial participant.[13] Such data is also called *source data* that encompasses all 'original records and certified copies of original records of clinical findings, observations, or other activities in a clinical trial necessary for the reconstruction and evaluation of the trial'.[14]

Upon the trial completion, 'raw' trial data is abstracted, coded, transcribed and 'locked' into analysable datasets.[15] The contents of a CSR include clinical overviews, the trial protocol and its amendments, sample case report form, description of the statistical methods,[16] reports on bio-pharmaceutical, pharmacokinetics and pharmacodynamics studies, efficacy and safety studies, etc.[17] A CSR is usually prepared and submitted to drug authorities to obtain drug marketing authorisation.

---

[11] Zarin and Tse (2016).

[12] EMA (21 Mar 2019) European Medicines Agency policy on publication of clinical data for medicinal products for human use. Policy/0070 [hereinafter EMA publication policy 0070], p. 3. Sometimes literature refers to CSRs as summary data. The present analysis follows the approach of the EU Clinical Trials Regulation and distinguishes between a CSR and a summary of a CSR; therefore, complete CSRs are considered non-summary data. Depending on the practice of a regulatory authority, CSRs can be treated separately from IPD. For instance, the EMA makes such distinction. See EMA publication policy 0070, p. 3. According to the International Council for Harmonisation of Technical Requirements for Pharmaceuticals for Human Use, patient data listings (including individual efficacy response data) constitute an appendix to a CSR that might or might not need to be submitted together with a CSR but should be provided upon a regulatory authority's request. ICH (30 Nov 1995) ICH Harmonised tripartite guideline. Structure and content of clinical study reports. E3, para 16.

[13] EMA (2 Mar 2016) External guidance on the implementation of the European Medicines Agency policy on the publication of clinical data for medicinal products for human use. EMA/90915/2016, p. 9.

[14] ICH (10 Jun 1996) ICH Harmonised tripartite guideline. Guideline for good clinical practice. E6, para 1.51.

[15] Institute of Medicine of the National Academies (2015), pp. 98–99; Zarin and Tse (2016).

[16] The trial protocol and statistical analysis plan are also known as 'meta-data'. Institute of Medicine of the National Academies (2015), pp. 92–93.

[17] CSRs are prepared following the structure specified in Directive 2001/83/EC, annex I, part I, module 5 that, in turn, follows the common format of the International Conference on Harmonisation of Technical Requirements for Registration of Pharmaceuticals for Human Use. See Reg 726/2004/EC, art 6. On the CSR contents particularities, see generally EMA (23 Jun 2004) Note for guidance on the inclusion of appendices to clinical study reports in marketing authorisation

Summary-level data refers to summary trial results and summaries of CSRs—both are subject to mandatory disclosure in the EU database.[18] A summary of a CSR outlines the trial background and its rationale, describes the study sample, design, type of intervention and main outcome measures, and provides the results on efficacy and safety.[19] The difference between a summary CSR and IPD is appreciable—the former usually ranges between 5 and 20 pages, while the latter can comprise thousands of pages providing a 'mostly hidden and untapped source of detailed and exhaustive data [...] well beyond the most extensive journal publications'.[20]

The main benefits and concerns associated with sharing data at various levels can be summarised as follows:

- access to summary trial results promotes general public awareness and can help detect publication biases, but it does not enable exploratory data analysis;
- access to CSRs can support transparency in regulatory decision making regarding drug marketing authorisation; however, CSRs may contain business-related information that can qualify as commercially confidential;
- access to *raw data* can support both confirmatory and exploratory data analyses; however, it can be 'overly burdensome' and put at risk personal data protection.[21]

3.1.3 The 'Life-Cycle' of Clinical Trial Data

The main stages of the clinical trial data 'life-cycle' are data acquisition, analysis, publication, archiving and storage. Data is collected throughout these phases according to the parameters predefined in a trial protocol. Such records comprise the baseline characteristics (e.g. gender, age, weight), disease factors, treatment response variables and other relevant characteristics.[22] Methods of the data acquisition include:

- electronic data capture (data collection by sensor-equipped devices);
- personal interviews;
- paper-based clinical report forms;

applications. CHMP/EWP/2998/03; ICH (30 Nov 1995) ICH Harmonised tripartite guideline. Structure and content of clinical study reports. E3.

[18] See Chap. 4 at Sect. 4.3.1.1.

[19] Reg 536/2014/EU, art 37(4) and annex IV.

[20] Doshi and Jefferson (2013). See EMA (20 Oct 2016) Opening up clinical data on new medicines. EMA provides public access to clinical reports. EMA/650519/2016 (reporting that the first CSRs related to two medicines released by the EMA on 20 October 2016 comprised 'approximately 260,000 pages of information'). See also Sudlow et al. (2016) (noting that a 'typical CSR (including all appendices) can be in excess of 1000 pages').

[21] Institute of Medicine of the National Academies (2015), p. 7 (emphasis added).

[22] Wang et al. (2009), p. 1056.

- self-administered diaries and questionnaires of study subjects;
- the transfer of data from other data systems (e.g. laboratory data).[23]

During a trial and upon its completion, clinical trial data is analysed according to the pre-specified statistical analysis plan and can be further used either for regulatory purposes or new research directed at questions and hypotheses that had not been addressed or even envisaged by the investigators of the original study.

3.2 Clinical Trial Data as a Source of Medical Knowledge

3.2.1 Clinical Trial Data as a Source of Scientific Knowledge

In science, data plays an evidentiary role. In clinical trials, ensuring the reliability and robustness of test data is the central principle[24] and the decisive factor for the trial authorisation.[25]

Clinical trials can be regarded as one of the major sources of pharmacological knowledge, in line with the definition of a clinical study as 'any investigation in relation to humans intended to *discover* or *verify* the clinical, pharmacological or other pharmacodynamic effects of one or more medicinal products'.[26] Trials generate evidence regarding the pharmacological properties of the tested substances, the molecule-target interaction and its impact on the disease progression, and the relationship between the desired and adverse effects of medical intervention.[27] Such insights play a paramount role in drug development.[28] However, it is worth emphasising that, before data can become scientific knowledge, primary ('raw') research data should be analysed, validated and systematised.[29]

A peculiar feature of primary data from clinical trials is that it can provide a valuable resource for exploratory research—that can take place in academic medical research and industry drug R&D—that can address questions far beyond the hypothesis investigated in the original study.[30]

---

[23] Trocky and Brandt (2009), p. 198.

[24] Reg 536/2014/EU, rec 1, art 3(a).

[25] Reg 536/2014/EU, rec 1, art 6(1)(b)(i) third indent.

[26] Reg 536/2014/EU, art 2 (2)(1)(a) (emphasis added).

[27] British Pharmacological Society. Pharmacology skills for drug discovery. https://thebiologist.rsb.org.uk/images/Pharmacology_Skills_for_Drug_Discovery.pdf. Accessed 26 Mar 2021.

[28] ibid.

[29] Dalrymple (2003), p. 35 (emphasising that scientific knowledge is systematic, testable and verifiable).

[30] Implications of secondary IPD analysis for cumulativeness of medical research and drug R&D are discussed more in detail in Chap. 8 at Sects. 8.2.3.1 and 8.2.3.2.

3.2.2 *The Types of Data Analyses*

3.2.2.1 Primary and Secondary Data Analyses

Primary data analysis is conducted according to the hypothesis, research questions and statistical analysis plan that are *prospectively* defined in a trial protocol.[31] When trials are carried out to obtain drug marketing authorisation, the primary analysis evaluates the benefit-risk balance of an investigational medicinal product. Secondary data analyses can be conducted by initial investigators and researchers who were not involved in the original study. Secondary data analyses can be directed at validating the conclusions of the primary analysis (*confirmatory* data analysis) and exploring research questions beyond those addressed in the original trial (*exploratory* data analysis).[32] When conducting secondary data analysis, a different methodology can be applied to verify the trial outcome, or other variables can be added to answer a different research question.[33]

Depending on the research objective, secondary data analysis can be performed by using individual datasets or aggregated data from multiple trials, summary-level data or IPD. Apart from the source data, it requires access to the trial protocol, statistical analysis plan and other meta-data[34] used in the initial trial.

3.2.2.2 Confirmatory Secondary Data Analysis

Validation of the trial results through confirmatory data analysis plays an important role in enabling the reproducibility of research.[35] Confirmatory analysis can use the same or different statistical methods.[36] At this point, it should be emphasised that, even though the present study focuses on access to data for *exploratory* purposes, the importance of confirmatory analysis cannot be underestimated. After all, for the subsequent research to build on the knowledge gained in earlier research, the validity of such knowledge should be confirmed in the first place.

[31] Institute of Medicine of the National Academies (2015), p. 24.

[32] Data from confirmatory trials can be used to generate and explore new hypotheses that would need to be confirmed in the subsequent studies. See ICH (5 Feb 1998) ICH Harmonised tripartite guideline. Statistical principles for clinical trials. E9, p. 4; Wang and Bakhai (2006), p. 4.

[33] Grady and Hearst (2007), p. 211.

[34] Koenig et al. (2015), p. 10; Institute of Medicine of the National Academies (2015), p. 104.

[35] The issue of reproducibility in clinical trials is discussed in more detailed in Chap. 6 at Sect. 6.4.2.

[36] Menikoff J (7 Mar 2017) Letter to the ICMJE secretariat. http://www.icmje.org/news-and-editorials/menikoff_icmje_questions_20170307.pdf. Accessed 26 Mar 2021.

3.2.2.3 Exploratory Secondary Analysis

Exploratory analysis is often known as hypothesis-generating as it can be carried out to generate[37] or evaluate research hypotheses not directly related to the initial trial.[38] For instance, a new hypothesis can investigate the homogeneity of response across different patient subgroups[39] or interaction patterns between treatments and health outcomes.[40] New insights can lead to new promising endpoints[41] and drug targets.[42]

For enabling meaningful exploratory analysis, data needs to be aggregated and managed as a research resource.[43] Exploratory data analysis is distinct from exploratory trials: exploratory data analysis can use data from *any* trials—not only from exploratory studies in the early phases of drug development—for purposes other than the original study. The outcomes of exploratory data analysis are probabilistic and need to be validated in the subsequent hypothesis-testing studies.[44] Notably, according to the statistics on access to data, researchers have a much higher interest in using IPD to explore new hypotheses than to validate the conclusions of the primary data analysis.[45]

3.2.2.4 Subgroup Data Analysis

A treatment effect can vary depending on the subgroup characteristics or other factors that can influence the trial outcome (covariates), e.g. it can increase or

[37] Exploratory analysis is sometimes referred to as analysis suggested by data. CONSORT. Glossary. http://www.consort-statement.org/resources/glossary#E. Accessed 26 Mar 2021.

[38] Institute of Medicine of the National Academies (2015), p. 20. See also Lauer (2010), p. 91 (noting that clinical trials 'can function as rich sources of observational data, useful for exploring questions that go beyond their original hypotheses'); Browner et al. (2007), p. 61 (referring to trial data as 'a fertile source of potential research questions for future studies').

[39] Selby et al. (2018), p. 285.

[40] Menikoff J (7 Mar 2017) Letter to the ICMJE Secretariat. http://www.icmje.org/news-and-editorials/menikoff_icmje_questions_20170307.pdf. Accessed 26 May 2021.

[41] USFDA (4 Jun 2013) Availability of masked and de-identified non-summary safety and efficacy data—request for comments. Fed. Reg. 78(107), p. 33422. https://www.govinfo.gov/content/pkg/FR-2013-06-04/pdf/2013-13083.pdf. Accessed 26 Mar 2021.

[42] Institute of Medicine of the National Academies (2015), p. 20.

[43] ibid pp. 164–167.

[44] See e.g. Selby et al. (2018), p. 285; Gustafsson et al. (2010), pp. 938–939 (discussing the example where secondary analysis of historical trial data generated two new hypotheses—one was successfully confirmed in the subsequent research and the other one could not be validated).

[45] Strom et al. (2016), p. 1608 (reporting that only three out of 177 requests for IPD submitted through the ClinicalStudyDataRequest.com portal between 7 May 2013 and 14 November 2015 were filed for confirmatory analysis purposes; the rest of the proposals were directed at new research questions).

decrease with age.[46] Therefore, it is recommended that the regular statistical analysis of clinical trial data 'should be extended by (exploratory) analysis if the existence of subgroups of patients for which the efficacy estimate differs, is suspected'.[47] Subgroup analysis refers to the *ad hoc* evaluation of the trial results in a trial population subset[48] that aims to identify the subgroup characteristics that can account for the observed difference(s) in the treatment effect and determine whether such differences can be considered homogeneous.[49]

Due to the small sample sizes, subgroup analysis usually lacks statistical power.[50] The results of the same subgroup should be compared across related studies to demonstrate consistency and homogeneity of subgroup effects.[51] For that, one needs to get access not only to the datasets from multiple trials eligible for the subgroup analysis but also to detailed data regarding the patient characteristics.[52]

The analysis of the subgroup effects can inform clinical practice (e.g. allow to tailor a treatment plan according to a particular patient category[53]) and subsequent research. Even though results of the subgroup analysis should be viewed as 'merely suggestive',[54] they can lead to new hypotheses that can be tested in the subsequent adequately powered studies.[55] The subgroup analysis is also applied in drug R&D directed at small populations, orphan diseases[56] and personalised medicine.[57]

3.2.2.5 Interaction Analysis

The term 'interaction' refers to the dependency of the treatment contrast—the difference between an investigational product and a control treatment—on third

[46] ICH (5 Feb 1998) ICH Harmonised tripartite guideline. Statistical principles for clinical trials. E9, p. 27.

[47] Cleophas et al. (2006), pp. 149–150.

[48] ICH (5 Feb 1998) ICH Harmonised tripartite guideline. Statistical principles for clinical trials. E9, pp. 26–27.

[49] Meinert (2012), p. 454.

[50] ibid.

[51] Song and Bachmann (2016).

[52] ibid.

[53] Selby et al. (2018), p. 285.

[54] Brody (2016), p. 88. See also ICH (5 Feb 1998) ICH Harmonised tripartite guideline. Statistical principles for clinical trials. E9, p. 29 (noting that the results of exploratory analysis 'should be interpreted cautiously [and] any conclusion of treatment efficacy (or lack thereof) or safety based solely on exploratory subgroup analyses are unlikely to be accepted').

[55] Biltaji et al. (2017), p. 2338; Cleophas et al. (2006), pp. 149–150 (noting that '[i]f such subgroups are identified, the exploratory nature of the regression analysis should be emphasized and the subgroup issue should be further assessed in subsequent independent and prospective data-sets').

[56] Koenig et al. (2015), p. 23.

[57] Song and Bachmann (2016).

factors.[58] By stratifying study subjects into subgroups, interaction analysis can expose how the associations between the predictor(s)[59] and the trial outcome can change depending on third factors.[60] For instance, the level of smoking can be the third factor that can modify the effect of the predictor (e.g. coffee drinking) on the outcome (e.g. myocardial infarction).[61] Therefore, interaction is also known as 'effect modification'.

Interaction analysis can be part of a subgroup analysis or meta-analysis.[62] In situations where the interactions between the variables can be anticipated, subgroup and interaction analyses are carried out as part of the confirmatory analysis. However, the biological plausibility and statistical significance of an interaction effect can be a challenge because the estimates of the associations between the predictor and outcome variables in different subgroups can diverge.[63] Thus, in most cases, interaction analysis is exploratory.[64]

3.2.2.6 Predictive Models and Prognostic Variables

According to the statistics on access to IPD through the CSDR platform, the most often indicated purpose of secondary IPD analysis is to develop predictive models.[65] Modelling disease progression can be based on the aggregated data from past studies of the same disease or indication.[66]

---

[58] ICH (5 Feb 1998) ICH Harmonised tripartite guideline. Statistical principles for clinical trials. E9, p. 34.

[59] A predictor is a variable (e.g. age, sex, smoking history, the intake of supplements) that can predict a health outcome (e.g. heart attack or quality of life). By analysing the associations between different variables predicting a health outcome, investigators can draw implications regarding causality. Cummings et al. (2007), p. 7.

[60] Newman et al. (2007b), pp. 137–138.

[61] ibid.

[62] ibid.

[63] ibid p. 138.

[64] ICH (5 Feb 1998) ICH Harmonised tripartite guideline. Statistical principles for clinical trials. E9, p. 27. See also Cleophas et al. (2006), p. 141 (noting that, 'when studying interactions, the results of the regression analysis are more valid when complemented by additional exploratory analyses within relevant subgroups of patients or within strata defined by the covariates').

[65] ibid. Other common purposes of secondary IPD analysis include meta-analysis, survival analysis and testing new analytical methods.

[66] USFDA (4 Jun 2013) Availability of masked and de-identified non-summary safety and efficacy data—request for comments. 78(107) Fed. Reg, p. 33423. https://www.govinfo.gov/content/pkg/FR-2013-06-04/pdf/2013-13083.pdf. Accessed 26 Mar 2021. See also USFDA (2011) Advancing regulatory science at FDA, p. 12 (stating that clinical trial data can be leveraged 'to develop quantitative models and measures of disease progression'). https://www.fda.gov/media/81109/download. Accessed 26 Mar 2021. Stewart and Tierney (2002), p. 89 (noting that IPD can be used in the exploratory analysis for constructing or validating prognostic indices).

IPD meta-analysis can form the basis for evaluating the prognostic effects of individual characteristics or defining particular risk groups according to multiple prognostic factors.[67] Such results can further inform the stratified randomisation in the subsequent trials.[68] Importantly, summary-level aggregate data used in the meta-analysis cannot reveal prognostic effects or the interactions between the treatment effect and individual participant characteristics.[69]

3.2.2.7 Meta-analysis and Systematic Reviews

Secondary data analysis can be carried out on data from one trial and data aggregated from multiple trials that addressed a specific issue. The latter type is known as meta-analysis and is defined as '[t]he formal evaluation of the quantitative evidence from two or more trials bearing on the same question'.[70]

Meta-analysis is often used for preparing *systematic reviews*[71] that systematise the body of knowledge regarding a particular subject[72] and are regarded as the basic principle of scientific research.[73] For instance, evidence synthesis can identify unmet needs or unresolved uncertainties ('knowledge gaps')[74] that can become a starting point of drug discovery programs.[75]

Meta-analysis can be based on the summary-level published results or IPD. IPD meta-analysis is considered to be the 'gold standard' for systematic reviews.[76] Its benefits vis-à-vis meta-analysis of summary-level data include replicating the primary analysis,[77] deriving more detailed and reliable conclusions regarding the treatment effect, enabling additional subgroup analyses,[78] assessing relative benefits and risks of treatments that have not been examined in earlier comparative

[67] Tierney et al. (2015), p. 1330.

[68] ibid.

[69] ibid. As observed by Tierney et al., 'IPD meta-analyses have played a role in the selection of participants, and in the conduct, analysis, and interpretation of trials, particularly in response to subgroup or prognostic factor analyses'. Tierney et al. (2015), p. 1332.

[70] ICH (5 Feb 1998) ICH Harmonised tripartite guideline. Statistical principles for clinical trials. E9, p. 34. See also Cooper and Patall (2009), pp. 165–166; Cochrane Methods Group. About IPD meta-analyses. https://methods.cochrane.org/ipdma/about-ipd-meta-analyses. Accessed 26 Mar 2021.

[71] Egger (1997), p. 1371.

[72] Engberg (2008), pp. 258–265; Haidich (2010), p. 29.

[73] Mulrow (1994), p. 597.

[74] Ioannidis (2004), p. 522.

[75] Tierney et al. (2015), p. 1325.

[76] Cochrane Methods Group. About IPD meta-analysis. https://methods.cochrane.org/ipdma/about-ipd-meta-analyses. Accessed 26 Mar 2021. See generally Tierney et al. (2015); Simmonds et al. (2015); Stewart and Parmar (1993).

[77] Cooper and Patall (2009), p. 167; Nevitt et al. (2017).

[78] Riley et al. (2010).

effectiveness studies,[79] etc.[80] IPD meta-analysis can identify differences in the treatment effect across the subgroups that might not be detected through a meta-analysis of the summary-level data.[81] Furthermore, IPD-based meta-analysis can play an important role in informing the design of new trials.[82]

However, conducting IPD meta-analysis is 'resource demanding, time consuming, and methodologically challenging'.[83] Furthermore, the importance of the totality of evidence should be emphasised, as conclusions and clinical recommendations can be reliable only if meta-analysis and systematic review are conducted based on *all* legitimate studies,[84] which is often a challenge in the case of both summary and non-summary clinical trial data.[85]

3.2.3 Fields of Research

Even a cursory look at the titles of articles based on secondary IPD analysis can give an idea of a broad range of medical disciplines and sub-disciplines.[86] The main fields of research and the relevance of IPD analysis are briefly outlined below.

3.2.3.1 Pharmacology

Data gathered in clinical trials is one of the major sources of pharmacological knowledge.[87] Clinical pharmacology is a field of medical research that concerned all aspects of the interaction between a human organism and a drug that can affect

---

[79] Naci et al. (2015), p. 58.

[80] Cooper and Patall (2009), p. 167.

[81] Institute of Medicine of the National Academies (2015), p. 12.

[82] Tierney et al. (2015), p. 1332 (noting that, even though IPD meta-analysis reports usually focus on providing recommendations for clinical practice, they can play 'an equally important role in informing subsequent clinical research'); Institute of Medicine of the National Academies (2015), pp. 61–62 (with further references). See also Stewart et al. (2011), p. 18:2 (definining situations where IPD analysis can be particularly important, such as conducting complex types of analyses (e.g. multivariate analysis) and exploring interactions between the intervention and patient-level characteristics).

[83] Nevitt et al. (2017).

[84] Council of Europe (2012), para 6.C.20.2.

[85] Bath and Gray (2009), p. 25 (stating that, '[u]nfortunately, if summary meta-analyses are complicated by missing trial data, this problem is magnified in analyses based on individual patient data').

[86] Clinical Study Data Request. Metrics. Published proposals. https://www.clinicalstudydatarequest.com/Metrics/Published-Proposals.aspx. Accessed 26 Mar 2021.

[87] British Pharmacological Society. Pharmacology skills for drug discovery. https://thebiologist.rsb.org.uk/images/Pharmacology_Skills_for_Drug_Discovery.pdf. Accessed 26 Mar 2021.

normal and abnormal biochemical functions.[88] Pharmacological knowledge plays a paramount role in drug R&D, guiding decision making throughout all stages of drug discovery, preclinical and clinical development.[89]

3.2.3.2 Epidemiology, Clinical Epidemiology and Pharmacoepidemiology

Epidemiology is a discipline that studies the distribution and determination of health-related states or events (including diseases) that can be applied to prevent a disease or control its progression.[90] Clinical pharmacology examines the effects of drugs in humans, while pharmacoepidemiology is a field of applied research that links the disciplines of epidemiology and clinical pharmacology by studying chronic diseases to understand the effects of drugs in a large number of people.[91]

Historically, case-control studies used to be designed as epidemiologic studies conducted to identify disease risk factors.[92] Nowadays, even though randomised clinical trials are sometimes used synonymously with experimental epidemiology, the difference between epidemiological studies and clinical trials lies in that the former are observational, while the latter involve medical intervention (e.g. administration of treatment).[93]

By definition, clinical trials are conducted to discover or verify the clinical, pharmacological or other pharmacodynamic effects of medicinal products[94] and to clarify the relationship between a disease and risk factors.[95] In epidemiology, a 'risk factor' is defined as 'a behavior, environmental exposure, or inherent human characteristic that is associated with an important health condition', while a 'cause' refers to 'a specific event, condition, or characteristic that precedes the health outcome and is necessary for its occurrence'.[96] Accordingly, clinical trial data analysis contributes to understanding the epidemiology of disease[97] and clinical pharmacology.

Furthermore, fundamental knowledge of disease determinants plays a paramount role in planning trials, especially in 'anticipating the important confounding factors

---

[88] Vallance and Smart (2006), p. 7.

[89] See e.g. Gallin et al. (2018), p. 649 ff; British Pharmacological Society. Pharmacology skills for drug discovery. https://thebiologist.rsb.org.uk/images/Pharmacology_Skills_for_Drug_Discovery.pdf. Accessed 26 Mar 2021.

[90] Merrill (2015), p. 3.

[91] Strom (2005), pp. 3–5.

[92] Newman et al. (2007a), p. 113.

[93] Meinert (2012), p. 69.

[94] Reg 536/2014/EU, art (2)(1) (emphasis added).

[95] Cummings et al. (2007), p. 21.

[96] Merrill (2015), p. 3.

[97] Institute of Medicine of the National Academies (2015), p. 31.

and background incidences of concurrent illnesses'.[98] Thus, courses in biostatistics and epidemiology form the core of training of clinical trialists in methods of design and analysis of scientific experiments.[99]

3.2.3.3 Aetiology, Pathology and Pathophysiology

As proclaimed under the Declaration of Helsinki, the principal objective of medical research in humans is 'to understand the causes, development and effects of diseases and improve preventive, diagnostic and therapeutic interventions'.[100] In the 2000 version, the provision had a slightly different wording stating that '[t]he primary purpose of medical research involving human subjects is to improve prophylactic, diagnostic and therapeutic procedures and the understanding of the *aetiology* and *pathogenesis* of disease'.[101] Aetiology is a discipline, which explains 'what sets the disease process in motion';[102] pathogenesis describes how a disease progresses.[103]

Furthermore, clinical trial data analysis can advance the understanding of the *physiology* and *pathophysiology* of disease states.[104] Physiology studies processes and mechanisms within a living organism, while pathophysiology examines structural and functional changes that occur in a living organism at a cellular and tissue level that are associated with a disease by combining knowledge of physiology and pathology.[105]

3.2.3.4 Research on Biomarkers

A biomarker ('biological marker') is defined as an objectively measurable indicator of normal physiological or pathogenic processes, disease progression or effects of a therapeutic intervention.[106] Biomarker discovery refers to the process of identifying chemical entities that are 'intimately associated [with] a biological/physiological process'.[107]

---

[98] CIOMS (2005), p. 67.

[99] Meinert (2012), pp. 424–425.

[100] Declaration of Helsinki, para (6).

[101] CIOMS (2005), p. 259 (emphasis added).

[102] Porth (2011), p. xix.

[103] ibid.

[104] Berlin et al. (2014).

[105] Braun and Anderson (2007), p. 2.

[106] National Institutes of Health, Biomarkers Definition Working Group (2001), p. 91. Biomarkers can be applied inter alia to indicate a disease prognosis or predict and monitor clinical response to medical intervention. ibid. See also Schulte and Mazzuckelli (1991), p. 239 (defining a biomarker as 'an indicator that signals events in biological systems or samples [...] generally taken to be any biochemical, genetic, or immunologic indicator that can be measured in a biological specimen').

[107] Laterza and Zhao (2016), p. 28.

IPD analysis can contribute to research on biomarkers. In particular, it can facilitate the discovery and validation of biomarkers to the extent that it can improve the understanding of a disease, its underlying mechanisms and associated factors.[108] Once identified and qualified, biomarkers play an important role in drug development,[109] especially personalised medicine.[110] Biomarkers can serve as predictive indicators of safety, efficacy and pharmacodynamic responses to treatment and, in so doing, reduce 'exposure to ineffective experimental treatments'[111] and 'improve the success rate and cost-effectiveness of rational drug development'.[112]

3.3 Exploratory Analysis of Clinical Trial Data in Drug R&D

3.3.1 'Data-Driven' Drug R&D

Clinical trial data analysis can be applied in different stages of drug discovery and development. Methods of data analysis can vary in complexity from data-mining[113] to the advanced forms of machine learning.[114] Over the past decades, several disciplines have emerged at the interface between biomedical sciences and informatics, such as clinical research informatics,[115] biomedical informatics[116] and

[108] ibid pp. 28–29.

[109] In the early stages of drug development, biomarkers can be applied, for instance, in the development of the proof of concept and safety studies. In later stages, they can aid in selecting the dose, understanding clinical efficacy in a subset of the study population or identifying risks. See Singh et al. (2016), p. 202; Wnek et al. (2016), p. 143; Ray (2016), p. 1 ff; USFDA (2011) Advancing regulatory science at FDA, p. 12. https://www.fda.gov/media/81109/download. Accessed 26 Mar 2021. On the types of clinical biomarkers and their role in drug development, see e.g. Lee (2016), p. 47.

[110] European Commission (25 Oct 2013) Use of '-omics' technologies in the development of personalised medicine. SWD(2013) 436 final, pp. 11–12; Singh et al. (2016), p. 202.

[111] Strimbu and Tavel (2010).

[112] Davis et al. (2016), p. 17 (with further references).

[113] Data-mining can be applied in identifying, selecting and prioritising potential disease target, diagnostic or prognostic markers. See Yang et al. (2009), pp. 150–151; Kilicoglu (2018).

[114] For instance, machine learning can be applied to the subgroup analysis. See e.g. Helal (2016), p. 561; Lipkovich et al. (2018).

[115] Clinical research informatics utilises methods of informatics to develop and manage medical knowledge and information. American Medical Informatics Association. Clinical research informatics. http://AMIA.org/applications-informatics/clinical-research-informatics. Accessed 26 Mar 2021.

[116] Biomedical informatics is defined as 'the interdisciplinary field that studies and pursues the effective uses of biomedical data, information, and knowledge for scientific inquiry, problem solving and decision making, motivated by efforts to improve human health'. Kulikowski et al.

translational bioinformatics.[117] The common feature shared by these disciplines is the application of computational methods to the collections of health-related data—including data from preclinical and clinical studies[118]—to discover unanticipated patterns and correlations between multiple variables.[119]

These relatively novel research methodologies complement rather than substitute the traditional hypothesis-driven approach.[120] They can be applied in all stages of drug discovery and development, including compound screening and lead selection, clinical trial design and management, prediction and detection of adverse events, as well as drug repurposing.[121] Computer models of cells, organs and systems that integrate historical data from trials can predict clinical risks and benefits of drug candidates in different patient populations.[122]

At the same time, it is important to emphasise that, without a sound methodological basis, the 'data-driven' exploratory analysis can amount to 'the "aimless" data dredging'.[123]

3.3.2 *The Application of Data Analytics in Drug Discovery*

Drug discovery starts with identifying a protein associated with a disease (a 'target') and a hypothesis that its inhibition or activation can cause a change in a disease state.[124] A potentially 'druggable' target can be found within a broad variety of moieties, including molecular entities, biological pathways and disease biomarkers.[125] A hypothesis can emerge from analysing data from diverse sources,

(2012), p. 933. Bioinformatic methods apply algorithms to analyse biological data, including DNA sequence, RNA expression and cells proteins. See Altman (2012), p. 994.

[117] Translational bioinformatics is defined as 'the development of storage, analytic, and interpretive methods to optimize the transformation of increasingly voluminous biomedical data into proactive, predictive, preventative, and participatory health'. Butte (2008), p. 709.

[118] Besides clinical trial data, data-driven drug discovery can integrate and utilise other types of health-related data such as genomic, epigenetic, transcriptomic and proteomic data. Xia (2017), p. 1709; Altman (2012), p. 994; Butte and Ito (2012), p. 949.

[119] Mayo et al. (2017) (noting that machine learning approaches 'can be used to leverage the wide range of data element categories contained in [data resource systems] to identify unanticipated interactions and dependencies that should be considered in the RCT design').

[120] Jones and Warren (2006), p. 253.

[121] Butte and Ito (2012), p. 950 (with further references).

[122] USFDA (2011) Advancing regulatory science at FDA, p. 9. https://www.fda.gov/media/81109/download. Accessed 26 Mar 2021.

[123] Biltaji et al. (2017), p. 2337; Elwood (2017), p. 254.

[124] Hughes et al. (2011), p. 1239.

[125] Yang et al. (2009), p. 147.

such as scientific publications, clinical records and biomedical data, including IPD from clinical trials.[126]

With the advent of 'big data' analytic techniques, leading pharmaceutical companies have been cooperating with data scientists, researchers in the field of artificial intelligence (AI) and AI-based platforms specialising in drug R&D[127] to leverage the synergies of large volumes of aggregated health-related data, advanced algorithms and computer-processing power in R&D programs, both in-house and through the partnerships.[128] Examples of industry-led data-pooling initiatives include the Project Data Sphere,[129] C-Path's Online Data Repository (CODR) for Alzheimer's disease, ITN Trial Share and the Pooled Resource Open-access ALS Clinical Trials (PRO-ACT). In the words of Sanofi's Chief Medical Officer in North America (the co-organiser of the Project Data Sphere), '[c]linical trial data sets can hold hidden gems of insight into disease and trial design',[130] and pooling trial results can 'increase the odds of finding these gems'.[131] Notably, such cooperative initiatives tend to be implemented in the therapeutic areas where uncertainty and risk of failure are particularly high.[132]

Various forms of machine learning have been integrated into different stages of drug R&D, promising to 'turn the drug-discovery paradigm upside down by using patient-driven biology and data to derive more-predictive hypotheses rather than the traditional trial-and-error approach'.[133] Even though the AI-based methods 'have yet to bear fruit in terms of drugs being progressed to market',[134] they have been

[126]This method is also known as a 'system approach' to drug discovery, whereby potential targets are selected by studying a disease using data gathered in earlier clinical trials. Yang et al. (2009), p. 147.

[127]Examples include the collaborative projects implemented by Pfizer and IBM Watson, Sanofi and Exscientia, Genentech and GNS Healthcare.

[128]Examples include Sanofi's 'Translational Medicines for Patients' program, Novartis' Institute for Biomedical research and Pfizer's 'Precision Medicine Analytics Ecosystem' initiative.

[129]The platform provides access to trial protocols, case report forms and comparator arm data from cancer clinical trials that can be used as a quasi-comparator arm in future studies or in the decease progression models. See Project Data Sphere. https://www.projectdatasphere.org/projectdatasphere/html/about. Accessed 26 Mar 2021.

[130]Pharma firms pool and share cancer trial data (2014).

[131]ibid.

[132]Critical Path Institute (10 Jul 2013) U.S. Food and Drug Administration and European Medicines Agency reach landmark decisions on Critical Path Institute's clinical trial simulation tool for Alzheimer's Disease. https://c-path.org/wp-content/uploads/2014/03/US-FDA-EMA-agency-reach-landmark-decisions-C-Path-clinical-trial-simulation-tool-for-alzheimers-disease.pdf. Accessed 26 Mar 2021.

[133]Fleming (2018), p. 56.

[134]Sellwood et al. (2018), p. 2027. Notably, AI-based methods have been used in medicinal chemistry since the 1960s. ibid 2025. See also WHO, IUPHAR, CIOMS (2012), p. 33 (concluding that 'high expectations of innovation models that involve combinatorial chemistry, high-throughput screening, rational drug design, pharmacogenomics, bioinformatics and disease and pathway modelling have not been met despite the high level of investment').

increasingly applied in designing new chemical structures, predicting the molecular property profile, synthesising compounds, etc.[135]

3.3.3 *Secondary IPD Analysis in Planning and Design of New Trials*

There are numerous ways of how secondary data analysis can benefit the decision-making during the planning and implementation of new studies.[136] Systematic reviews and meta-analysis have traditionally been used as an integral part of designing randomised trials,[137] especially phase III studies.[138] For some trial parameters, only IPD-based meta-analysis can provide the necessary information, which otherwise cannot be derived from the summary-level data.[139] For instance, IPD meta-analysis can reveal patient-level effect modifiers that should be controlled for in the subsequent studies,[140] establish the consistency of the treatment benefits across the related trial outcomes, provide with 'extra reassurance that a certain intervention should be used'[141] and a better understanding of the mode of action.[142] The study by Tierney et al. shows that IPD meta-analysis can be particularly useful for choosing the comparator, frequency and dose of treatment, the recruitment of trial participants, the sample size calculation and the analysis and interpretation of the results.[143] Data from the past exploratory or pilot studies can be especially valuable as it can allow researchers to formulate reasonable assumptions, e.g. concerning the size of the treatment effect to be investigated in a trial.

Modelling and simulation are often employed in drug R&D to configure how a study should be designed to answer the research question of interest. Computer models can integrate various types of data, including pharmacokinetic, pharmacodynamic, safety and efficacy data, to predict risk-benefit outcomes for different options of the trial design and safety in different patient populations,[144] forecast

---

[135] Sellwood et al. (2018).

[136] For an overview, see Clayton et al. (2017).

[137] Jones et al. (2013).

[138] Cleophas et al. (2006), p. 205.

[139] Tierney et al. (2015), p. 1331.

[140] Sutton et al. (2007), p. 2496 (pointing out that, 'if patient level covariates are identified as explaining heterogeneity, designing future studies controlling such effects through study design and pooling sub-grouped data defined by such covariate would seem sensible').

[141] ibid.

[142] ibid.

[143] Tierney et al. (2015), p. 1325.

[144] USFDA (2011) Advancing regulatory science at FDA, p. 9. https://www.fda.gov/media/81109/download. Accessed 26 Mar 2021.

the treatment efficacy ('expected efficacy') or simulate the trial outcome by asking the 'what if' questions (e.g. what if a different dosage is used).[145]

3.3.4 *Secondary Analysis of Data from Unsuccessful Trials*

A layperson would perhaps associate 'failed' clinical trials mainly with adverse effects, as the problem of non-reported and non-published adverse trial outcomes received much publicity.[146] However, most trials fail to confirm not safety but efficacy.[147]

The prompt dissemination of negative results is essential to inform the ongoing and planned research directed at the same or related problems and, thus, reduce 'experimental dead ends'.[148] Even where the original hypothesis regarding the treatment effect could not be supported, an investigational product might still be found effective under the modified conditions, e.g. in combination therapies or populations with different baseline characteristics.[149] In this regard, IPD availability

---

[145] A notable example is the clinical trial simulation tool for Alzheimer's Disease developed by the Critical Path Institute's (C-Path), which the EMA found to be 'suitable [...] for use in drug development as a longitudinal model for describing changes in cognition in patients with mild and moderate [Alzheimer's Disease], and for use in assisting in trial designs in mild and moderate [Alzheimer's Disease], as defined by the context of use'. EMA (19 Sep 2013) Qualification opinion of a novel data driven model of disease progression and trial evaluation in mild and moderate Alzheimer's disease. EMA/CHMP/SAWP/567188/2013, p. 50. The model can estimate inter alia a quantitative rationale for the selection of inclusion criteria and compare the results of the *post hoc* analyses with historical controls to reduce the risk of false positives. ibid p. 2. The model was developed based on the de-identified IPD provided by the research-based pharmaceutical companies. Critical Path Institute (10 Jul 2013) US Food and Drug Administration and European Medicines Agency Reach Landmark Decisions on Critical Path Institute's Clinical Trial Simulation Tool for Alzheimer's Disease, pp. 1–2. https://c-path.org/wp-content/uploads/2014/03/US-FDA-EMA-agency-reach-landmark-decisions-C-Path-clinical-trial-simulation-tool-for-alzheimers-disease.pdf. Accessed 26 Mar 2021.

[146] See e.g. Ioannidis and Lau (2001); Golder et al. (2016); Prayle et al. (2012); Law et al. (2011); Zarin et al. (2011).

[147] For the statistics, see e.g. Harrison (2016), pp. 817–818.

[148] Nightingale and Mahdi (2006), p. 81 (with further references). See also British Pharmacological Society. Pharmacology skills for drug discovery. https://thebiologist.rsb.org.uk/images/Pharmacology_Skills_for_Drug_Discovery.pdf. Accessed 26 Mar 2021; Multi-Regional Clinical Trials Center at Harvard University (2014) Overview of data disclosure initiatives: current and ongoing data transparency activities in the pharmaceutical industry (observing that 'access to data from discontinued programs, including early, exploratory trials, could inform future research and potentially reduce the risk to subjects in new clinical trials'). https://www.regulations.gov/comment/FDA-2013-N-0271-0031. Accessed 26 Mar 2021.

[149] A failure to confirm a hypothesis should be distinguished from 'uninformative' results. The latter means that a trial might have had inadequate power but does not necessarily imply the lack of association between the treatment and the intended health benefit. See Altman and Bland (1995), p. 485.

is crucial as its analysis can yield further insights about disease progression factors and individual responses to the treatment that can guide subsequent research.[150]

3.4 Secondary Data Analysis by Drug Regulators

3.4.1 Advancing Regulatory Science

Exploratory analysis of aggregated data submitted for drug marketing authorisation can be conducted by a drug authority to address healthcare questions beyond the benefit-risk assessment of a drug candidate. Such potential was recognised by the USFDA. While the Agency generally acknowledges that data from clinical and preclinical studies has 'a tremendous potential to [...] provide new opportunities for innovation in medical product development',[151] it also intends to use aggregated data submitted for drug marketing authorisation—in particular, data related to drug action, clinical outcomes, safety, and biomarkers—to promote innovation in regulatory decision making and advance 'regulatory science'.[152] The latter is defined as 'the science of developing new tools, standards, and approaches to assess the safety, efficacy, quality, and performance of all FDA-regulated products'.[153] Towards that goal, the FDA's strategic plan envisages that the analysis of pooled clinical trial datasets by the Agency's experts will 'explore differences in specific populations and subpopulations [...] different subsets of diseases, improve understanding of relationships between clinical parameters and outcomes, and evaluate clinical utility of potential biomarkers'.[154]

[150] See e.g. Gustafsson et al. (2010), p. 938 (discussing the examples of cardiovascular trials). In this regard, the Restored Trials Initiative should be mentioned. See Doshi et al. (2013) (proposing a system for independent analysis of clinical study reports on abandoned and non-reported trials. Even though secondary analyses under the initiative intended to follow the analyses specified in the original trial protocols, as acknowledge by Professor Doshi, exploratory analysis of IPD from abandoned and non-reported trials can be highly valuable. E-mail correspondence of 14 Jul 2018 (on file with the author).

[151] USFDA (4 Jun 2013) Availability of masked and de-identified non-summary safety and efficacy data—request for comments. 78(107) Fed. Reg, p. 33422. https://www.govinfo.gov/content/pkg/FR-2013-06-04/pdf/2013-13083.pdf. Accessed 26 Mar 2021. In particular, the FDA envisages that the analysis of aggregated safety and effectiveness data can be used to address 'key hurdles in drug development[,] identify potentially valid endpoints for clinical trials, understand the predictive value of preclinical models, clarify how medical products work in different diseases, and inform development of novel clinical designs and endpoints to the benefit of patients'.

[152] ibid.

[153] USFDA (2011) Advancing regulatory science at FDA, p. 12. https://www.fda.gov/media/81109/download. Accessed 26 Mar 2021. The specific mandate to promote public health innovation by advancing regulatory science is vested in the USFDA by the Administration Safety and Innovation Act (FDASIA). FDASIA, Pub. L. 112-144, sec 1124.

[154] USFDA (2011) Advancing regulatory science at FDA, p. 12. https://www.fda.gov/media/81109/download. Accessed 26 Mar 2021.

To date, the USFDA's example of leveraging non-summary efficacy and safety data for advancing 'regulatory science' is perhaps unique, given the Agency's significant expertise in biomedical statistics and experience in IPD analysis.[155] The EMA expressed interest in exploring applications of 'big data' analysis in drug discovery and clinical development and opportunities to use data to supplement prospective randomised clinical trials (RCTs).[156] In 2016, it announced its intention to 'continue to engage with stakeholders and [...] develop the skills and regulatory processes needed to ensure [that] big data is harnessed to facilitate robust medicines assessments and complement clinical trial data'.[157] While the EMA has not followed up with a strategy and measures on a scale comparable to the USFDA,[158] some developments are worth mentioning, such as extrapolation based on clinical trial data.

3.4.2 *Extrapolation*

The concept of extrapolation reflects the idea of using data from past trials in planning new ones. As defined by the EMA, extrapolation means

> extending information and conclusions available from studies in one or more subgroups of the patient population (source population(s)) or in related conditions or with related medicinal products, in order to make *inferences* for another subgroup of the population (target population), or condition or product, thus *reducing* the amount of, or general need for, additional information (types of studies, design modifications, number of patients required) needed to reach conclusions.[159]

The underlying rationale is 'to avoid unnecessary studies in the target population for ethical reasons, for efficiency, and to allocate resources to areas where studies are

[155] USFDA (4 Jun 2013) Availability of masked and de-identified non-summary safety and efficacy data—request for comments. 78(107) Fed. Reg, p. 33422. https://www.govinfo.gov/content/pkg/FR-2013-06-04/pdf/2013-13083.pdf. Accessed 26 Mar 2021. In this regard, the FDA's approach differs in principle from that of the EMA. See Koenig et al. (2015), pp. 10–11. On the EMA's data analysis practice, see also Chap. 6 at Sect. 6.5.1.2.

[156] In November 2016, the EMA held a workshop intending to identify the possibilities and challenges of how the potential of 'big data' could be leveraged to support drug R&D and regulatory decision-making. See EMA (2017) Identifying opportunities for 'big data' in medicines development and Regulatory Science. Report from a workshop held by EMA on 14–15 November 2016. EMA/740359/2016.

[157] ibid.

[158] At least, not at the time of writing. For the background information and related documents, see Workshop on identifying opportunities for 'big data' in medicines development and regulatory science. https://www.ema.europa.eu/en/events/workshop-identifying-opportunities-big-data-medicines-development-regulatory-science. Accessed 26 Mar 2021.

[159] EMA (9 Oct 2017) Reflection paper on the use of extrapolation in the development of medicines for paediatrics. EMA/199678/2016, p. 3 (emphasis added).

the most needed'.[160] The main benefits of extrapolation stem from 'the more targeted generation of evidence'[161] and the elimination of redundant research in situations where no uncertainty regarding a treatment exists.[162]

Extrapolation starts with a systematic review of all available and relevant data.[163] Based on the synthesis and analysis of the existing evidence, researchers can identify knowledge gaps and outstanding uncertainties,[164] make inferences and formulate predictions,[165] define specific research questions, assumptions and objectives that need to be addressed in new studies.[166] In this manner, the design of non-clinical and clinical studies can be informed by the existing evidence on the mechanism of action, pharmacokinetics, pharmacodynamics and clinical efficacy.

The extent to which earlier findings can be extrapolated from a 'source population' (a subgroup of the patient population examined in the original trial) to a 'target population' (a subgroup of the population, to which the results are extended) depends on the degree of *ceKrtainty* in understanding the disease mechanism and clinical pharmacology of the medical intervention in question. To a large extent, the strength of conclusions would hinge upon the totality of the accessible evidence. In some situations, extrapolation might 'not be justifiable [because] the disease is completely different in [other] subgroups [...] or the understanding of the drug's pharmacology is insufficient'.[167] In other situations, it would be '*unethical* not to extrapolate since the understanding of the disease and drug pharmacology is [...] well established'.[168] Many cases would likely fall somewhere within this spectrum and would require further studies.[169]

---

[160]EMA (19 Mar 2013) Concept paper on extrapolation of efficacy and safety in medicine development. EMA/129698/2012, p. 2.

[161]EMA (9 Oct 2017) Reflection paper on the use of extrapolation in the development of medicines for paediatrics. EMA/199678/2016, p. 4 (emphasis added).

[162]ibid p. 6 (pointing out that no new data needs to be generated to confirm the relationship between a treatment and efficacy if such relationship is well established and quantified and can be extrapolated to the target population, and if no gap in knowledge remains). Extrapolation can be of crucial importance for paediatric studies, where the extrapolation of data from adults to children could provide a basis for regulatory decision making, e.g. when planning and developing paediatric investigation plans. ibid pp. 9–10.

[163]ibid p. 12.

[164]ibid p. 6.

[165]ibid p. 7.

[166]ibid pp. 8, 12.

[167]EMA (19 Mar 2013) Concept paper on extrapolation of efficacy and safety in medicine development. EMA/129698/2012, p. 4.

[168]ibid (emphasis added).

[169]ibid.

3.5 Conclusion on Chapter 3

This chapter presented an overview of how secondary IPD analysis can be applied in scientific research, drug R&D and regulatory practice. While the list of examples is by no means exhaustive, it can give an idea that the research potential of aggregated non-summary clinical trial data is much greater than the benefit-risk assessment of an investigational product.

Several points are worth highlighting. First, the spectrum of applications of secondary IPD analysis is expansive and diverse. Potential benefits include a better informed clinical practice, validated and systematised medical knowledge and more targeted and optimised drug R&D. Exploratory analysis of aggregated IPD from multiple trials can generate new hypotheses regarding structure-activity correlations. Notably, insights drawn from the reviewed literature contrast the opinion that clinical trial data 'is unlikely [to] provide insights or generate new research that will significantly affect public welfare',[170] or that '[n]o evidence has been put forward [. . .] where information contained in clinical-trial data reveals details of what other molecules might be developed'.[171]

Second, the distinction between the use of data in basic (non-commercial) and applied (commercial) research is notional. Knowledge derived through data analysis in the 'upstream' medical research forms the basis for the 'downstream' drug development. In this view, the purpose limitation 'for non-commercial research'[172] stipulated under the contractual terms of access might not be meaningful and enforceable in practice.

Third, it should be emphasised that, while access to data is an indispensable prerequisite, its research potential can be realised only through data analysis. For that, various practical, methodological[173] and legal issues need to be clarified and resolved.

[170] Atholl J, Pugatch MP and Taylor D (2013) Clinical trials and data transparency – the public interest case. A briefing paper, p. 10. http://www.pugatch-consilium.com/reports/PC_Report_Clinical_Trials_OnLine_Version.pdf. Accessed 26 Mar 2021.

[171] EMA (30 Apr 2013) Advice to the European Medicines Agency from the Clinical Trial Advisory Group on Legal Aspects (CTAG5) – final advice, lines 89–90. https://www.ema.europa.eu/en/documents/other/ctag5-advice-european-medicines-agency-clinical-trial-advisory-group-legal-aspects-final-advice_en.pdf. Accessed 26 Mar 2021. Notably, this statement seems to contradict the very purpose of the EMA policy for data publication, namely, to promote drug research and innovation through secondary data analysis. See EMA publication policy 0070, pp. 3–4.

[172] EMA publication policy 0070, annex 2.

[173] See e.g. Cleophas et al. (2006), p. 205; Biltaji et al. (2017), p. 2338; Mayo et al. (2017).

References

Altman RB (2012) Translational bioinformatics: linking the molecular world to the clinical world. Clin Pharmacol Ther 91(6):994–1000. https://doi.org/10.1038/clpt.2012.49

Altman DG, Bland JM (1995) Absence of evidence is not evidence of absence. BMJ 311 (7003):485. https://doi.org/10.1136/bmj.311.7003.485

Bath PMW, Gray LJ (2009) Systematic reviews as a tool for planning and interpreting trials. Int J Stroke 4(1):23–27. https://doi.org/10.1111/j.1747-4949.2009.00235.x

Berlin JA et al (2014) Bumps and bridges on the road to responsible sharing of clinical trial data. Clin Trials 11(1):7–12. https://doi.org/10.1177/1740774513514497

Biltaji E et al (2017) Can *ad hoc* analyses of clinical trials help personalize treatment decisions? Br J Clin Pharmacol 83(11):2337–2338. https://doi.org/10.1111/bcp.13377

Braun CA, Anderson CM (2007) Pathophysiology: functional alterations in human health. Lippincott Williams & Wilkins, Baltimore, Philadelphia

Brody T (2016) Clinical trials: study design, endpoints and biomarkers, drug safety, and FDA and ICH Guidelines, 2nd edn. Elsevier, Amsterdam

Browner WS, Newman TB, Hulley SB (2007) Getting ready to estimate sample size: hypotheses and underlying principles. In: Hulley SB et al (eds) Designing clinical research, 3rd edn. Wolters Kluwer Health, Philadelphia, pp 51–64

Butte AJ (2008) Translational bioinformatics: coming of age. J Am Med Inform Assoc 15 (6):709–714. https://doi.org/10.1197/jamia.M2824

Butte AJ, Ito S (2012) Translational bioinformatics: data-driven drug discovery and development. Clin Pharmacol Ther 91(6):949–952. https://doi.org/10.1038/clpt.2012.55

CIOMS (2005) Management of safety information from clinical trials. Report of CIOMS working group VI. CIOMS, Geneva

Clayton GL et al (2017) The INVEST Project: investigating the use of evidence synthesis in the design and analysis of clinical trials. Trials 18(1):219. https://doi.org/10.1186/s13063-017-1955-y

Cleophas TJ, Zwinderman AH, Cleophas TF (2006) Statistics applied to clinical trials, 3rd edn. Springer, Dordrecht

Cooper H, Patall EA (2009) The relative benefits of meta-analysis conducted with individual participant data versus aggregated data. Psychol Methods 14(2):165–176. https://doi.org/10.1037/a0015565

Council of Europe (2012) Guide for research ethics committee members. Council of Europe

Cummings SR, Browner WS, Hulley SB (2007) Conceiving the research question. In: Hulley SB et al (eds) Designing clinical research, 3rd edn. Wolters Kluwer Health, Philadelphia, pp 17–26

Dalrymple D (2003) Scientific knowledge as a global public good: contributions to innovation and the economy. In: Esanu JM, Uhlir PF (eds) The role of scientific & technical data and information in the public domain: proceedings of a symposium. National Academy of Sciences, Washington DC, pp 35–51

Davis RA, Mayer AP, Bowsher RR (2016) Biomarkers in drug discovery and development: pre-analytical and analytical considerations. In: Weiner R, Kelley M (eds) Translating molecular biomarkers into clinical assays. Techniques and applications, vol 21. Springer, Cham, pp 17–25

Doshi P, Jefferson T (2013) Clinical study reports of randomised controlled trials: an exploratory review of previously confidential industry reports. BMJ Open 3:e002496. https://doi.org/10.1136/bmjopen-2012-002496

Doshi P et al (2013) Restoring invisible and abandoned trials: a call for people to publish the findings. BMJ 346:f2865. https://doi.org/10.1136/bmj.f2865

Egger M (1997) Meta-analysis: potentials and promise. BMJ 315(7119):1371–1374. https://doi.org/10.1136/bmj.315.7119.1371

Elwood M (2017) Critical appraisal of epidemiological studies and clinical trials, 4th edn. OUP, Oxford

Engberg S (2008) Systematic reviews and meta-analysis: studies of studies. J Wound Ostomy Continence Nurs 35(3):258–265. https://doi.org/10.1097/01.WON.0000319122.76112.23
Fleming N (2018) Computer-calculated compounds. Nature 557:55–57
Gallin JI, Ognibene FP, Johnson LL (2018) Principles and practice of clinical research, 4th edn. Elsevier, London
George SL, Wang X, Pang H (2016) Endpoints for cancer clinical trials. In: George SL, Wang X, Pang H (eds) Cancer clinical trials: current and controversial issues in design and analysis. Taylor & Francis Group, Boca Raton, pp 3–36
Goffin J (2009) Introduction to clinical trials. In: Cox Gad S (ed) Clinical trials handbook. Wiley, Hoboken, pp 1–22
Golder S et al (2016) Reporting of adverse events in published and unpublished studies of health care interventions: a systematic review. PLoS Med 13(9):e1002127. https://doi.org/10.1371/journal.pmed.1002127
Grady D, Hearst N (2007) Utilizing existing databases. In: Hulley SB et al (eds) Designing clinical research, 3rd edn. Wolters Kluwer Health, Philadelphia, pp 207–224
Gustafsson F et al (2010) Maximizing scientific knowledge from randomized clinical trials. Am Heart J 159(6):937–943. https://doi.org/10.1016/j.ahj.2010.03.002
Haidich AB (2010) Meta-analysis in medical research. Hippokratia 14(Suppl 1):29–37
Harrison RK (2016) Phase II and phase III failures: 2013-2015. Nat Rev Drug Discov 15 (12):817–818. https://doi.org/10.1038/nrd.2016.184
Helal S (2016) Subgroup discovery algorithms: a survey and empirical evaluation. J Comput Sci Technol 31:561–576. https://doi.org/10.1007/s11390-016-1647-1
Hughes JP et al (2011) Principles of early drug discovery. Br J Pharmacol 162(6):1239–1249. https://doi.org/10.1111/j.1476-5381.2010.01127.x
Huque M, Röhmel J (2010) Multiplicity problems in clinical trials: a regulatory perspective. In: Dmitrienko A, Tamhane AC, Bretz F (eds) Multiple testing problems in pharmaceutical statistics. Taylor & Francis Group, Boca Raton, pp 1–34
Institute of Medicine of the National Academies (2015) Sharing clinical trial data: maximizing benefits, minimizing risk. The National Academies Press, Washington DC
Ioannidis JP (2004) Systematic review of medical evidence. J Law Policy 12:509–535
Ioannidis JP, Lau J (2001) Completeness of safety reporting in randomized trials: an evaluation of 7 medical areas. JAMA 285(4):437–443. https://doi.org/10.1001/jama.285.4.437
Jones SD, Warren P (2006) Proteomics and drug discovery. In: Chorghade MS (ed) Drug discovery and development, vol I. Wiley, Hoboken, pp 233–272
Jones AP et al (2013) The use of systematic reviews in the planning, design and conduct of randomised trials: a retrospective cohort of NIHR HTA funded trials. BMC Med Res Methodol 13:50. https://doi.org/10.1186/1471-2288-13-50
Kilicoglu H (2018) Biomedical text mining for research rigor and integrity: tasks, challenges, directions. Brief Bioinform 19(6):1400–1414. https://doi.org/10.1093/bib/bbx057
Koenig F et al (2015) Sharing clinical trial data on patient level: opportunities and challenges. Biom J 57(1):8–26. https://doi.org/10.1002/bimj.201300283
Kulikowski CA et al (2012) AMIA Board White Paper: definition of biomedical informatics and specification of core competencies for graduate education in the discipline. J Am Med Inform Assoc 19(6):931–938. https://doi.org/10.1136/amiajnl-2012-001053
Laake P, Breien HB (2015) Research strategies, planning, and analysis research. In: Laake P, Benestad HB, Olsen BR (eds) Medical and biological sciences: from planning and preparation to grant application and publication. Elsevier, London, San Diego
Laterza OF, Zhao X (2016) Biomarker discovery. In: Weiner R, Kelley M (eds) Translating molecular biomarkers into clinical assays. Techniques and applications, vol 21. Springer, Cham, pp 27–36
Lauer MS (2010) Data primarily collected for new insights. In: Grossmann C et al (eds) Clinical data as the basic staple of health learning: creating and protecting a public good. National Academy of Sciences, Washington DC, pp 90–99

Law MR, Kawasumi Y, Morgan SG (2011) Despite law, fewer than one in eight completed studies of drugs and biologics are reported on time on ClinicalTrials.gov. Health Aff (Millwood) 30 (12):2338–2345. https://doi.org/10.1377/hlthaff.2011.0172

Lee JW (2016) Biomarkers in discovery and preclinical phase during drug development. In: Weiner R, Kelley M (eds) Translating molecular biomarkers into clinical assays. Techniques and applications, vol 21. Springer, Cham, pp 47–56

Lipkovich I et al (2018) Multiplicity issues in exploratory subgroup analysis. J Biopharm Stat 28 (1):63–81. https://doi.org/10.1080/10543406.2017.1397009

Mayo CS et al (2017) Big data in designing clinical trials: opportunities and challenges. Front Oncol 7:187. https://doi.org/10.3389/fonc.2017.00187

Meinert CL (2012) Clinical trials: design, conduct and analysis, 2nd edn. OUP, Oxford, New York

Merrill RM (2015) Introduction to epidemiology, 7th edn. Jones & Bartlett Publishers, Burlington

Moyé LA (2003) Multiple analyses in clinical trials: fundamentals for investigators. Springer, New York, Berlin, Heidelberg

Mulrow CD (1994) Rationale for systematic reviews. BMJ 309(6954):597–599. https://doi.org/10.1136/bmj.309.6954.597

Naci H, Cooper J, Mossialos E (2015) Timely publication and sharing of trial data: opportunities and challenges for comparative effectiveness research in cardiovascular disease. Eur Heart J Qual Care Clin Outcomes 1(2):58–65. https://doi.org/10.1093/ehjqcco/qcv012

National Institutes of Health, Biomarkers Definition Working Group (2001) Biomarkers and surrogate endpoints: preferred definitions and conceptual framework. Clin Pharmacol Ther 69 (3):89–95. https://doi.org/10.1067/mcp.2001.113989

Nevitt SJ et al (2017) Exploring changes over time and characteristics associated with data retrieval across individual participant data meta-analyses: systematic review. BMJ 357:j1390. https://doi.org/10.1136/bmj.j1390

Newman TB et al (2007a) Designing cross-sectional and case-control studies. In: Hulley SB et al (eds) Designing clinical research, 3rd edn. Wolters Kluwer Health, Philadelphia, pp 109–126

Newman TB, Browner WS, Hulley SB (2007b) Enhancing causal inference in observational studies. In: Hulley SB et al (eds) Designing clinical research, 3rd edn. Wolters Kluwer Health, Philadelphia, pp 127–146

Nightingale P, Mahdi S (2006) The evolution of pharmaceutical innovation. In: Mazzucato M, Dosi G (eds) Knowledge accumulation and industry evolution: the case of pharma-biotech. CUP, Cambridge, pp 73–111

Pharma firms pool and share cancer trial data (2014) Nat Rev Drug Discov 13:323. https://doi.org/10.1038/nrd4331

Porth CM (2011) Preface. In: Porth CM (ed) Essentials of pathophysiology: concepts of altered health states, 3rd edn. Wolters Kluwer, Philadelphia, pp ix–xii

Prayle AP, Hurley MN, Smyth AR (2012) Compliance with mandatory reporting of clinical trial results on ClinicalTrials.gov: cross sectional study. BMJ 344:d7373. https://doi.org/10.1136/bmj.d7373

Ray C (2016) Fit-for-purpose validation. In: Weiner R, Kelley M (eds) Translating molecular biomarkers into clinical assays. Techniques and applications, vol 21. Springer, Cham, pp 1–15

Riley RD, Lambert PC, Abo-Zaid G (2010) Meta-analysis of individual participant data: rationale, conduct, and reporting. BMJ 340:c221. https://doi.org/10.1136/bmj.c221

Schulte P, Mazzuckelli LF (1991) Validation of biological markers for quantitative risk assessment. Environ Health Perspect 90:239–246. https://doi.org/10.1289/ehp.90-1519476

Selby JV et al (2018) Using large data sets for population-based health research. In: Gallin JI, Ognibene FP, Johnson LL (eds) Principles and practice of clinical research, 4th edn. Elsevier, London, pp 269–292

Sellwood MA et al (2018) Artificial intelligence in drug discovery. Future Med Chem 10 (17):2025–2028. https://doi.org/10.4155/fmc-2018-0212

Simmonds M, Stewart G, Stewart L (2015) A decade of individual participant data meta-analyses: a review of current practice. Contemp Clin Trials 45:76–83. https://doi.org/10.1016/j.cct.2015.06.012

Singh U, Dolled-Filhart M, Wu D (2016) *In situ* hybridization in clinical biomarker development. In: Weiner R, Kelley M (eds) Translating molecular biomarkers into clinical assays. Techniques and applications, vol 21. Springer, Cham, pp 201–210

Song F, Bachmann MO (2016) Cumulative subgroup analysis to reduce waste in clinical research for individualised medicine. BMC Med 14:197. https://doi.org/10.1186/s12916-016-0744-x

Stewart LA, Parmar MKB (1993) Meta-analysis of the literature or of individual patient data: is there a difference? Lancet 341(8842):418–422. https://doi.org/10.1016/0140-6736(93)93004-k

Stewart LA, Tierney JF (2002) To IPD or not to IPD? Advantages and disadvantages of systematic reviews using individual patient data. Eval Health Prof 25(1):76–97. https://doi.org/10.1177/0163278702025001006

Stewart LA, Tierney JF, Clarke M (2011) Reviews of individual patient data. In: Higgins JPT, Green S (eds) Cochrane handbook for systematic reviews of interventions. Wiley, Hoboken, pp 18:1–18:9

Strimbu K, Tavel J (2010) What are biomarkers? Curr Opin HIV AIDS 5(6):463–466. https://doi.org/10.1097/COH.0b013e32833ed177

Strom BL (2005) What is pharmacoepidemiology. In: Strom BL (ed) Pharmacoepidemiology. Wiley, Hoboken, pp 3–16

Strom BL et al (2016) Data sharing – is the juice worth the squeeze? N Engl J Med 375(17):1608–1609. https://doi.org/10.1056/NEJMp1610336

Sudlow R et al (2016) EFSPI/PSI working group on data sharing: accessing and working with pharmaceutical clinical trial patient level datasets – a primer for academic researchers. BMC Med Res Methodol 16(73). https://doi.org/10.1186/s12874-016-0171-x

Sutton AJ et al (2007) Evidence-based sample size calculations based upon updated meta-analysis. Stat Med 26(12):2479–2500. https://doi.org/10.1002/sim.2704

Tierney JF et al (2015) How individual participant data meta-analyses have influenced trial design, conduct, and analysis. J Clin Epideiol 68(11):1325–1335. https://doi.org/10.1016/j.jclinepi.2015.05.024

Trocky N, Brandt C (2009) Process of data management. In: Cox Gad S (ed) Clinical trials handbook. Wiley, Hoboken, pp 185–202

Vallance P, Smart TG (2006) The future of pharmacology. Br J Pharmacol 147(Suppl 1):304–307. https://doi.org/10.1038/sj.bjp.0706454

Wang D, Bakhai A (2006) Clinical trials: a practical guide to design, analysis, and reporting. Remedica, London

Wang D, Bakhai A, Maffulli N (2009) Statistical methods for analysis of clinical trials. In: Cox Gad S (ed) Clinical trials handbook. Wiley, Hoboken, pp 1053–1080

WHO, IUPHAR, CIOMS (2012) Clinical pharmacology in health care, teaching and research. CIOMS, Geneva

Wnek R, Tseng M, Wu D (2016) Current flow cytometry methods for the clinical development of immunomodulatory biologics. In: Weiner R, Kelley M (eds) Translating molecular biomarkers into clinical assays. Techniques and applications, vol 21. Springer, Cham, pp 141–151

Xia X (2017) Bioinformatics and drug discovery. Curr Top Med Chem 17(15):1709–1726. https://doi.org/10.2174/1568026617666161116143440

Yang Y, Adelstein SJ, Kassis AI (2009) Target discovery from data mining approaches. Drug Discov Today 14(3–4):147–154. https://doi.org/10.1016/j.drudis.2008.12.005

Zarin DA, Tse T (2016) Sharing individual participant data (IPD) within the context of the trial reporting system (TRS). PLoS Med 13(1):e1001946. https://doi.org/10.1371/journal.pmed.1001946

Zarin DA et al (2011) The ClinicalTrials.gov results database – update and key issues. N Engl J Med 364(9):852–860. https://doi.org/10.1056/NEJMsa1012065

Part II
Analysis *De Lege Lata*

Chapter 4
Legal Sources of Control Over and Access to Clinical Trial Data Under the EU Applicable Framework

Abstract This chapter examines the applicable legal framework at EU level that currently determines the conditions and scope of access to clinical trial data. First, the main sources of EU primary and secondary law applicable to clinical trial data throughout its lifecycle and particularly relevant for this study are outlined. Next, the legal determinants of trial sponsors' control over and third-party access to individual patient-level trial data are surveyed and analysed. Finally, uncertainties and limitations of the current regulatory approach regarding the post-trial accessibility of non-summary data are identified.

4.1 The EU Legal and Regulatory Framework Applicable to Clinical Trial Data

4.1.1 Relevant Provisions Under Primary Law

The pharmaceutical sector is intensely regulated along 'the entire value chain including R&D activities'.[1] Given the overarching safety concerns related to public health, regulation of clinical trials constitutes an area of shared competence of the Union and Member States.[2] Article 5 of the TEU, which lays down the principles of the use of Union competencies, is explicitly referenced by the EU Clinical Trials Regulation[3] in relation to the objective of guaranteeing that, 'throughout the Union, clinical trial data are reliable and robust while ensuring respect for the rights, safety, dignity and well-being of subjects'.[4] Since such objective 'cannot be sufficiently achieved by the Member States but, by reason of its scale, can rather be better

[1]European Commission (8 Jul 2009) Pharmaceutical sector inquiry final report, p. 95. https://ec.europa.eu/competition/sectors/pharmaceuticals/inquiry/staff_working_paper_part1.pdf. Accessed 26 Mar 2021.

[2]TFEU, art 4(2)(k).

[3]Reg 536/2014/EU.

[4]ibid rec 85.

D. Kim, *Access to Non-Summary Clinical Trial Data for Research Purposes Under EU Law*, Munich Studies on Innovation and Competition 16,
https://doi.org/10.1007/978-3-030-86778-2_4

achieved at Union level',[5] the Union can adopt measures in accordance with the principles of subsidiarity and proportionality as set out in Article 5 of the TEU.[6]

Articles 114 and 168(4)(c) of the TFEU constitute a double legal basis for the EU Clinical Trials Regulation[7] and correspond to two complementary policy goals: to achieve an internal market as regards clinical trials and medicinal products and to ensure high standards of quality and safety for medicinal products.[8] The reliability and robustness of trial data can be viewed as an integral element of both objectives, given that such data serves to prove the quality and safety of medicinal products that are authorised for marketing in the EU.[9]

The authorisation and conduct of clinical trials shall be carried out in accordance with the rights, freedoms and principles recognised under the Charter of Fundamental Rights (CFR) of the European Union, in particular, the right to human dignity, the integrity of the person, the rights of the child, respect for private and family life, the protection of personal data, and the freedom of art and science.[10] In line with the CJEU case law, free and informed consent of medical and biological research subjects is part of the fundamental human right to integrity and Union law.[11] The protection of the rights, safety, dignity and well-being of subjects and their prevalence over all other interests constitute a general principle of conducting clinical trials in the EU.[12]

Furthermore, Article 15(3) of the TFEU and Article 42 of the CFR that guarantee the right of access to documents are of particular relevance for this study as they form the legal basis of the EMA transparency policies.[13]

4.1.2 Relevant Sources of Secondary Law

4.1.2.1 The EU Regulation on Clinical Trials

General Aspects

The EU Clinical Trials Regulation contains the main substantive and procedural provisions that govern different aspects of interventional studies conducted in the

[5] ibid.

[6] ibid.

[7] ibid rec 82.

[8] ibid.

[9] ibid.

[10] ibid.

[11] See Explanations relating to the Charter of Fundamental Rights (2007) OJ C 303, C 303/18; Reg 536/2014/EU, rec 27.

[12] Reg 536/2014/EU, art 3(a).

[13] For a detailed discussion, see below at Sect. 4.3.3.

EU.[14] It was adopted on 16 April 2014, entered into force on 16 June 2014 and will be applicable six months after the European Commission issues the notice confirming the full functionality of the EU portal and database.[15] Before the adoption of the EU Clinical Trials Regulation, the European Commission revised the Clinical Trials Directive[16] in pursuit of the following objectives:

- to modernise the regulatory framework for submission, assessment, and regulatory follow-up of applications for clinical trials;
- to adapt regulatory requirements to practical considerations and needs;
- to address the global dimension of clinical trials when ensuring compliance with good clinical practice.[17]

These objectives relate to three problem issues that prompted the revision of the Clinical Trials Directive, namely: (i) separate submission, diverging assessments and regulatory follow-up of applications for clinical trials; (ii) regulatory requirements that are not adapted to practical considerations and needs; (iii) reliability of clinical trial data in a globalised research environment.[18] In line with these objectives, the main novelties introduced by the EU Clinical Trials Regulation are:

- a streamlined application procedure through the EU portal;[19]
- the concept of 'low-intervention' studies, for which 'administrative burdens and operational costs [can be reduced] considerably without compromising on patient safety';[20]
- mandatory trial registration in a public register[21] and inspections of non-EU countries' regulatory systems for clinical trials.[22]

---

[14] Reg 536/2014/EU, art 1.

[15] Reg 536/2014/EU, art 99. On 21 April 2021, the EMA's Management Board confirmed that the Clinical Trials Information System is 'fully functional and meets the functional specifications, following an independent, successful audit'; at the time of writing, the European Commission is expected to consider 'if the conditions set by the [EU Clinical Trials] Regulation are met and, once confirmed, [to] publish a notice in the Official Journal of the European Union'. EMA. Clinical trial regulation. http://www.ema.europa.eu/ema/index.jsp?curl=pages/regulation/general/general_content_000629.jsp. Accessed 29 Jun 2021.

[16] Dir 2001/20/EC.

[17] European Commission, SWD(2012) 200 final, vol. I, pp. 17–27.

[18] ibid.

[19] Reg 536/2014/EU, rec 4, art 5.

[20] European Commission, SWD(2012) 200 final, vol. I, pp. 35–38. Low-intervention studies are trials that involve investigational medicinal products that have been already authorised for marketing. See also Reg 536/2014/EU, art 2(2)(3).

[21] Reg 536/2014/EU, rec 67; art 25(6).

[22] Reg 536/2014/EU, rec 64; art 79. See also European Commission, SWD(2012) 200 final, vol. I, pp. 70–71 (concluding that the objective of addressing the global dimension of clinical trials and ensuring compliance with GCP would be 'achieved reasonably well' by combining the measures of mandatory trial registration and the inspections of non-EU countries' regulatory systems for clinical trials).

Overall, the EU Clinical Trials Regulation aims to achieve 'an internal market as regards clinical trials and medicinal products for human use'[23] and to ensure 'high standards of quality and safety for medicinal products to meet common safety concerns as regards these products'.[24] As a general principle, it proclaims that a clinical trial can be conducted only if:

- the rights, safety, dignity and well-being of subjects are protected and prevail over all other interests; and
- it is designed to generate reliable and robust data.[25]

Data Reliability and Robustness

While the EU Clinical Trials Regulation does not define 'data reliability' and 'data robustness', these terms form the key components of *good clinical practice* (GCP), understood as

> ethical and scientific quality requirements for designing, conducting, performing, monitoring, auditing, recording, analysing and reporting clinical trials ensuring that the rights, safety and well-being of subjects are protected, and that the data generated in the clinical trial are *reliable* and *robust*.[26]

The wording is a slight modification of the definition under the ICH Guideline for Good Clinical Practice, which uses the terms 'credible and accurate'[27] instead of 'reliable and robust' in relation to trial results. The notions of data credibility, accuracy, reliability and robustness pertain to the overarching concept of *quality of a body of evidence*, 'the extent to which one can be confident that an estimate of [treatment] effect is near the true value for an outcome, across studies'.[28]

Data reliability and robustness are closely related but not synonymous. Reliability implies that data generated in a trial can be relied upon when a drug is authorised for marketing if such data shows that a medicinal product meets the quality and safety standards. Data robustness relates to the internal validity of a trial, which means that a trial is designed to answer the research question in a credible (unbiased) manner.[29]

---

[23] Reg 536/2014/EU, rec 82 (emphasis added).

[24] ibid.

[25] Reg 536/2014/EU, art 3.

[26] Reg 536/2014/EU, art 2(2)(30) (emphasis added).

[27] ICH (10 Jun 1996) ICH Harmonised tripartite guideline. Guideline for good clinical practice. E6, para 1(24). Furthermore, the ICH guideline lists the types of documents that 'individually and collectively' allow to evaluate the conduct of a trial and the quality of the generated data. ibid paras 8.1–4.8.

[28] Higgins et al. (2017), p. 8:4.

[29] ibid p. 8:3. See also ICH (10 Jun 1996) ICH Harmonised tripartite guideline. Guideline for good clinical practice. E6, para 6.4 (stating that 'scientific integrity of the trial and the credibility of the data from the trial depend substantially on the trial design').

In particular, data robustness is evaluated vis-à-vis the trial endpoints and the statistical methods applied.[30]

The EU Database for Clinical Trial Data

Among the novelties introduced by the EU Clinical Trials Regulation was the setup of a database intended to 'ensure a sufficient level of transparency in the clinical trials' by making publicly accessible 'all relevant information as regards the clinical trial submitted through the EU portal'.[31] The EMA shall establish and maintain the database in collaboration with the Member States and the European Commission.[32]

Further, the new EU database for clinical trials should be distinguished from the EudraCT (the European Union Drug Regulating Authorities Clinical Trials) and the EudraVigilance databases. The EudraVigilance database was established to disseminate information regarding adverse reactions to medicinal products authorised for the European market.[33] The EudraCT database was launched under the Clinical Trials Directive in 2004.[34] As a general rule, data submitted to the EudraCT database is accessible only to national competent authorities (NCA),[35] while limited information can be made publicly accessible or available for registered users.[36]

[30] European Commission, SWD(2012) 200 final, vol. I, p. 13.

[31] Reg 536/2014/EU, rec 67.

[32] Reg 536/2014/EU, art 81(4)(b). The EMA publication policy 0070 and the EU database are distinct instruments: the former applies to trial data submitted to the EMA through the centralised marketing authorisation procedure after 1 January 2015; the latter will apply to data generated in clinical trials approved after the EU Clinical Trials Regulation becomes applicable. EMA (21 Mar 2019) Questions and answers on the European Medicines Agency policy on publication of clinical data for medicinal products for human use. EMA/357536/2014, rev. 2, pp. 3, 6–7. https://www.ema.europa.eu/en/documents/report/questions-answers-european-medicines-agency-policy-publication-clinical-data-medicinal-products_en.pdf. Accessed 26 Mar 2021. Data disclosed under EMA's publication policy 0070 can be accessed through the website launched and operated by the EMA: https://clinicaldata.ema.europa.eu/web/cdp/home. Accessed 26 Mar 2021.

[33] Reg 726/2004/EC, art 57(1)(d). The EudraVigilance database is accessible to health-care professionals, marketing authorisation holders and the public, provided that personal data protection is guaranteed. ibid.

[34] For further details regarding the functionalities of the EudraCT database, see https://eudract.ema.europa.eu/help/Default.htm. Accessed 26 Mar 2021.

[35] Dir 2001/20/EC, art 11(1).

[36] EudraCT. Overview of the EudraCT public and secure applications. https://eudract.ema.europa.eu/help/Default.htm#eudract/overview_eudract_public_secure.htm. Accessed 26 Mar 2021.

4.1.2.2 The EU Drug Authorisation Regulation

The EU Drug Authorisation Regulation[37] lays down the procedures for the authorisation and supervision of medicinal products for human and veterinary use. It also establishes the EMA[38] responsible inter alia for coordinating the scientific evaluation of the quality, safety, and efficacy of medicinal products applied for the EU-wide marketing authorisation.[39]

The EU Clinical Trials Regulation and the EU Drug Authorisation Regulation pursue distinct objectives: the former intends to ensure the rights and safety of the trial participants and the reliability and robustness of trial data; the latter aims to guarantee the safety and efficacy of the marketed medicinal products. As emphasised by the European Commission, the conditions of the authorisation of clinical trials are unrelated to the regulation and authorisation of medicinal products.[40] One aspect where two bodies of rules intersect is the quality of clinical trial data that provides the evidentiary basis for decision making regarding drug marketing authorisation.[41]

4.1.2.3 The EMA's Guidance on the Implementation of the Publication Policy

The European Commission issued numerous guidance documents intending to achieve the uniform application of the substantive and procedural provisions under the EU Clinical Trials Regulation.[42] Among the guidance documents adopted by the EMA, particularly relevant for this study is the 'External guidance on the implementation of the European Medicines Agency policy on the publication of clinical data for medicinal products for human use' adopted in 2016.[43] The document provides a detailed methodology for identifying and redacting commercially confidential information (CCI) from the dossiers submitted to the EMA and

[37] Regulation (EC) No 726/2004 of the European Parliament and of the Council of 31 March 2004 laying down Community procedures for the authorisation and supervision of medicinal products for human and veterinary use and establishing a European Medicines Agency (30 Apr 2004) OJ L 136.

[38] Reg 726/2004/EC, art 55.

[39] Reg 726/2004/EC, art 57(1)(a).

[40] European Commission, SWD(2012) 200 final, vol. I, p. 13.

[41] ibid.

[42] For a compilation of guidance documents applicable to clinical trials, see EudraLex – Volume 10 – Clinical trials guidelines. https://ec.europa.eu/health/documents/eudralex/vol-10_en#fragment0. Accessed 26 Mar 2021.

[43] EMA (15 Oct 2018) External guidance on the implementation of the European Medicines Agency policy on the publication of clinical data for medicinal products for human use. EMA/90915/2016, version 1.4. https://www.ema.europa.eu/en/documents/regulatory-procedural-guideline/external-guidance-implementation-european-medicines-agency-policy-publication-clinical-data_en-3.pdf. Accessed 26 Mar 2021. Since its adoption in 2016, the procedural aspects were subsequently clarified in the revised versions of the guidance.

de-identifying personal data. It emphasises that the ultimate purpose of the EMA's transparency policies is 'to retain a *maximum* of *scientifically useful* information on medicinal products for the benefit of the public while achieving adequate anonymisation'.[44]

Having sketched the core components of the regulatory framework applicable to clinical trials, let us now identify key legal determinants of control over and access to individual patient-level trial data.

4.2 Legal Sources of Control of Trial Sponsors Over Individual Patient-Level Clinical Trial Data

4.2.1 Do Drug Sponsors 'Own' Clinical Trial Data?

4.2.1.1 Competing Claims of Data Ownership

Drug companies claimed that clinical trial data constitutes their 'proprietary information'[45] and refer in contracts to all trial data as their 'exclusive property'.[46] Some medical commentators raised the question of who 'owns' clinical trial data[47] and contemplated that data 'ownership' can potentially be claimed by trial participants, investigators, trial sponsors and the general public.[48] Data 'ownership' was also reported to be 'a critical obstacle to data sharing efforts, particularly for *industry funded* research, with sponsors often explicitly retaining *ownership* rights'.[49]

---

[44] ibid p. 43 (emphasis added).

[45] See e.g. Case C-389/13 *EMA v AbbVie* [2013] ECLI:EU:C:2013:794, para 18; Case T-33/17 *Amicus Therapeutics v EMA* [2018] ECLI:EU:T:2018:595, para 26; Case T-235/15 *Pari Pharma v EMA* [2018] ECLI:EU:T:2018:65, para 63. See also EMA (30 Apr 2013) Advice to the European Medicines Agency from the Clinical Trial Advisory Group on Legal Aspects (CTAG5) – final advice, lines 160–161 (referring to the arguments of the participants of the workshop on access to clinical trial data and transparency held by the EMA in 2012 that 'clinical-trial data are commercially confidential [...] as they contain information such as [...] *proprietary* information regarding efficacy and safety measurements and statistical analysis' (emphasis added)). https://www.ema.europa.eu/en/documents/other/ctag5-advice-european-medicines-agency-clinical-trial-advisory-group-legal-aspects-final-advice_en.pdf. Accessed 26 Mar 2021. See also EFSPI, European Federation of Statisticians in the Pharmaceutical Industry (EFSPI) position on European Medicines Agency (EMA) access to clinical trial data initiative (referring to trial sponsors as 'the owners of the data' throughout the text). https://www.efspi.org/documents/publications/efspipositiononema250413.pdf. Accessed 26 Mar 2021.

[46] On the contractual practice of obtaining exclusive control over clinical trial data, see below at Sect. 4.2.5 in this chapter.

[47] See e.g. Drazen (2002); Vickers (2006).

[48] Drazen (2002), p. 409.

[49] Rathi (2012) (emphasis added) (with further references).

The Directive 2001/83/EC indeed uses the terms 'owner' and 'ownership': in particular, it refers to 'sponsor or *other owner* of the data',[50] who shall retain all documentation related to the trial as long as the corresponding medicinal product is authorised for marketing. Further, it stipulates that the final trial report 'shall be retained by the sponsor or subsequent *owner*, for five years after the medicinal product is no longer authorized',[51] and that '[a]ny change of *ownership* of the data shall be documented'.[52]

The EU Clinical Trials Regulation contains an analogous obligation, which states:

> Any transfer of *ownership* of the *content* of the clinical trial master file shall be documented. The new *owner* shall assume the responsibilities set [therein].[53]

While 'data ownership' is mentioned in these provisions only in passing, the question arises: Can the content of CSRs, trial master files and, most importantly, IPD be 'owned' in the sense of property law?[54]

4.2.1.2 De facto Control But Not de jure Ownership of IPD

The Obligation to Protect Data Against Unauthorised Access as the Source of de facto Exclusive Control

The EU Clinical Trials Regulation vests in trial sponsors and investigators an obligation to record, process, handle, and store *all* clinical trial information 'in such a way that it can be accurately reported, interpreted and verified while the confidentiality of records and the personal data of the subjects remain protected in accordance with the applicable law on personal data protection'.[55] Further, the Regulation requires that

> [a]ppropriate technical and organisational measures shall be implemented to protect information and personal data processed against *unauthorised* or unlawful access, disclosure, dissemination, alteration, or destruction or accidental loss [. . .].[56]

Even though this provision explicitly mentions the entities in charge of implementing the said measures, it follows from its positioning that the obligation

---

[50] Dir 2001/83/EC, annex I, pt 4(B)(2)(c) (emphasis added).

[51] Dir 2001/83/EC, annex I, pt 4(B)(2)(d) (emphasis added).

[52] ibid (emphasis added).

[53] Reg 536/2014/EU, art 58 (emphasis added). A clinical trial master file 'shall [. . .] contain the essential documents relating to that clinical trial which allow verification of the conduct of a clinical trial and the quality of the data generated'. Reg 536/2014/EU, art 57.

[54] The question is posed in view of the civil law tradition, given the study's focus on EU law. On the fitness of the concept of 'data ownership' in civil law jurisdictions, see Drexl (2018), p. 89 ff.

[55] Reg 536/2014/EU, art 56(1).

[56] Reg 536/2014/EU, art 56(2) (emphasis added).

is vested in trial sponsors and trial investigators. By this obligation and given that trial sponsors usually stipulate the transfer of all data gathered in a trial in contracts with investigators and trial subjects, trial sponsors become capable of exercising de facto exclusive control over any data generated in the course of a trial.

Factual control, however, should not be equated with the right of ownership, which is regarded as 'the most extensive [. . .] property right'.[57] While, from a legal perspective, de facto control is distinguished from the right to own,[58] control and ownership may appear synonymous to a layperson.[59] Such confusion may, in part, stem from the notion of control as a definitional feature of legal ownership.[60]

The requirement that trial data shall be recorded, handled and stored 'adequately' is rationalised on the grounds of 'ensuring subject rights and safety, the robustness and reliability of the data generated in the clinical trial, accurate reporting and interpretation, effective monitoring by the sponsor and effective inspection by Member States'.[61] In other words, this duty is mainly motivated by the considerations for protecting trial subjects and public health, including by way of pharmacovigilance. In favour of that also speaks that the duration of the obligation for data archival is linked to the duration of the corresponding drug marketing authorisation.[62]

The obligation to store all trial data and protect it *against destruction* by its very nature rejects the idea of property rights in such data. Such obligation contradicts a property law conception of the right of ownership, which presupposes an 'absolute dominion' over 'a thing', in particular, by exercising the rights to use, transfer, exclude third-party access to and use of, and destroy the object of property rights.[63] In this view, the terms 'data owner' and 'data ownership' under the EU Clinical Trials Regulation and Directive 2001/83/EC[64] cannot be interpreted as vesting fully-fledged property rights in IPD. More appropriate would be to refer to trial sponsors as 'data holders'.

As personal data protection can significantly limit the discretion of trial sponsors over IPD, let us consider next the nature and scope of trial subjects' individual rights in data collected in trials.

---

[57] Hoof (2010), p. 27. See also van Erp and Akkermans (2010), p. 33.

[58] See e.g. European Commission, SWD(2017) 2 final, pp. 34–35 (emphasising that the concept of ownership should be distinguished from de facto possession).

[59] See e.g. OECD (2015), p. 195 (pointing out that the concept of ownership is 'often misunderstood and/or misused' by businesses).

[60] See e.g. Coleman and Kraus (1989), p. 1339 (observing that '[a] perfectly natural way of characterizing what it means to have a right to a resource or to property is in terms of autonomy or control'). It goes beyond the scope of this study to examine jurisdictional differences regarding 'ownership'.

[61] Reg 536/2010/EU, rec 51.

[62] Dir 2001/83/EC, annex I, pt 4(B)(2)(c).

[63] Penner (1996), p. 743.

[64] Above (nn 50–53) and the accompanying text.

No Property Rights in IPD as Personal Data

The need to ensure that trial-related data and information of personal nature are treated as confidential is emphasised by various international norm-setting instruments in the area of medical research, including the Declaration of Helsinki,[65] the ICH Guideline for Good Clinical Practice[66] and the Oviedo Convention.[67] In the EU, the right to personal data protection is guaranteed under Article 8 of the CFR and Article 16(1) of the TFEU; the General Data Protection Regulation (the GDPR) harmonises the main rules in this regard.[68] Quite obviously, IPD qualifies as personal data within the meaning of the GDPR as 'information relating to an *identified* or *identifiable* natural person ("data subject")'.[69] Furthermore, IPD falls within a special category of personal data, namely, 'data concerning health'.[70] Such data merits a higher level of protection and, as a general rule, can be processed only upon obtaining from a data subject explicit consent to the processing of personal data for the predefined specific purpose[71] or where other conditions can apply.[72]

The European legal tradition approaches personal data protection from a human rights perspective, which implies that 'the concept of (commercial) property may not be vested in privacy because privacy is attached to individuals by virtue of their personhood and, as such, this right cannot be waived or transferred to others'.[73] Thus, personal data protection is conceived as a negative, *autonomy-based* right of non-interference into the personal sphere.[74]

---

[65] Declaration of Helsinki, paras 9 and 24.

[66] ICH (10 Jun 1996) ICH Harmonised tripartite guideline. Guideline for good clinical practice. E6, para 4.8.10(o), (p).

[67] Additional protocol to the Oviedo Convention concerning biomedical research (25 Jan 2005) CETS No. 195, arts 25 and 26.

[68] Regulation (EU) 2016/679 of the European Parliament and of the Council of 27 April 2016 on the protection of natural persons with regard to the processing of personal data and on the free movement of such data, and repealing Directive 95/46/EC (General Data Protection Regulation) (4 May 2016) OJ L 119 [hereinafter the GDPR].

[69] GDPR, art 4(1) (emphasis added).

[70] GDPR, art 9(1). Article 4(15) defines 'data concerning health' as 'personal data related to the physical or mental health of a natural person, including the provision of health care services, which reveal information about his or her health status'.

[71] GDPR, art 9(2)(a).

[72] GDPR, art 9(2)(b)-(j).

[73] Prins (2006), p. 234.

[74] ibid. See also European Commission, SWD(2017) 2 final, p. 33 (pointing out that protection of personal data 'is a fundamental right in itself'; therefore, the new ownership right 'would not be conceivable' with regard to personal data). But see Purtova (2011), p. 47 (arguing that the EU approach to personal data protection does not 'adequately grasp the new structure of relationships within data flow'). She concludes that property rights in personal data could better manage the complexity of the relationships between the numerous actors involved in processing personal data due to their *erga omnes* effect and fragmentation. ibid p. 61.

Under the GDPR, the protection of natural persons in relation to the processing of personal data is regarded as a fundamental right.[75] Protection is granted not in the form of a fully-fledged right of ownership,[76] but as specific, inalienable[77] rights that allow the data subject to exercise control over third-party access to and use of data qualified as personal, namely:

(a) the right to access to personal data;[78]
(b) the right to restrict processing of personal data;[79]
(c) the right to data portability;[80]
(d) the right to object to processing of personal data;[81]
(e) the right not to be subject to a decision based solely on automated processing, including profiling;[82]
(f) the right of erasure of personal data (the 'the right to be forgotten').[83]

Furthermore, personal data protection is reinforced under the pharmaceutical sector legislation, including the EU Clinical Trial Regulation,[84] Drug Marketing Authorisation Regulation[85] and Directive 2001/83/EC.[86]

Pseudonymisation or anonymisation can change the legal status of IPD, as personal data protection does not apply to 'information which does not relate to an identified or identifiable natural person or to personal data rendered anonymous in such a manner that the data subject is not or no longer identifiable'.[87] Importantly, as long as personal data that has undergone pseudonymisation could still be attributed to a natural person with the help of additional information, it 'should be considered

[75] GDPR, rec 1 (emphasis added) (invoking Article 8(1) of the CFR and Article 16(1) of the TFEU).

[76] Recital 4 of the GDPR states that '[t]he right to the protection of personal data is not an absolute right; it must be considered in relation to its function in society and be balanced against other fundamental rights, in accordance with the principle of proportionality'. Recital 7 of the GDPR holds that natural persons 'should have control of their own personal data'.

[77] Lynskey (2015), p. 241.

[78] GDPR, art 15.

[79] GDPR, art 18.

[80] GDPR, art 20.

[81] GDPR, art 21.

[82] GDPR, art 22.

[83] GDPR, art 17.

[84] Reg 536/2014/EU, art 56(1) (mandating that, when trial data is recorded, processed, handled and stored by trial sponsors and investigators, the confidentiality of records and the personal data of the subjects remain protected according to the applicable law on personal data protection).

[85] Reg 726/2004/EC, art 57(d) (vesting in the EMA the responsibility to guarantee personal data protection while ensuring the dissemination of information regarding the adverse reactions to medicinal products authorised for marketing within the European Community).

[86] Dir 2001/83/EC, art 54a(3)(a).

[87] GDPR, rec 26.

to be information on an *identifiable* natural person'.[88] An 'identifiable' natural person is

> one who can be identified, directly or indirectly, in particular by reference to an identifier such as a name, an identification number, location data, an online identifier or to one or more factors specific to the physical, physiological, genetic, mental, economic, cultural or social identity of that natural person.[89]

Accordingly, the data status as personal or non-personal cannot be fixed permanently because, first, identification depends on the additional information allowing for the identification and, second, identification techniques continue to be improved. In the case of clinical trial data, pseudonymisation or anonymisation might be challenging, especially for data related to rare diseases.[90] The persisting risk of re-identification can restrain trial sponsors from transferring IPD,[91] which is arguably another indication that IPD cannot be susceptible to property rights owned by trial sponsors.[92]

In situations where the threshold of pseudonymisation can be met, or data anonymisation can be accomplished, the question arises whether such IPD could be subject to the right of ownership.

No *in rem* Rights in Sensor-Generated Data

Clinical trial data can represent qualitative or quantitative characteristics of health states (e.g. the measurements of the physiological process). In many cases, such data is obtained through the sensors of medical devices. In the EU, the issue of ownership of sensor-generated data was debated intensely in the context of the Digital Single Market strategy and the data-driven economy. According to the European Commission, 'raw' sensor-generated data *de lege lata* is not protected under property or IP

[88] ibid (emphasis added).

[89] GDPR, art 4(1).

[90] See e.g. CIOMS (2016), p. 51; Institute of Medicine of the National Academies (2015), p. 13. But see Cavoukian A, El Emam K (2011) Dispelling the myths surrounding deidentification: anonymization remains a strong tool for protecting privacy, p. 7 (finding that the re-identification of individuals in medical databases by using sophisticated de-identification tools and public information happens rarely). https://www.ipc.on.ca/wp-content/uploads/2016/11/anonymization.pdf. Accessed 26 Mar 2021. The authors point out that the re-identification is 'a difficult and time-consuming task' that requires technical skills. ibid.

[91] See e.g. PhRMA and EFPIA (18 Jul 2013) Principles for responsible clinical trial data sharing. Our commitment to patients and researchers, p. 4 (stating that drug companies would not share patient-level data 'when there is a reasonable likelihood that individual patients could be re-identified'). https://www.efpia.eu/media/25189/principles-for-responsible-clinical-trial-data-sharing.pdf. Accessed 26 Mar 2021.

[92] See also above (nn 61–63) and the accompanying text.

rights.[93] The policy and academic debate evolved around the idea of introducing a 'producer's right' in the machine-generated data in the form of *in rem* rights, including the exclusive rights to use data and licence its usage.[94]

Machine- or sensor-generated data is defined by the European Commission as

> [data] created *without the direct intervention* of a human by computer processes, applications or services, or by sensors processing information received from equipment, software or machinery, whether virtual or real.[95]

Clinical trial data—even if obtained by sensors of medical devices—is unlikely to satisfy this definition, given that trial data is gathered under the direct supervision of trial investigators, who are obliged to ensure the compliance with a protocol and the GCP standards,[96] as well as the protection of safety and well-being of study subjects during the trial conduct.[97] In this view, the debate over 'industrial data ownership' is of limited relevance for IPD.

Data Fixed and Stored in a Material Medium

As pointed out by the European Commission, 'intangible data' is a non-rivalrous good and 'as such cannot be captured by the traditional definitions of property law'.[98] While, conceptually, one can understand data as *abstract* (intangible) content, in practice, such content would be in most cases fixed and stored in a *material* carrier. The ownership of a data carrier as a physical and movable object is subject to property law.[99] However, protection of property rights in a material carrier can be of

---

[93]European Commission, SWD(2017) 2 final, p. 19. With regard to the eligibility for the trade secret protection, the Commission doubted whether 'individual data generated by interconnected machines and devices could be regarded as "trade secrets" in the sense of [the Trade Secrets] Directive, mostly because of its lack of commercial value as individual data; however, combination of data (datasets) can be trade secrets under [the Trade Secrets] Directive if all the criteria are met'.

[94]ibid p. 33.

[95]European Commission, COM(2017) 9 final, p. 9 (emphasis added). As pointed out by the Commission, sensor-generated data can be personal or non-personal, whereby the differentiation depends on whether a natural person can be identified. ibid.

[96]Reg 536/2014/EU, art 47.

[97]Reg 536/2014/EU, art 48.

[98]European Commission, SWD(2017) 2 final, p. 19. See also Max Planck Institute for Innovation and Competition (2017) Arguments against data ownership: ten questions and answers, p. 1 (pointing out that ownership rights 'can only be recognised and provided by law' and that 'a "data ownership right" does not currently exist either at EU or Member State level, or any other industrialised country'). https://www.ip.mpg.de/fileadmin/ipmpg/content/forschung/Argumentarium-Dateneigentum_eng.pdf. Accessed 26 Mar 2021.

[99]van Erp S (2017) Ownership of data and the *numerus clausus* of legal objects. Maastricht European Private Law Institute Working Paper No. 2017/6 1-23, p. 13. https://cris.maastrichtuniversity.nl/en/publications/ownership-of-data-the-numerus-clausus-of-legal-objects. Accessed 26 Mar 2021. The question of how property law of EU Member States applies to clinical trial data goes beyond the scope of this study.

limited relevance[100] for data, especially in an environment where data sharing becomes instantaneous, rampant and ubiquitous.

Property law is not the only legal basis for the protection against unauthorised access to data. Legislation on cybersecurity provides remedies against interference with technological measures of protection in a virtual space.[101] Furthermore, specialised regimes of IP protection such as trade secrets,[102] database protection and test data exclusivity need to be considered.[103]

4.2.2 The Applicability of the EU Trade Secrets Directive to Non-summary Clinical Trial Data

4.2.2.1 The Legal Definition of a Trade Secret

The EU Trade Secrets Directive[104] protects information against unauthorised acquisition, use, and disclosure[105] if it meets the following cumulative criteria. The information shall

(a) be secret in the sense that it is not, as a body or in the precise configuration and assembly of its components, generally known among or readily accessible to persons within the circles that normally deal with the kind of information in question;
(b) have commercial value because it is secret;

[100] See Drexl (2018), p. 90 (noting that 'reliance on property in the physical carrier will not provide sufficient protection where the person who has a legitimate interest in protection is different from the person owning the physical career').

[101] At EU level, criminal penalties for offences against information systems were harmonised by the Directive 2013/40/EU of the European Parliament and of the Council of 12 August 2013 on attacks against information systems and replacing Council Framework Decision 2005/222/JHA (14 Aug 2013) OJ L 218.

[102] While the relationship between trade secrets and IP has been debated, this study adheres to the view that trade secrets can be regarded as a subset of IP law, albeit they are not protected by exclusive rights. See European Commission (28 Nov 2013) Impact assessment accompanying the document proposal for a Directive of the European Parliament and of the Council on the protection of undisclosed know-how and business information (trade secrets) against their unlawful acquisition, use and disclosure, SWD(2013) 471 final, p. 66.

[103] As for other fields of IP law, copyright protection can apply to the scientific publications based on the IPD analysis, but not to 'raw' IPD as such; patents can be obtained for inventions that might be developed on the basis of IPD analysis (e.g. therapeutic applications). See Chap. 8 at Sect. 8.1.4.3 (discussing how secondary analysis of historical IPD might contribute to drug development).

[104] Directive 2016/943/EU of the European Parliament and of the Council of 8 June 2016 on the protection of undisclosed know-how and business information (trade secrets) against their unlawful acquisition, use and disclosure (15 Jun 2016) OJ L 157/1.

[105] Dir 2016/943/EU, art 4.

(c) be subject to reasonable steps under the circumstances, by the person lawfully in control of the information, to keep it secret.[106]

4.2.2.2 The Applicability of Trade Secret Protection to CSRs

According to the EMA Guidance on the redaction of CCI, the mere statement that clinical trial data includes trade secrets is insufficient and falls into the rejection category 'irrelevant justification'.[107] Meanwhile, the EMA's practice of granting access to CSRs has been contested by pharmaceutical companies alleging inter alia that CSRs contain trade secrets. In particular, in *Pari Pharma v EMA*, protection for similarity and superiority reports against disclosure by the EMA was claimed based on Article 39(2) TRIPS and the EU Trade Secrets Directive.[108] The CJEU held that the information contained in those reports was *generally known* since it was, in part, publicly available and, in part, 'could be obtained without difficulty and without any particular *inventiveness*'.[109] The claim for trade secret protection, in that case, was denied, given that the reports at issue contained information 'generally known among or readily accessible to persons within the circles that normally deal with the kind of information in question'.[110] The main implication of that ruling is that the qualification of the content of a CSR as a trade secret is context-dependent and, thus, as such, cannot be excluded.

While the judgement in *Pari Pharma v EMA* concerned CSRs, the question of whether trade secret protection applies to IPD requires further consideration.[111]

---

[106] Dir 2016/943/EU, art 2(1).

[107] EMA, External guidance on the publication of clinical data (n 43), pp. 53–54.

[108] Case T-235/15 *Pari Pharma v EMA* [2018] ECLI:EU:T:2018:65, para 57. In that case, the applicant failed to show that, 'in itself, the compilation of all the information consisted, for example, of new scientific conclusions or considerations relating to an inventive strategy which gives the undertaking a commercial advantage over its competitors and that it therefore provided added value'. ibid para 107.

[109] ibid para 113 (emphasis added). As a side note, it is worth noting that the criterion of 'inventiveness' is irrelevant for the question of the eligibility for trade secret protection. The reports at issue contained the results of the market survey conducted by Pari Pharma. See also EMA, External guidance on the publication of clinical data (n 43), p. 49 (stating that, in many cases, clinical study reports build on 'logic and common sense in line with the content of publicly available documents', including scientific and regulatory guidelines and guidance documents and, therefore, should be deemed as common knowledge).

[110] ibid paras 113–114.

[111] At the time of writing, the EMA does not disclose IPD.

4.2.2.3 The Applicability of Trade Secret Protection to IPD

The qualification of data as personal does not eliminate the applicability of trade secret protection.[112] Thus, it is important to consider whether IPD can meet the qualifying criteria under the EU Trade Secrets Directive.

The Secrecy Requirement

To fulfil the secrecy criterion, information or data should not be 'generally known among or readily accessible to persons within the circles that normally deal with the kind of information in question'. In the case of IPD, the obligation on trial sponsors and trial investigators to protect all data from trials against unauthorised access and disclosure[113] effectively excludes medical researchers from 'normally dealing with' IPD, notwithstanding the desirability of secondary IPD analysis.[114] Where drug authorities might be viewed as those who 'normally deal with' IPD when assessing applications for drug marketing authorisation, they are likely to be bound by the obligation of professional secrecy.[115] Where test data might be shared on a contractual basis, the party acquiring access to data would usually be bound by confidentiality obligations.[116] In other words, the secrecy of IPD would still be preserved, even if not in the absolute sense.

The situation is somewhat paradoxical, given a strong normative impetus for the accessibility of primary research findings as a principle of scientific research in general and medical research in humans in particular.[117] Besides, it should be

---

[112]The coexistence of data protection and trade secrets is not excluded. For instance, Recital 35 of the EU Trade Secrets Directive emphasises that the trade secret holder, when taking measures to protect a trade secret, should respect the right to protection of personal data of individuals whose personal data may be processed. Recital 63 of the GDPR states that, where possible, 'the controller should be able to provide remote access to a secure system which would provide the data subject with direct access to his or her personal data' and that such right 'should not adversely affect the rights or freedoms of others, including trade secrets'. See also Article 29 Data Protection Working Party (13 Dec 2016) Guidelines on the right to data portability. WP 242 rev.01, p. 10 (emphasising that '[a] potential business risk cannot [...] in and of itself serve as the basis for a refusal to answer the portability request and data controllers can transfer the personal data provided by data subjects in a form that does not release information covered by trade secrets or intellectual property rights'). https://ec.europa.eu/information_society/newsroom/image/document/2016-51/wp242_en_40852.pdf. Accessed 26 Mar 2021. On the relationship between trade secret protection and personal data protection, see e.g. Graef et al. (2018).

[113]Above at Sect. 4.2.1.2, subheading 'The Obligation to Protect Data Against Unauthorised Access as the Source of de facto Exclusive Control' in this chapter.

[114]As discussed in Chap. 3.

[115]The duty of confidentiality of EU institutions—thus, the EMA—is guaranteed under Article 339 of the TFEU.

[116]See Chap. 5 at Sect. 5.1.3.

[117]See CIOMS (2016), p. 51 (emphasising that researchers, sponsors and research ethics committees 'must share data for further research where possible'); Krumholz (2015).

emphasised that the regulatory requirement to store trial data 'in such a way that it can be [...] verified'[118] is rationalised precisely on the grounds of enabling 'effective monitoring' and pharmacovigilance.[119] Nevertheless, medical practitioners and researchers are rarely able to access IPD held by drug companies.[120] The very objective of the EMA's publication policy—to enable 'the wider scientific community to make use of detailed clinical data to develop new knowledge in the interest of public health' and to 'allow third parties to verify the original analysis and conclusions'[121]—indicates that, as a matter of the status quo, primary data is not accessible to those who could conduct secondary IPD analysis.

The Requirement of Commercial Value Due to Secrecy

The second prong requires establishing a causal relationship between the secrecy of information and its commercial value.[122] At the outset, it may not be clear what commercial value might result from the secrecy of clinical trial data. As noted earlier, trial data has, first and foremost, social and scientific value in terms of informing clinical practice and medical research.[123] Methods commonly used in the valuation of intangible assets could be applied to define the potential commercial value of trial data.[124]

First, one might be inclined to relate the economic value of clinical trial data with the investment made into conducting trials. Such approach would be in line with the method based on the *historical costs* of acquiring an asset.[125] However, to satisfy the commercial value requirement, the causal link between secrecy and commercial value should be established. The costs-based method does not relate the sunk costs of generating data to confidentiality protection. Hence, the costs incurred in conducting trials cannot be relevant for the eligibility for trade secrets protection.

A *market-based* approach estimates the value of an asset by comparing actual arm's-length market transactions involving similar assets in the existing market.[126] Such valuation method can be applied where information on the (comparable) actual market transactions is available. However, it can be of limited utility for intangible assets if neither such assets nor comparable goods are traded.[127] A variation of a market-based method is the *licensing analogy*, which determines the value of an

---

[118] Above (n 55) and the accompanying text.

[119] Above (nn 61–62) and the accompanying text.

[120] For a review of evidence, see Chap. 6 at Sect. 6.3.2.

[121] EMA publication policy 0070, p. 4.

[122] Dir 2016/943/EU, art 2(1)(b).

[123] See Chap. 2 at Sect. 2.1.2.

[124] Anson (2005), p. 32 ff.

[125] ibid p. 33 (noting that historical costs provide 'an absolute minimum value for the asset').

[126] Anson (2005), p. 34.

[127] ibid.

asset based on its exploitation through licensing.[128] While the licensing analogy can help determine the value of intellectual assets such as technological know-how, it might be of limited relevance for clinical trial data. The main 'business function' of efficacy and safety data subsists in supporting drug marketing authorisation—after that, it is usually not marketed as a commodity.[129] In situations where IPD is shared for 'legitimate research' purposes, access is usually granted without payment obligations under confidentiality agreements.[130]

The next method presupposes that an intangible asset can have 'an *implied* value with no tangible proof that the value would have or could have been realized'.[131] Such value can be 'expressed as *competitive advantage* value, and it is measured by imputing a company's competitive position or competitive value with and without the trade secret'.[132] Such approach echoes the notion of 'actual or potential commercial value' under the EU Trade Secrets Directive 'where its unlawful acquisition, use or disclosure is likely to harm the interests of the person lawfully controlling it, in that it undermines that person's *scientific* and *technical potential*, business or financial interests, strategic positions or *ability to compete*'.[133] Accordingly, the commercial value of IPD could be defined based on the assessment of actual or potential competitive effects of its disclosure. As will be discussed more in detail later, disclosure can entail two types of competitive effects. First, one needs to consider whether the competitors' use of data can facilitate *price competition*, e.g. whether clinical dossiers can be used for marketing approval of a generic product.[134] Second, it should be considered whether exploratory IPD analysis by competitors might facilitate the development of a new product and, thus, diminish the advantage of the trial sponsor (the initial holder of the re-analysed data) in competition in innovation. In such situations, the value of data cannot be determined in abstract terms. Given that the second requirement of the trade secret definition is context-dependent,[135] its fulfilment would depend on how the third-party use of data

---

[128] Jager (2002), p. 135 ff.

[129] See below (Chap. 9, nn 29–32) and the accompanying text.

[130] See e.g. Data use agreement, para 1.2. http://yoda.yale.edu/data-use-agreement. Accessed 26 Mar 2021. See also CSDR standard contract template for clinical trial data sharing (10 Apr 2017). https://www.clinicalstudydatarequest.com/Documents/CSDR%20DATA%20SHARING%20AGREEMENT%20Version%201%204.10.2017.pdf. Accessed 26 Mar 2021.

[131] Anson et al. (2005), p. 84 (emphasis added).

[132] ibid (emphasis added).

[133] Dir 2016/943/EU, rec 14 (emphasis added).

[134] For an analysis of how IPD disclosure might affect regulatory incentives that delay the marketing authorisation of generic products, Chap. 5 at Sect. 5.4.

[135] As the CJEU held in *EMA v AbbVie*:

> The extent to which the disclosure of [clinical study reports] causes serious and irreparable harm depends on a combination of *circumstances*, such as, inter alia, the professional and *commercial importance* of the information for the undertaking which provides it and the utility of *that* information for other undertakings which are liable to examine and use it subsequently.

Case C-389/13 *EMA v AbbVie* [2013] ECLI:EU:C:2013:794, para 42 (emphasis added).

would impact competitive dynamics under the particular market conditions. As it will be shown later, control over third-party secondary IPD analysis can play a more prominent role in protecting competitive advantage in R&D[136] rather than in originator-generic competition.[137]

The Requirement of Protection Measures to Preserve Secrecy

The fulfilment of the requirement that 'the person lawfully in control' of the information or data needs to undertake 'reasonable steps [. . .] to keep it secret'[138] is quite straightforward in the case of IPD. Trial sponsors and investigators can be deemed as 'the persons lawfully in control' of trial data. As pointed out above, they have a duty under the EU Clinical Trials Regulation to implement appropriate technical and organisational measures to protect all information and data obtained in trials, including against unauthorised access, disclosure, dissemination or accidental loss.[139] Due to this obligation and the duty to protect the personal data of trial participants,[140] the threshold of 'reasonableness' of the measures undertaken to protect trial data against disclosure can be easily met.

On balance, *de lege lata* IPD would likely qualify for protection under the EU Trade Secrets Directive. From a *de lege ferenda* perspective, such outcome appears contentious. The secrecy condition is incompatible with the very purpose of conducting medical research in humans—gaining validated medical knowledge, which, by definition, requires unconditional access to primary data for confirmatory analysis. Given that the qualifying criteria under Article 2(1) of the EU Trade Secrets Directive apply cumulatively, the situation in the EU can change once the second phase of the EMA's publication policy 0043 is launched and IPD becomes 'readily accessible' within a broader circle of medical researchers.[141]

---

[136] For a more detailed discussion of implications of IPD disclosure for competition in R&D, see Chap. 8 at Sect. 8.1.4.

[137] On the potential re-use of IPD for obtaining marketing approval for generic drugs, see Chap. 5 at Sect. 5.4.1.

[138] Dir 2016/943/EU, art 2(1)(c) (emphasis added).

[139] Reg 536/2014/EU, art 56(2).

[140] Reg 536/2014/EU, arts 56, 81(4) and 93.

[141] On the implementation of the EMA publication policy 0070, see Chap. 2 at Sect. 2.3.1.2.

4.2.3 The Applicability of the EU Database Directive to IPD

4.2.3.1 IPD from an Individual Trial

The EU Database Directive[142] sets the conditions for protecting databases, defined as collections of 'independent works, data or other materials arranged in a systematic or methodical way and individually accessible by electronic or other means'.[143] There is little doubt that IPD is gathered and arranged systematically and methodically, particularly according to the methodology described in a trial protocol[144] and following the structure of a patient case report form.[145]

Can IPD obtained in a trial be deemed as a collection of 'independent' data? The basic unit of IPD can be seen as a single quantitative measurement or a qualitative characteristic of the health state obtained according to the safety and efficacy variables pre-specified in a trial protocol.[146] One could argue that such measurements present a meaningful whole only in their *entirety*, which has to do with the statistical power[147] and the sample size[148] of a trial. The number of trial participants—hence, the aggregate number of measurements characterising the state of health—is based on the statistical power considerations and needs to be 'large enough to provide a reliable answer to the question addressed'.[149] In turn, the research question corresponding to the trial outcome of interest[150] is statistically described as variables allowing to assess an effect size of medical intervention.[151] On the one hand, one could argue that all data elements can be viewed as semantically dependent since their analytical utility subsists in the totality of evidence. On the other hand, IPD is

[142] Directive 96/9/EC of the European Parliament and of the Council of 11 March 1996 on the legal protection of databases (27 Mar 1996) OJ L 77.

[143] Dir 96/9/EC, art 1(2).

[144] A trial protocol describes the methodology for answering a research question in a credible way. Reg 536/2014/EU, art (2)(22) and annex I(D).

[145] See ICH (10 Jun 1996) ICH Harmonised tripartite guideline. Guideline for good clinical practice. E6, para 1.11 (defining a case report form as a 'document designed to record all of the protocol required information to be reported to the sponsor on each trial subject').

[146] ICH (5 Feb 1998) ICH Harmonised tripartite guideline. Statistical principles for clinical trials. E9, p. 7 ff.

[147] The power of a trial means the statistical significance of the trial evidence. More specifically, it refers to the statistical probability that the null hypothesis will be rejected if it is not true. Day (2007), p. 160.

[148] The trial sample size refers to the number of trial participants and is estimated based inter alia on calculating the power of a trial.

[149] ICH (5 Feb 1998) ICH Harmonised tripartite guideline. Statistical principles for clinical trials. E9, p. 16.

[150] An 'outcome of interest' is defined as a measurable effect of a medicinal intervention on the disease clinical manifestations.

[151] The 'effect size' of a health intervention refers to the measure that correlates with the outcome of interest and statistically characterises the trial hypothesis.

recorded separately for each patient and can be interpreted individually for each study subject and, thus, have autonomous meaning.

In *Verlag Esterbauer*,[152] the CJEU interpreted the concept of 'independent materials' rather 'generously',[153] holding that, as long as an individual piece of—in that case, geographical—information can be extracted, it has 'autonomous informative value' and, thus, constitutes an independent material in the sense of the Database Directive.[154] Whether an individual piece of information extracted from a collection has 'autonomous informative value' is assessed 'in the light of the value of the information not for a typical user of the collection concerned, but for each third party interested by the extracted material'.[155] In light of this interpretation, IPD as a collection of measurements is likely to fulfil the definitional criterion of data elements independence, given that each measurement can characterise the state of health of individual study participants.

Once the definitional prerequisites are met, the eligibility for two types of protection under the EU Database Directive should be considered, namely, copyright and the *sui generis* database right.

4.2.3.2 The Applicability of the Copyright Type of Database Protection

Copyright protection can apply if a database constitutes 'the author's own intellectual creation' due to the selection or arrangement of its contents.[156] According to the established case law of the CJEU, this standard is met if, 'through the selection or arrangement of the data which [a database] contains, its author expresses his *creative ability* in an *original manner* by making *free* and *creative choices* [...] and thus stamps his "*personal touch*"'.[157] The exercise of freedom and creativity can hardly be contemplated in the case of IPD. The arbitrary selection of IPD is a thorny issue as it points in the direction of biases in the presentation and reporting of evidence that can distort the interpretation of the trial outcomes.[158] As far as the arrangement of IPD is concerned, individual case reports are compiled and stored according to the *standardised* predetermined structure and format, notwithstanding whether IPD is obtained from a single trial or whether it can be aggregated from different trials.[159] In this regard, trial investigators, trial sponsors or an entity willing to undertake data

---

[152] Case 490-14 *Freistaat Bayern v Verlag Esterbauer* [2015] ECLI:EU:C:2015:735.

[153] Drexl (2018), p. 74.

[154] Case 490-14 *Freistaat Bayern v Verlag Esterbauer* [2015] ECLI:EU:C:2015:735, para 24 ff.

[155] ibid para 27.

[156] Dir 96/9/EC, art 3(1).

[157] Case C-604/10 *Football Dataco and Others* [2012] ECLI:EU:C:2012:115, para 38 (and case law cited) (emphasis added).

[158] For a detailed discussion, see Chap. 6 at Sects. 6.4.2.2 and 6.4.2.3.

[159] CSRs follow the structure specified in Directive 2001/83/EC, annex I, part I, module 5. See also Chap. 3 at Sect. 3.1.2.

pooling would normally be constrained by technical considerations that 'leave no room for creative freedom'.[160] Consequently, aggregated IPD is unlikely to qualify for copyright protection for databases.

4.2.3.3 The Applicability of the sui generis Database Right

The eligibility for protection under the *sui generis* database right is contingent on whether a qualitatively and (or) quantitatively substantial investment is made in obtaining, verifying, or presenting the database contents.[161] The investment made in either of these activities has to be independent of the resources spent creating the database materials.[162] In the case of IPD, this excludes the costs of conducting trials.

The difficulty in drawing a line between 'creating' and 'obtaining' data has been pointed out by legal scholars[163] and reportedly encountered in practice.[164] Such distinction can be especially problematic for 'sole source' databases, where it might be difficult to show that an independent investment was made in obtaining the database contents.[165] Clinical trial data can easily fall within the category of 'sole source' databases, given that trials are designed to address novel questions[166] and, hence, generate unique data.[167] Accordingly, in situations where IPD from past trials is shared contractually, there is no act of 'obtaining' data—separate from generating IPD—on the part of a trial sponsor (data holder). However, on the part of an entity acquiring IPD for its subsequent arrangement as a database, the corresponding expenditure can qualify as an independent investment in obtaining the database

---

[160]Case C-604/10 *Football Dataco and Others* [2012] ECLI:EU:C:2012:115, para 39 (and case law cited).

[161]Dir 96/9/EC, art 7(1). Recital 40 specifies that 'such investment may consist in the deployment of financial resources and/or the expending of time, effort and energy'.

[162]Established case law of the CJEU: Case C-203/02 *BHB Horseracing* [2004] ECLI:EU: C:2004:695, para 35; Case C-444/02 *Fixtures Marketing* [2004] ECLI:EU:C:2004:697, para 45. As held by the CJEU, the 'purpose of the protection by the *sui generis* right provided for by the [database] directive is to promote the establishment of storage and processing systems for existing information and not the creation of materials capable of being collected subsequently in a database'. ibid para 40.

[163]See e.g. Drexl (2018), p. 70 ff; Leistner (2017), p. 28.

[164]See Fisher et al. (2018), p. 61 (reporting that it is mainly database producers, who often encounter difficulty in making the distinction between creating and obtaining data as it is 'sometimes impossible for database producers to separate [the costs] and, therefore, to prove the investment in these two types of effort').

[165]See Drexl (2018), p. 70 (noting that, '[b]y excluding investment in the creation of data, the case law [of the CJEU] considerably reduces the likelihood that a sui generis right will exist in so-called "sole source" databases').

[166]Cummings et al. (2007), p. 20.

[167]This is likely to be the case with scientific data in general. See Beunen (2007), p. 233 (noting that scientists discovering factual data become single-source producers of information holding 'a *de facto* monopoly on their materials').

contents.[168] Such investment can give rise to protection under the *sui generis* database right, provided that the substantiality criterion is met.[169]

As far as the investment in the verification of the database contents is concerned, the requirement was interpreted by the CJEU as referring to 'the resources used, with a view to ensuring the reliability of the information contained in that database, to monitor the accuracy of the materials collected when the database was created and during its operation'.[170] In the case of IPD, substantial resources, in principle, should not be necessary to verify the reliability and accuracy of data. The need for such efforts would indicate a systematic violation of the principle of reliability and robustness of trial data.[171] As a rule, resources related to IPD reliability and robustness would be expended during the design and conduct of clinical trials.[172] As clarified by the CJEU, such investment would not fall within the scope of Article 7(1) of the Database Directive.[173]

The presentation of aggregated IPD as the database contents can be a plausible candidate to fulfil the requirement of an independent investment where IPD is prepared for secondary use, which is unrelated to the generation of data. For instance, IPD anonymisation involves considerable costs[174] and can constitute a standalone activity required for enabling secondary data analysis. Next, one would think that expenditures related to the setup of the data-sharing platforms, such as the Clinical Study Data Request (CSDR),[175] could qualify as an independent investment. However, the CSDR portal—the industry's major initiative for IPD sharing to

[168] See Drexl (2018), p. 71 (pointing out that 'the distinction between creating and obtaining data can easily be applied where the underlying data were created by a person or entity that is different from the database maker').

[169] The substantiality criterion is unlikely to present a hurdle, given that the case law set a rather low threshold. Leistner (2017), p. 30.

[170] Case C-203/02 *BHB Horseracing* [2004] ECLI:EU:C:2004:695, para 42.

[171] Reg 536/2014/EU, art 3(b). On the concepts of reliability and robustness of clinical trial data, see this chapter at Sect. 4.1.2.1, subheading 'Data Reliability and Robustness'. On systematic errors in design and methodology of clinical trials that can impair the reliability and robustness of trial data, see Chap. 6 at Sects. 6.4.2.2–6.4.2.4.

[172] Reg 536/2014/EU, art 48 (providing inter alia for the obligation on trial sponsors to ensure reliability and robustness of trial data).

[173] Case C-203/02 *BHB Horseracing* [2004] ECLI:EU:C:2004:695, para 42.

[174] See e.g. Institute of Medicine of the National Academies (2015), p. 68 (reporting that the costs of de-identifying a data-set can range from 10,000 to 100,000 US dollars, and the costs of developing the in-house capacity to automate data de-identification range from 100,000 to 500,000 US dollars). Such costs can easily meet the 'substantiality' threshold, which is viewed to be rather low. See Leistner (2017), p. 30; Fisher et al. (2018), p. 53 (noting that 'commentators are split on the issue [of the threshold of the investment eligible for the sui generis protection] but generally, national courts have been generous and granted protection for relatively low-level investments').

[175] https://clinicalstudydatarequest.com. Accessed 26 Mar 2021.

date—functions only as a data broker facilitating the negotiations between IPD holders and parties interested in undertaking secondary data analysis.[176]

In sum, primary ('raw') data as such directly obtained in clinical trials, even if it can meet the definitional criteria of a database under the EU Database Directive, is unlikely to qualify for protection under the *sui generis* database right. At the same time, the applicability of the *sui generis* database protection to aggregated IPD cannot be excluded in the context of data sharing where substantial investment might be made into the obtaining (from third parties), verification or presentation of the pre-existing IPD as a database. However, the current lack of IPD databases can call into question the effectiveness of this type of incentive in the case of clinical trial data.[177]

4.2.4 *Data Exclusivity Protection*

Test data exclusivity refers to a *sui generis* type of protection adopted by many jurisdictions with varying modalities.[178] By delaying generic competition, it intends to promote innovation incentives of research-based drug companies.[179] In the EU, this type of protection was introduced by Directive 87/21/EEC.[180] Since test data exclusivity presents a derogation from the abridged marketing authorisation procedure, it is helpful to start by explaining what the latter means.

4.2.4.1 The Abridged Procedure for Drug Marketing Authorisation

As a rule, applicants for drug marketing authorisation are required to submit the results of pharmaceutical (chemical, biological or microbiological) tests, preclinical

[176] Once a data-sharing agreement is signed, 'a research area is created in the secure data access system' where the trial sponsor can upload anonymised data and supporting documentation. Frequently asked questions. https://clinicalstudydatarequest.com/Help/Help-FAQS.aspx. Accessed 26 Mar 2021.

[177] On the restrictive data-sharing policies and practices of drug companies, see Chap. 5 at Sect. 5.1.3.

[178] For an overview of data exclusivity regimes in 44 jurisdictions, see IFPMA (2011) Data exclusivity: encouraging development of new medicines. https://www.ifpma.org/wp-content/uploads/2016/01/IFPMA_2011_Data_Exclusivity__En_Web.pdf. Accessed 26 Mar 2021. The international proliferation of test data exclusivity can be attributed to the protection obligations under Article 39(3) of the TRIPS Agreement and the FTAs. For a detailed account, see de Carvalho (2018) and Shaikh (2016).

[179] For a critical account, see Chap. 5 at Sect. 5.2.4.2.

[180] Council Directive 87/21/EEC of 22 December 1986 amending Directive 65/65/EEC on the approximation of provisions laid down by law, regulation or administrative action relating to proprietary medicinal products (17 Jan 1987) OJ L 15.

(toxicological and pharmacological) tests, and clinical trials.[181] The abridged authorisation procedure means that this requirement can be waived for generic applicants, who can instead obtain marketing approval based on the bioequivalence studies demonstrating the interchangeability of a generic product with the originator drug.

A *generic* medicinal product is defined as

> a medicinal product which has the same qualitative and quantitative composition in active substances and the same pharmaceutical form as the reference medicinal product, and whose *bioequivalence* with the reference medicinal product has been demonstrated by appropriate *bioavailability* studies.[182]

The abridged procedure is also known as approval by referencing the originator's clinical trial data or the 'referential use' of the originator's data.[183] This pathway was introduced by the Directive 65/65/EEC.[184] The underlying policy rationale is to avoid 'repetitive tests on humans or animals without over-riding cause'.[185]

4.2.4.2 Data Exclusivity as a Temporary Derogation from the Abridged Procedure

In the EU, the duration of test data exclusivity was harmonised under the Directive 2004/27/EC[186] amending the Directive 2001/83/EC. The protection term is often expressed as a summation '8+2+1', which means that

---

[181] Dir 2001/83/EC, art 8(3)(i). Annex I lists the particularities of an application dossier.

[182] Dir 2001/83/EC, art 10(2)(b) (emphasis added). Furthermore, the provision clarifies that

> different salts, esters, ethers, isomers, mixtures of isomers, complexes or derivatives of an active substance shall be considered to be the same active substance, unless they *differ significantly* in properties with regard to safety and/or efficacy. In such cases, additional information providing proof of the safety and/or efficacy of the various salts, esters or derivatives of an authorised active substance must be supplied by the applicant (emphasis added).

[183] The 'reference medicinal product' means the originator drug authorised on the basis of full efficacy and safety data. Dir 2001/83/EC, art 10(2)(a).

[184] Council Directive 65/65/EEC of 26 January 1965 on the approximation of provisions laid down by Law, Regulation or Administrative Action relating to proprietary medicinal products (9 Feb 1965) OJ 022.

[185] As affirmed by the CJEU, the purpose of the abridged procedure is 'to relieve applicants for marketing authorisation of the obligation to carry out pharmacological and toxicological tests and clinical trials' and, in so doing, avoid the repetition of tests on humans or animals, 'unless absolutely necessary'. Case C-368/96 *The Queen v The Licensing Authority* [1998] ECLI:EU:C:1998:583, paras 69, 71. See also Dir 2001/83/EC, rec 10; Dir 87/21/EEC, rec 4.

[186] Directive 2004/27/EC of the European Parliament and of the Council of 31 March 2004 amending Directive 2001/83/EC on the Community code relating to medicinal products for human use (30 Apr 2004) OJ L 136.

- during eight years after the originator's drug approval, the originator's clinical data cannot be used (referenced) for generic drug approval (i.e. test data exclusivity);
- during the subsequent two years, an application for a generic product can be reviewed and approved based on the referenced data, but the product cannot be launched on the market until the ten-year term, calculated from the date of the reference product approval, expires (i.e. market exclusivity);
- an additional, one (non-cumulative) year of protection can be granted if, during the first eight out of ten years of protection, the holder of the marketing authorisation succeeds in obtaining authorisation for one or more new therapeutic indications (provided that such indications represent 'a significant clinical benefit in comparison with existing therapies'[187]).[188]

Let us consider next the scope of test data exclusivity and the nature of protection.[189]

4.2.4.3 The Nature and Scope of Test Data Exclusivity Protection

Test data exclusivity should be distinguished from the duty of confidentiality of government authorities.[190] As explained earlier, test data exclusivity provides for a *sui generis* type of protection, which is implemented in the form of the time-limited derogation from the abbreviated drug marketing authorisation pathway.[191] In this regard, test data in and of itself is a symbolic object of protection—rather than data as such, protection is directed at the economic interests of trial sponsors in earning returns on R&D[192] and, thus, at facilitating innovation incentives.

In the 1996 judgement, the CJEU held that protection under Article 4(8)(a)(iii) of Directive 65/65/EEC, as amended by Directive 87/21/EEC (the 'precursor' of Article 10(a)(iii) of the Directive 2001/83/EC),

> confers on the owner of [the original authorised medicinal] product an *exclusive right to make use* of the results of the pharmacological and toxicological tests and clinical trials

---

[187] Dir 2001/83/EC, art 10(1), para 4.

[188] Dir 2001/83/EC, art 10. For a lay summary of these rules, see European Commission (2018) Notice to applicants, vol. 2A, rev 9, p. 42 ff.

[189] On the potential impact of CSRs and IPD disclosure on this type of protection, see Chap. 5 at Sect. 5.4.1.

[190] In the case of the EMA, the primary source of the duty 'not to disclose information of the kind covered by the obligation of professional secrecy' is Article 339 of the TFEU. See also below at Sect. 4.3.3 in this chapter.

[191] See above at Sect. 4.2.4.1 in this chapter.

[192] Dir 2001/83/EC, rec 9 (pointing out the need to ensure that 'innovative firms are not placed at a disadvantage').

> placed in the file on that product for a period of 6 or 10 years from the grant of the first marketing authorisation for that product in the Community.[193]

The balance of interest concerning the provision at issue was articulated by the Court as follows:

> the Community legislature took account of the interests of innovating firms in its approach to the *right to property relating* to pharmacological, toxicological and clinical data and, to a certain extent, ensured the protection of innovation, while pursuing the aim of avoiding repetition of tests on humans or animals unless absolutely necessary.[194]

A few points need to be clarified regarding the Court's pronouncement of data exclusivity as a property right.[195] First of all, while the Court recognised the right to property in relation to pharmacological, toxicological and clinical data, it is assumed that it referred to the *dossiers* submitted for drug marketing authorisation and not to *any* data gathered in clinical studies. Second, the wording 'the right to property *relating* to data' indicates that the right does not subsist in data as such but in *making use* of the results of the pharmacological, toxicological and clinical studies, as proof of efficacy and safety of a drug candidate, within the limited period.[196] Such right is exclusive in the sense that subsequent applicants are excluded from 'relying' on the originator's test data, while the data holder can still consent to the 'referential' use of the safety and efficacy data.[197] Accordingly, test data exclusivity shall be viewed as affording a context-specific, time- and purpose-bound type of regulatory protection and not as vesting property-type rights in the *source* data (IPD) as such. Furthermore, the requirements and practice of IPD analysis differ among drug authorities, and, in many cases, IPD is not even submitted within an application dossier.[198]

4.2.5 Contractually Obtained Exclusive Control

4.2.5.1 Contractual Practice Related to Obtaining Clinical Trial Data

Two categories of contracts related to obtaining clinical trial data can be distinguished: agreements between trial sponsors and trial participants and collaborative agreements between trial sponsors and other parties involved in conducting trials (such as trial investigators and funders).

---

[193] Case C-368/96 *The Queen v The Licensing Authority* [1998] ECLI:EU:C:1998:583, para 81 (emphasis added).

[194] ibid para 83.

[195] It is worth pointing out that the decision in *The Queen v The Licensing Authority* has been criticised for not addressing the issue of test data protection form a 'proper dogmatic' perspective. Cottier et al. (2000), p. 60.

[196] Case C-368/96 *The Queen v The Licensing Authority* [1998] ECLI:EU:C:1998:583, para 81.

[197] Dir 2001/83/EC, art 10(1)(a)(i) (emphasis added).

[198] See Chap. 6 at Sect. 6.5.1.2.

(i) Contracts for informed consent

Under the GDPR, individual patient data falls within one of the special categories of personal data: data concerning health.[199] Processing such data is generally prohibited unless 'the data subject has given explicit consent to the processing of those personal data for one or more specified purposes'.[200] Thus, a trial can be conducted only if all study subjects have given their informed consent,[201] which is defined as

> a free and voluntary expression of willingness to participate in a particular clinical trial, after having been informed of all aspects of the clinical trial that are relevant to the subject's decision to participate [. . .] in the clinical trial.[202]

The EU Clinical Trials Regulation envisages that, without prejudice to personal data protection rights, sponsors may also ask the study subjects for their prior consent for 'the use of his or her data *outside* the protocol of the clinical trial *exclusively for scientific purposes*'.[203]

(ii) Collaborative agreements

Collaborative agreements can be concluded between the trial sponsors, co-sponsors and medical research institutions with expertise and facilities to carry out a trial (contract research organisations). Notwithstanding the absence of statutory property rights in personal data, one can come across a contractual definition of all data gathered in trials as 'exclusive property' of trial sponsors. For instance, the clause on 'Ownership and Intellectual Property' extracted from an agreement between a US sponsor and an EU-based investor for conducting provided by Osborne Clarke LLP[204] reads:

> All data [. . .] of the Client provided to Party A by and/or on behalf of the Client in connection with the Services, including all data and information from the Clinical Trial existing prior to and after the Effective Date of this Agreement, irrespective of whether provided in paper, oral, electronic or other form (including, but not limited to original case

[199] GDPR, art 9(1). 'Data concerning health' is defined as 'personal data related to the physical or mental health of a natural person, including the provision of health care services, which reveal information about his or her health status'. GDPR, art 4(15).

[200] GDPR, art 9(2)(a).

[201] Reg 536/2014/EU, art 28(1)(c).

[202] Reg 536/2014/EU, art (2)(21).

[203] Reg 536/2014/EU, art 28(2) (emphasis added). Furthermore, Recital 76 states that, to ensure 'the robustness and reliability of data from clinical trials used for *scientific* purposes', the withdrawal of informed consent 'should not affect the results of activities already carried out, such as the storage and use of data obtained on the basis of informed consent *before* withdrawal' (emphasis added). For an analysis of the GDPR provisions relevant for clinical trial data processing for primary and secondary research purposes, see European Data Protection Board (23 Jan 2019) Opinion 3/2019 concerning the questions and answers on the interplay between the Clinical Trials Regulation (CTR) and the General Data Protection Regulation (GDPR) (art. 70.1.b)).

[204] Osborne Clarke LLP (2016).

> report forms, and including dictionaries and data entry copies of case report forms), shall be the *exclusive property* of Client.[205]

Furthermore, the contract states that

> any and all data and other information (tangible and intangible) resulting from, and/or generated or made in the performance of the Services or in support of the Clinical Trial prior to and after the Effective Date of this Agreement, including without limitation writings (irrespective of whether in written, oral or electronic form) [...] shall be the *exclusive property* of the Client.[206]

4.2.5.2 Implications of Contractually Defined 'Exclusive Property' in Trial Data

Property law in modern legal systems—both civil law and common law traditions—tends to be 'heavily codified'.[207] In civil law jurisdictions, the *numerus clausus* principle prevents the creation of property rights 'outside of the typical scheme provided by the Code'.[208] Assuming that there are neither statutory property rights in personal data,[209] nor can data, as an intangible subject matter, 'be captured by the traditional definitions of property law',[210] the contractual characterisation of trial data as 'exclusive property' can be reduced to a material carrier, in which data can be stored. Hence, protection under national property law can be applied to an extent to which third-party unauthorised access to and use of data interferes with property rights in such material carrier.[211]

To summarise, none of the legal regimes surveyed above appears to provide for property-type rights in individual patient-level data as such. The legal status quo of IPD can be defined as de facto exclusive control of trial sponsors due to the obligation to store and protect all data gathered in trials against unauthorised use.[212] The next section examines legal determinants of access to clinical trial data under the applicable regimes at EU level.

---

[205] ibid p. 154 (emphasis added).

[206] ibid (emphasis added).

[207] Mattei (1998), p. 160.

[208] ibid p. 161. See also von Bar and Drobnig (2004), p. 320 ff.

[209] Above at Sect. 4.2.1.2, subheading 'No Property Rights in IPD as Personal Data'.

[210] European Commission, SWD(2017) 2 final, p. 19 (emphasis added).

[211] See above (nn 98–100) and the accompanying text.

[212] Above (n 56–60) and the accompanying text.

4.3 Access Regimes Applicable to Clinical Trial Data

4.3.1 Regulatory Requirements for Clinical Trial Data Disclosure

4.3.1.1 What Data Is (Supposed to Be) in the Public Domain?

As mentioned earlier, several databases provide information about trials conducted in the EU.[213] Study descriptions and summary results can be accessed via the EU Clinical Trials Register.[214] Summary results and summary to laypersons[215] shall also be accessible via the EU clinical trial database established under the EU Clinical Trials Regulation.[216] The obligation applies irrespective of the trial outcomes, which is an important novelty compared to the requirement under the Clinical Trials Directive to submit the SUSARs to the database accessible only by the Commission, the NCAs and the EMA.[217]

Furthermore, where a trial is conducted in view of obtaining drug marketing authorisation, applicants for marketing authorisation are also required to submit CSRs to the EU database.[218] According to some policymakers, this obligation is 'a huge step forward [that] means that Europe is now leading the way globally in clinical trial transparency'.[219] However, 'full' clinical study reports might not be available, given the reservation for CCI. Some scholars find that the EU Clinical Trials Regulation 'takes a modest step towards transparency'[220] as it lacks the mandatory obligations to share full data and publish CSRs from unsuccessful clinical trials.[221] Nevertheless, the importance of the new transparency requirements should not be downplayed, considering that the Clinical Trials Directive did not contain provisions addressing public accessibility of clinical trial data.

As far as IPD is concerned, the EU Clinical Trials Regulation states that only if the sponsor decides to share 'raw' data voluntarily, the European Commission 'shall produce guidelines for the formatting and sharing of those data'.[222] In this regard,

---

[213] See generally Chap. 3.

[214] EMA. EU Clinical Trials Register. About the EU Clinical Trials Register. https://www.clinicaltrialsregister.eu/about.html.

[215] Reg 536/2014/EU, rec 39; art 37(4). The contents of both types of summaries are specified in Annexes VI and V, respectively.

[216] Above at Sect. 4.1.2.1, subheading 'The EU Database for Clinical Trial Data'.

[217] Dir 2001/20/EC, art 17(3)(a); rec 9.

[218] Reg 536/2014/EU, art 37(4).

[219] European Commission (2015), p. 27 (citing Glenis Willmott, Member of the European Parliament and Rapporteur for the EU Clinical Trials Regulation).

[220] Hervey and McHale (2015), p. 319.

[221] ibid 320.

[222] Reg 536/2014/EU, art 37(4) (emphasis added).

the EMA's publication policy 0070 envisaging the disclosure of anonymised IPD goes beyond what is mandated by the regulator.[223]

4.3.1.2 Uncertainty Regarding the Definition of CCI

Publication of clinical trial data in the EU database is subject to the exceptions for protecting personal data[224] and CCI.[225] In the latter case, the status of the drug marketing authorisation needs to be taken into account when deciding whether confidentiality can be justified.[226] Protection of CCI will not apply if there is an overriding public interest in disclosure.[227]

The EU Clinical Trials Regulation does not specify what particular elements of marketing authorisation dossiers can qualify as CCI. Highlighting 'the lack of a legal definition', the EMA views CCI as 'any information which is not in the public domain or publicly available and where disclosure may *undermine the economic interest or competitive position* of the owner of the information'.[228] The Agency presumes that, in general, 'clinical data [comprised of clinical reports and IPD[229]] cannot be considered CCI [except for] in limited circumstances'.[230] Notably, before the policy implementation, the EMA's Advisory Group on Legal Aspects did not 'manage to reach an agreement [as to] whether or not clinical trial data contain CCI'.[231] To clarify such circumstances, the EMA lists the elements of CSRs that might be considered as CCI, such as the product development rationale, overviews of biopharmaceutics and clinical pharmacology, benefits and risks conclusions, summaries of biopharmaceutical and clinical pharmacology studies,[232] primary and secondary endpoints, including biomarkers and exploratory endpoints.[233]

Furthermore, to assist drug companies in preparing CSRs for publication, the EMA published guidance on identifying and redacting CCI.[234] The document

---

[223] EMA publication policy 0070, p. 7.

[224] Reg 536/2014/EU, art 81(7).

[225] Reg 536/2014/EU, art 81(4)(b).

[226] Reg 536/2014/EU, art 81(4)(b).

[227] ibid.

[228] EMA' access policy 0043, p. 4 (emphasis added).

[229] EMA publication policy 0070, p. 3.

[230] ibid p. 4. The following appendices of a CSR are subject to mandatory publication: the protocol and its amendments, statistical methods and a sample case report form. EMA, External guidance on the publication of clinical data (n 43), p. 88.

[231] EMA (30 Apr 2013) Advice to the European Medicines Agency from the Clinical Trial Advisory Group on Legal Aspects (CTAG5) – final advice, lines 8–10, 21 (putting forward arguments for and against). https://www.ema.europa.eu/en/documents/other/ctag5-advice-european-medicines-agency-clinical-trial-advisory-group-legal-aspects-final-advice_en.pdf. Accessed 26 Mar 2021.

[232] EMA publication policy 0070, pp. 17–18.

[233] ibid pp. 17, 19.

[234] EMA, External guidance on the publication of clinical data (n 43).

specifies four categories of information that shall not be considered as CCI: information that is already in the public domain, information that does not bear any innovative features (common knowledge), additional information the disclosure of which would be in the public interest, and information lacking sufficient or relevant justification.[235] The following claims would be considered by the EMA irrelevant if not supported by further explanation.

- Information is commercially confidential, competitively sensitive [...] and includes intellectual property, including trade secret[s].[236]
- The analytical methods are [the company]'s intellectual property, which [was] developed by expending a significant amount of time, and human, financial and commercial resources.[237]

As a general rule, the EMA requires the applicants for drug marketing authorisation to submit a detailed specification of how disclosure would affect their commercial interests, while the last word as to whether the justification is sufficient and adequate remains with the EMA.[238] In sum, the qualification of the contents of marketing authorisation dossiers as CCI is not category-specific but fact-dependent.

4.3.2 The Relevance of the Right of Access to Personal Data

As long as individual trial participants can be identified, clinical trial data shall qualify as personal data. EU law on data protection provides data subjects with specific rights, including the right of access.[239] Obviously, in the case of clinical trial data, this right can be exercised only by trial participants and does not constitute a legal basis for accessing IPD for secondary analysis in medical research and drug R&D.

---

[235] ibid p. 52.

[236] ibid p. 54 (emphasis added).

[237] ibid p. 53 (emphasis added). See also EMA publication policy 0070, annex III (listing specific sections of CSRs that can be redacted as CCI).

[238] EMA, External guidance on the publication of clinical data (n 43), p. 47.

[239] GDPR, art 15.

4.3.3 *Access to IPD Under the Right of Access to Documents*

4.3.3.1 The Right of Access to Documents Held by Public Authorities

Guaranteed under Article 15(3) of the TFEU and Article 42 of the CFR,[240] the right of access to documents is regarded as the 'core aspect' of the principles of transparency, democracy and good administration of the EU political system.[241] It can be exercised by any citizen of the Union and any natural or legal person residing or having its registered office in a Member State. The right applies to the documents developed by or submitted to the Union's institutions, bodies, offices and agencies to ensure transparency of their proceedings,[242] as specified under the EU Transparency Regulation.[243]

At EU level, the right of access to documents can provide a relevant basis for accessing clinical trial data held by the EMA.[244] The EMA access policy 0043 explicitly references Article 15 of the TFEU and the EU Transparency Regulation[245] and emphasises that openness and transparency are the 'paramount values enshrined in the TEU and in the TFEU [that] strengthen the principles of democracy and good administration'.[246] The EMA publication policy 0070 was adopted to 'extend [the Agency's] approach to transparency' and 'without prejudice' to the EU Clinical Trials Regulation[247] and the EU Transparency Regulation.[248] Thus, both policies can be viewed as implementing the EMA's duty to 'adopt rules to ensure the availability to the public of regulatory, scientific or technical information concerning the authorisation or supervision of medicinal products which is not of a confidential nature'.[249]

---

[240] Explanations relating to the Charter of Fundamental Rights (2007) OJ C 303, C 303/28.

[241] See e.g. Curtin and Mendes (2014), p. 1104; Case C-52/05P *Sweden and Turco v Council* [2008] ECLI:EU:C:2008:374, para 45.

[242] TFEU, art 15 (3).

[243] Council Regulation (EC) 1049/2001 regarding public access to European Parliament, Council and Commission documents (31 May 2001) OJ L145/43.

[244] Reg 726/2004/EC, art 73. National laws usually have analogous rules on transparency and access to documents held by public authorities. For an overview of transparency regimes and their application to clinical trial data in several EU Member States, see Pugatch Consilum (2015). Overall, the study finds that transparency regulations vary greatly among the examined EU Member States with some considerable 'grey areas'. Even though drug authorities in Germany, Italy, Spain and Sweden 'are committed to [...] increased transparency, none of them have a policy of proactively publishing submitted clinical [trial data]'. ibid p. 4.

[245] EMA publication policy 0070, p. 4.

[246] ibid p. 1.

[247] ibid p. 2.

[248] ibid p. 1.

[249] Reg 726/2004/EC, art 80.

4.3.3.2 The Exception for Protection of Commercial Interests in the Case of Clinical Trial Data

The Exception for Protection of Commercial Interests

While the EU Transparency Regulation intends to give 'the fullest possible effect to the right of public access to documents',[250] its exercise is subject to exceptions and limitations[251] necessary to protect personal data[252] and commercial interests.[253] Furthermore, where access is petitioned regarding the documents submitted by third parties, the institution concerned has to consult the document submitter and assess whether the exceptions are applicable, 'unless it is clear that the document shall or shall not be disclosed'.[254] The assessment must be 'specific in nature'[255] and account for the competing interests at stake.[256]

The EU Clinical Trials Regulation articulates the balance of interests as follows:

> the publicly available information contained in the EU database should contribute to protecting *public health* and fostering the *innovation capacity* of European medical research, while recognising the *legitimate* economic interests of sponsors.[257]

What appears unclear is how the scope of 'legitimate economic interests' concerning IPD should be interpreted when applying the exception envisaged under the Transparency Regulation. In principle, restrictions on the exercise of fundamental rights can be justified only if they 'correspond to objectives of general interest pursued by the Community and do not constitute, with regard to the aim pursued, disproportionate and unreasonable interference undermining the very substance of those rights'.[258] The 'very substance' of the right of access to documents subsists in making administrative practices and regulatory decision-making transparent.[259] In the case of clinical trial data, this implies validating regulatory decisions and procedures concerning drug authorisation and supervision through secondary

---

[250] Reg 1049/2001/EC, rec 4.

[251] C-365/12 P *Commission v EnBW* [2014] EU:C:2014:112, para 85.

[252] Reg 1049/2001/EC, art 4(1)(b).

[253] Reg 1049/2001/EC, art 4(2), first indent.

[254] Reg 1049/2001/EC, art 4(4). In general, access can be granted to any document irrespective of whether it is drawn up or received by an institution in any area of its activity. Reg 1049/2001/EC, art 2(3).

[255] Joined cases T-355/04 and T-446/04 *Co-Frutta v Commission* [2010] ECLI:EU:T:2010:15, para 123.

[256] Case C-365/12 P *Commission v EnBW* [2014] ECLI:EU:C:2014:112, para 63.

[257] Reg 536/2014/EU, rec 67 (emphasis added).

[258] Explanations relating to the Charter of Fundamental Rights (2007) OJ C 303, C 303/32-33 (and case law cited).

[259] Above at Sect. 4.3.3.1.

confirmatory analysis. The purpose of re-use of IPD for secondary *exploratory* analysis seemingly exceeds the scope of the right of access to documents.[260]

The reservation 'unless it is *clear* that the document shall or shall not be disclosed'[261] is arguably a matter of perspective of the institution concerned. As noted earlier, the EMA presumes that clinical data comprising CSRs and IPD[262] generally does not constitute CCI, except in limited circumstances.[263] Meanwhile, this view was challenged by pharmaceutical companies in several cases brought before the CJEU.[264] When contesting the EMA's decision to grant third-party access to CSRs, they invoked inter alia the exception for CCI and claimed that CSRs constitute CCI[265] that qualifies for the general presumption of confidentiality.[266]

The Existence of the General Presumption of Confidentiality for CSRs

In some situations, the obligation on the institution to carry out the individual examination of the contents of documents submitted by third parties can be waived. So far, the CJEU acknowledged the existence of a general presumption of confidentiality only for few categories of documents, namely, 'documents in an administrative file relating to a procedure for reviewing State aid, the pleadings lodged by an institution in court proceedings, documents exchanged in the course of merger control proceedings, documents concerning infringement proceedings during the pre-litigation stage, and documents in administrative proceedings under Article 101 TFEU'.[267]

In *Amicus v EMA* and *PTC Therapeutics International v EMA*, the CJEU considered whether CSRs in their *entirety*[268] ought to be subject to the general

---

[260] As argued below at Sect. 4.3.3.4.

[261] Reg 1049/2001/EC, art 4(4) (emphasis added).

[262] EMA publication policy 0070, p. 3.

[263] ibid p. 4. The following appendices of a CSR are subject to mandatory publication: the protocol and its amendments, statistical methods and a sample case report form. EMA, External guidance on the publication of clinical data (n 43), p. 88.

[264] For a detailed analysis of these cases, see Kim (2017).

[265] Case T-33/17 *Amicus Therapeutics v EMA* [2018] ECLI:EU:T:2018:595, para 28; Case T-73/13 R *InterMune v EMA* [2013] ECLI:EU:T:2013:222, paras 14, 16; Case T-44/13 R *AbbVie v EMA* [2013] ECLI:EU:T:2013:221, paras 18, 24; Case T-235/15 *Pari Pharma v EMA* [2018] ECLI:EU:T:2018:65, para 34.

[266] Case T-33/17 *Amicus Therapeutics v EMA* [2018] ECLI:EU:T:2018:595, paras 21, 22; Case T-718/15 *PTC Therapeutics International v EMA* [2018] ECLI:EU:T:2018:66, para 26; Case T-235/15 *Pari Pharma v EMA* [2018] ECLI:EU:T:2018:65, paras 34, 36, 38. In the latter case, the dispute arose over the CHMP assessment report, which comprised the similarity and superiority reports submitted by Pari Pharma.

[267] Case C-513/16 *EMA v PTC Therapeutics International* [2018] ECLI:EU:C:2017:148, para 45.

[268] As claimed by the pharmaceutical companies in Case T-33/17 *Amicus Therapeutics v EMA* [2018] ECLI:EU:T:2018:595, para 7; Case T-718/15 *PTC Therapeutics International v EMA* [2018] ECLI:EU:T:2018:66, paras 4, 26.

presumption of confidentiality. The following criteria for recognising the existence of the general presumption of confidentiality, when applying the exceptions under Article 4 of the Transparency Regulation, were elucidated.

(i) All previously recognised general presumptions of confidentiality concerned the documents submitted within the files related to the *ongoing* administrative or judicial proceedings.[269]
(ii) The application of a general presumption is '*essentially dictated* by the overriding need to ensure that the procedures at issue operate correctly and to guarantee that their objectives are not jeopardised'.[270] In other words, third-party access to documents involved in certain administrative procedures would be '*incompatible* with the proper conduct of those procedures'.[271] Therefore, the general presumption of confidentiality is necessary to 'ensure that the integrity of the conduct of the procedure can be preserved by limiting intervention by third parties'.[272]
(iii) The legislation provides specific rules governing the procedures conducted by an EU institution, for which the requested documents were produced and submitted.[273]

Given that the request for access to CSRs in both *Amicus v EMA* and *PTC Therapeutics International v EMA* cases was made after the corresponding marketing authorisation was granted, the first factor was not relevant. As for the second criterion, upon the review of the sector-specific legislation on the publication of information related to drug marketing authorisation, the CJEU took the view that 'the EU legislature took the implicit view that the integrity of the marketing authorisation procedure is not undermined in the absence of a presumption of confidentiality'.[274] As for the third factor, the CJEU held that no provisions under the EU Drug Authorisation Regulation could be interpreted 'as evidence of the intention of the EU legislature to set up a system of restricted access to documents by means of a general presumption of confidentiality of documents'.[275] Quite to the contrary, the principle of transparency of drug marketing authorisation procedures[276]

[269] Case T-33/17 *Amicus Therapeutics v EMA* [2018] ECLI:EU:T:2018:595, para 36; Case T-718/15 *PTC Therapeutics International v EMA* [2018] ECLI:EU:T:2018:66, para 40.

[270] Case T-33/17 *Amicus Therapeutics v EMA* [2018] ECLI:EU:T:2018:595, para 35 (emphasis added).

[271] ibid (emphasis added).

[272] ibid. See also Case T-718/15 *PTC Therapeutics International v EMA* [2018] ECLI:EU:T:2018:66, para 39.

[273] Case T-33/17 *Amicus Therapeutics v EMA* [2018] ECLI:EU:T:2018:595, para 37 (and case law cited); Case T-718/15 *PTC Therapeutics International v EMA* [2018] ECLI:EU:T:2018:66, para 41.

[274] Case T-33/17 *Amicus Therapeutics v EMA* [2018] ECLI:EU:T:2018:595, para 46.

[275] Case T-718/15 *PTC Therapeutics International v EMA* [2018] ECLI:EU:T:2018:66, para 48; Case T-33/17 *Amicus Therapeutics v EMA* [2018] ECLI:EU:T:2018:595, para 43.

[276] Case T-718/15 *PTC Therapeutics International v EMA* [2018] ECLI:EU:T:2018:66, para 50; Case T-33/17 *Amicus Therapeutics v EMA* [2018] ECLI:EU:T:2018:595, para 45.

and public access to the related information should prevail, while the CCI exception should be interpreted strictly.[277] Accordingly, in the absence of a general presumption of confidentiality, the release of trial data held by the EMA remains subject to the case-by-case and element-by-element assessment of the dossier contents.[278]

4.3.3.3 Implications for the Publication of Clinical Trial Data in the EU Database

The interpretation of the exception for the protection of commercial interests under the EU Transparency Regulation bears directly on the scope of obligations regarding data publication in the EU database, given a systematic relationship between the right of access to documents and the transparency provisions under the sector regulations, as shown in the Table 4.1.

As evident, the reservations for CCI under the sector regulations are consistently linked to the exception for protecting commercial interests under the right of access. Consequently, the interpretation of the CCI exception under the EU Transparency Regulation would determine the scope of clinical trial data that can be accessed through the EU clinical trials database or lawfully released by the EMA.[279]

4.3.3.4 Limitations of the Right of Access to Documents as an Instrument of Access to IPD for Research Purposes

The importance of fostering transparency in the procedures and decision making concerning drug marketing authorisation is beyond doubt. However, as far as supporting the objectives pursued by the EMA publication policy is concerned, the transparency framework might be limited in several aspects. First, the scope of the right of access to documents by definition is restricted to the documents held by EU public authorities, which leaves out a considerable amount of clinical trial data—most importantly, IPD[280] and data from unsuccessful trials, which present high value for both scientific research and drug R&D.[281]

Second, access to IPD can be subject to the exceptions for protecting the personal data of trial participants and the commercial interests of trial sponsors. The scope and conditions of the application of both exceptions are uncertain. As for personal data,

---

[277] Case T-718/15 *PTC Therapeutics International v EMA* [2018] ECLI:EU:T:2018:66, para 52; Case T-33/17 *Amicus Therapeutics v EMA* [2018] ECLI:EU:T:2018:595, para 70.

[278] Case T-235/15 *Pari Pharma v EMA* [2018] ECLI:EU:T:2018:65, para 57; Case T-718/15 *PTC Therapeutics International v EMA* [2018] ECLI:EU:T:2018:66, para 53; T-33/17 *Amicus Therapeutics v EMA* [2018] ECLI:EU:T:2018:595, para 47.

[279] Above at Sect. 4.1.2.1, subheading 'The EU Database for Clinical Trial Data'.

[280] Such as in the case of the EMA; see Chap. 6 at Sect. 6.5.1.2.

[281] As discussed in Chap. 8 at Sect. 8.2.3.2.

Table 4.1 The relationship between the right of access to documents and reservations for CCI under the sector regulations

Legal provisions concerning clinical trial data disclosure (emphasis added)	
The TFEU	Article 15(3) Any citizen of the Union, and any natural or legal person residing or having its registered office in a Member State, *shall* have a right of access to documents of the Union institutions, bodies, offices and agencies, whatever their medium, *subject to the principles and the conditions* to be defined in accordance with this paragraph.
The CFR	Article 42 Any citizen of the Union, and any natural or legal person residing or having its registered office in a Member State, has a right of access to documents of the institutions, bodies, offices and agencies of the Union, whatever their medium.
Regulation 1049/2001/EC	Recital 4 The purpose of this Regulation is to give the fullest possible effect to the right of public access to documents and to lay down the general principles and limits on such access in accordance with Article 255(2) of the EC Treaty. Article 4 2. The institutions shall refuse access to a document where disclosure would undermine the protection of: – commercial interests of a natural or legal person, including intellectual property. [...] 3. As regards third-party documents, the institution shall consult the third party with a view to assessing whether an exception in paragraph 1 or 2 is applicable, unless it is *clear* that the document shall or shall not be disclosed.
Regulation 726/2004/EC	Article 13 3. The Agency shall immediately publish the assessment report on the medicinal product for human use drawn up by the Committee for Medicinal Products for Human Use and the reasons for its opinion in favour of granting authorisation, *after deletion of any information of a commercially confidential nature*. The European Public Assessment Report (EPAR) shall include a *summary* written in a manner that is understandable to the public. The summary shall contain in particular a section relating to the conditions of use of the medicinal product. Article 73 Regulation (EC) No 1049/2001 of the European Parliament and of the Council of 30 May 2001 regarding public access to European Parliament, Council and Commission documents shall apply to documents held by the Agency. Article 80 To ensure an appropriate level of transparency, the Management Board, on the basis of a proposal by the Executive Director and in agreement with the Commission, shall adopt rules to ensure the availability to the public of regulatory, scientific or technical information concerning the authorisation or supervision of medicinal products which is not of a confidential nature.

(continued)

Table 4.1 (continued)

Legal provisions concerning clinical trial data disclosure (emphasis added)	
Regulation 536/2014/EU	Recital 68 For the purposes of this Regulation, *in general* the data included in a clinical study report *should not be considered commercially confidential* once a marketing authorisation has been granted, the procedure for granting the marketing authorisation has been completed, the application for marketing authorisation has been withdrawn. In addition, the main characteristics of a clinical trial, the conclusion on Part I of the assessment report for the authorisation of a clinical trial, the decision on the authorisation of a clinical trial, the substantial modification of a clinical trial, and the clinical trial results including reasons for temporary halt and early termination, *in general*, should not be considered confidential. Article 81 1. The Agency shall, in collaboration with the Member States and the Commission, set up and maintain a EU database at Union level. [. . .] The EU database shall contain the data and information submitted in accordance with this Regulation. [. . .] 4. The EU database shall be publicly accessible unless, for all or part of the data and information contained therein, confidentiality is justified on any of the following grounds: [. . .] (b) protecting *commercially confidential information*, in particular through taking into account the status of the marketing authorisation for the medicinal product, unless there is an overriding public interest in disclosure.

uncertainty stems from the risk of re-identification of trial subjects. As for the protection of commercial interests, it is unclear to what extent competitive concerns should be taken into account. Pharmaceutical companies argued that the blanket disclosure of data would impede their competitive advantage, including in competition in innovation. In this regard, the EMA's intention to 'establish[] a level playing field'[282] for all drug developers by disclosing non-summary data appears questionable. If IPD constitutes a unique resource for exploratory research and its analysis may, eventually, lead to the development of new products, the sponsor of the original trial may have a legitimate interest in having the priority to carry out exploratory data analysis.[283]

Third, the underlying rationale of the right of access to documents is to improve *transparency* in policymaking and administrative practices of the public authorities. Even though there is no purpose-limitation in the sense that the requestor has to provide the reasons justifying the grant of access, the right of access to documents is meant to guarantee transparency in the public institutions' decision making within the scope of their mandate and responsibilities. In the case of clinical trial data, this

[282] EMA publication policy 0070, p. 4.

[283] For a detailed analysis, see Chap. 5 at Sect. 5.3.2 and Chap. 8 at Sect. 8.1.4.

implies that a third party can gain access to data held by the EMA for understanding and verifying regulatory decisions concerning the evaluation, marketing authorisation, supervision or pharmacovigilance of medicinal products[284] through confirmatory data analysis. The objective of promoting medical research and making drug development more *efficient* through *exploratory* data analysis[285]—albeit highly important and desirable—would exceed the scope of the right of access to documents. In other words, the right of access to documents does not appear to provide a relevant legal basis for fostering scientific knowledge and promoting drug R&D through secondary IPD analysis. In this regard, clinical trial data presents a rather unusual case where a public authority holds scientific data that can be used for purposes unrelated to the initial data collection or regulatory decision making.[286]

4.3.4 *Competition Law as an (Unsuitable) Instrument of Access to IPD*

The rules on access to data reviewed in the previous section can apply to IPD held by the public authorities. As for the data held by the drug companies, the only instrument applicable across sectors that might enable access is competition law. In particular, compulsory access to information or data held by business undertakings can, under certain conditions, be imposed as a remedy against anti-competitive conduct, as an exception from the general principle that companies should be free to choose trading partners and dispose of their property.[287]

4.3.4.1 Access to IPD as a Hypothetical Case on a Refusal to Deal

One could contemplate the following scenario. Companies A and B are research-based pharmaceutical companies. Company A sponsored clinical trials and commercialised drug X. Company B seeks access to IPD from the trials related to drug X to aggregate it with other datasets for conducting exploratory research. After A refuses to share data, B lodges a complaint with the European Commission alleging A' anti-competitive behaviour.

---

[284] Such responsibilities are vested in the EMA under Reg 726/2004/EC, art 55.

[285] EMA publication policy 0070, pp. 3–4.

[286] As shown in Chap. 3, exploratory IPD analysis can address questions beyond the benefit-risk balance of an investigational product.

[287] European Commission (24 Feb 2009) Communication from the Commission – Guidance on the Commission's enforcement priorities in applying Article 82 of the EC Treaty to abusive exclusionary conduct by dominant undertakings. OJ C 45/7 [hereinafter Enforcement priorities guidance], para 75.

At the outset, such scenario appears highly improbable. First, it is unlikely that a drug company would seek access to data held by a competitor, given that corporate data-sharing policies explicitly deny access to IPD to actual or potential competitors.[288] Second, the company intending to undertake exploratory IPD analysis might not be willing to negotiate access directly with the data holder (a competitor) not to disclose the purpose of secondary analysis—the strategic information related to the planned or ongoing research project.[289] Third, the transaction costs could be dissuasive, given that exploratory data analysis requires aggregating IPD from different trials that multiple data holders might hold.[290] Forth, resorting to competition law enforcement might be unattractive since the remedy of compulsory access to data would be applied *ex post,* while the proceedings can last for years.[291]

At the same time, an overview of the legal instruments of access to data would be incomplete without considering the role of competition law. Notwithstanding the impracticality and low probability of the above-outlined scenario, it presents an apt opportunity to contemplate 'how far upstream' intervention under competition law might reach to protect competition beyond an existing product market. This question has been of particular relevance for a data-driven economy where 'big data' constitutes a valuable R&D resource while only a limited number of undertakings can amass it.[292] The remainder of this part considers to what extent the current methodology of assessing refusals to deal can apply when access to (anonymised) IPD held by trial sponsors is sought for exploratory research.

4.3.4.2 The Assessment of a Refusal to Grant Access to IPD for Exploratory Analysis Under Article 102 of the TFEU

The Standard of Intervention Where Access to Information or Data Is Refused

Under EU competition law, compulsory access to data can be granted as a remedy in refusal-to-supply cases under Article 102 of the TFEU.[293] The provision lays down a general prohibition of abusing a dominant position and exemplifies several categories of abusive conduct. While holding a dominant position in itself is not abusive,[294] the assessment should consider pro- and anti-competitive effects of a refusal,

---

[288] On data-sharing policies of pharmaceutical companies, see Chap. 5 at Sect. 5.1.3 and Chap. 6 at Sect. 6.3.2.

[289] On the reverse 'information paradox', see Chap. 9 at Sect. 9.3.2.5.

[290] For a discussion, see Chap. 9 at Sect. 9.3.2.3.

[291] For instance, in *Microsoft* the proceedings last over 14 years and in *Magill* almost 10 years.

[292] On the role of competition law as an instrument of access to data in the context of data-driven innovation, see Drexl (2017b, c, d); Crémer et al. (2019), p. 73 ff.

[293] The refusal-to-supply category encompasses IP-related cases (e.g. refusal to license IP rights) and refusal to grant access to an essential facility or a network. Enforcement priorities guidance (n 287), para 78.

[294] ibid para 1.

including actual and potential effects on innovation.[295] The question is thus two-fold: whether the IPD holding company is dominant in a relevant market; if so, whether the refusal to grant access to IPD constitutes an abuse of dominance.

In a series of cases—*Magill*,[296] *IMS Health*,[297] and *Microsoft*[298]—the CJEU elaborated specific criteria of applying the general liability standard under Article 102 of the TFEU in situations where a refusal concerns access to information and data. These criteria became known as the 'exceptional circumstances' test.[299] While the test originated in case law where exclusive IP rights protected the refused subject matter,[300] the assessment in *Microsoft* proceeded under a presumption that trade secrets 'must be treated as equivalent to intellectual property rights'.[301] Given that IPD might qualify *de lege lata* for trade secrets protection,[302] *Microsoft* presents the 'most suitable precedent'.[303] The CJEU restated the criteria of the 'exceptional circumstances' test as follows.

- The refusal relates to a product or service *indispensable* for exercising a particular activity on a neighbouring market.
- The refusal is of such a kind as to *exclude any effective competition* on that neighbouring market.
- The refusal prevents the appearance of a new product for which there is potential consumer demand.[304]

Once these circumstances are established, the assessment should check for an objective justification before a refusal by a dominant undertaking can be deemed violating Article 102 of the TFEU.[305] The next section discusses the analytical

[295] ibid paras 75, 87.

[296] Joined Cases C-241/91 and C-242/91 [1995] *RTE and ITV v Commission* ('Magill') ECLI:EU:C:1995:98.

[297] Case C-418/01 *IMS Health* [2004] ECLI:EU:C:2004:257.

[298] Case T-201/04 *Microsoft v Commission* [2007] ECLI:EU:T:2007:289.

[299] ibid para 332.

[300] See Drexl J, Hilty RM et al. (2016) Data ownership and access to data – position statement of the Max Planck Institute for Innovation and Competition as of 16 August 2016 on the current European debate. Max Planck Institute for Innovation & Competition Research Paper No. 16-10 [hereinafter MPI position statement on data ownership and access to data], para 34 (emphasising that the exceptional circumstances test was developed 'under the assumption of IP protection for the subject matter at issue[, and] whether and how these findings can be applied to situations involving unprotected raw data is yet to be clarified'). https://www.ip.mpg.de/fileadmin/ipmpg/content/stellungnahmen/positionspaper-data-eng-2016_08_16-def.pdf. Accessed 26 Mar 2021.

[301] *Microsoft* (n 298), para 289.

[302] As concluded at Sect. 4.2.2 in this chapter.

[303] Drexl (2017a), para 125 (further noting that the *Magill* case 'laid the foundations for dealing with the issue of information-based dominance').

[304] *Microsoft* (n 298), para 332 (emphasis added).

[305] ibid para 333.

challenges of applying these factors to a case where access to IPD sought for exploratory research purposes might be refused.

The Dominance of the IPD Holder

Refusal-to-deal cases usually concern constellations with two interrelated markets: a primary market constituted by the requested product or service where an undertaking refusing to deal holds a dominant position and a secondary (neighbouring) market constituted by a product or service, which manufacturing or supply might require the requested product or serviced.[306] In a typical scenario, the firm dominant in the primary market would be vertically integrated and compete with the potential buyer (the petitioner for access to a resource) in a secondary market.[307]

While control over 'big data' can be assumed to constitute a source of market power,[308] the methodology of assessing the dominance of an undertaking exercising control over such data has not been settled. For one reason, it might not be clear how the relevant market for assessing dominance should be defined.[309] For another reason, it might be difficult to prove that market dominance stems from control over data.[310]

Where access is sought to information, dominance can be established if an undertaking is the 'only source' of the requested information.[311] On the one hand, IPD can be viewed as unique as clinical trials are normally designed to address questions and hypotheses novel for scientific research and clinical practice.[312] On the other hand, whether the IPD holder is the 'only source' of data should be assessed from the data user's perspective, which raises the question of data *substitutability* for the intended purpose.[313] For a researcher or a drug developer, the main value of IPD subsists in the knowledge that might be revealed through exploratory data analysis.

---

[306] *Microsoft* (n 298), para 335.

[307] Enforcement priorities guidance (n 287), para 76. See also Drexl (2017a), para 137 (noting that, in light of the CJEU's decision in *Huawei*, 'the question may be asked whether a refusal to license or a refusal to deal can also be considered abusive if the dominant frm is not vertically integrated').

[308] ibid para 36. See also Drexl (2017a), paras 117, 130 (arguing that 'control over big data should play a more prominent role in assessing market power and dominance', and that 'the collection of datasets for the purpose of enabling big data analysis may [. . .] enhance market power of the firm that controls access to the larger dataset').

[309] Crémer et al. (2019), p. 100. MPI position statement on data ownership and access to data (n 300), para 33 (noting that 'it is by no means clear how the relevant market for data should be defined when access concerns not individual data, but large data sets for data-mining purposes, and under what conditions different data sets can be considered as substitutable').

[310] MPI position statement on data ownership and access to data (n 300), para 33.

[311] In *Magill*, the CJEU found for the dominant position of the TV stations, which 'by force of circumstance' enjoyed 'a *de facto* monopoly over the information used to compile listings for the television programmes'. *Magill* (n 296), para 47.

[312] Cummings et al. (2007), p. 20.

[313] Drexl (2017a), para 128.

Whether other datasets or knowledge resources might substitute the requested IPD for that purpose is unclear because one cannot know *ex ante* what knowledge would be gained before undertaking the analysis.[314] Such uncertainty distinguishes access to 'big data' from cases where access is sought to 'identifiable and delineable'[315] information or data. The analysis becomes even more puzzling where exploratory analysis requires aggregating IPD held by multiple trial sponsors, who would need to be argued as the 'only sources' of IPD.

The argument that IPD is not substitutable for exploratory research might not be convincing as it would imply that no drug could have ever been developed without companies being able to access and analyse historical IPD held by other undertakings in hypothesis-generating research. As long as researchers can generate hypotheses regarding novel treatment effects relying on other knowledge resources or make assumptions[316] based on the background knowledge, the IPD holding company cannot be deemed the 'only source' of information for that purpose.

For the sake of analysis, let us consider how the assessment under competition law would proceed if dominance could be established.

IPD Indispensability for Exercising an Activity on a Neighbouring Market

When assessing indispensability, one first needs to define a neighbouring market where the activity is exercised for which the requested product or service is indispensable. Such market might not be straightforward in the case of IPD. Exploratory IPD analysis is usually conducted in preclinical research. The market for which IPD is needed as input would arguably be a market where the results of IPD analysis can be eventually commercialised—that is, a drug market. The output of exploratory analysis most relevant for drug development would be a hypothesis concerning a treatment effect—it would take years before a product developed on its basis can be launched on the market.[317] Drug markets are conventionally defined by the therapeutic fields.[318] Hypothetically, exploratory analysis of historical IPD can contribute

[314] See Drexl (2017a), para 128 (observing that substitutability of datasets 'will depend on the concrete circumstances, including the very nature of the information contained in the data'). Applying the concept of substitutability in the assessment of dominance 'in a world of big datasets [...] remains a most difficult task, since even the petitioner for access, such as a big data analyst, will often only have a vague understanding about the kind of data contained in the dataset and about which data will produce the most valuable new information based on observable correlations'. ibid para 129. See also MPI position statement on data ownership and access to data (n 300), para 33.

[315] See Drexl (2017a), para 127. See also MPI position statement on data ownership and access to data (n 300), para 37.

[316] Below (nn 330–331) and the accompanying text.

[317] For an analysis, see Chap. 8 at Sects. 8.1.4.3 and 8.1.4.4.

[318] To define a relevant market, competition law analysis of pharmaceutical cases often applies the Anatomical Therapeutic Chemical (ATC) classification, an international classification system developed by the WHO based on the chemical, pharmacological and therapeutic properties of medicines and organs or systems that they affect. WHO Collaborating Centre for Drug Statistics

to developing (i) a follow-on product within the same therapeutic field as the drug, which historical IPD is reanalysed,[319] and (ii) a drug in a different therapeutic field, especially if large volumes of aggregated historical data are analysed.[320] Given that it can hardly be predicted *ex ante* whether the prospective research might result in a viable hypothesis, the relevant drug market might not be identified when access to data is sought. In this regard, the case of access to IPD for exploratory research differs markedly from *Magill*, *IMS Health* and *Microsoft*, concerned with access to information necessary for entering an identifiable, existing market.

Even if the outcome of IPD analysis could be predicted, how could the indispensability of IPD be assessed? First, if the input indispensability is assessed by its substitutability, the same reasons against the data holder as the 'only source' of IPD[321] would apply. Second, the threshold of indispensability under EU competition law is high: the requested product or service would be deemed indispensable if 'no actual or potential substitute' exists,[322] or if technical, legal or economic obstacles make its recreation 'impossible, or even unreasonably difficult'.[323] The presence of technical and economic obstacles to IPD reproduction cannot be claimed, given that conducting clinical trials is a regular activity of the research-based drug companies. Legal obstacles cannot be argued either because duplicative trials are not outlawed.[324]

As noted above, claiming that IPD is indispensable for exploratory analysis due to its uniqueness would imply that no drug could have ever been developed without sharing IPD from past trials for hypothesis-generating research. Rather, what could rationalise the necessity of access to historical IPD are potential gains in efficiency at the sector level due to the cost-saving effect of utilising the existing knowledge resources. Exploratory analysis of (aggregated) IPD has an untapped potential to guide drug discovery and validate new drug targets and molecules more efficiently. Medical researchers point out that many clinically relevant research questions can be answered satisfactorily based on data from earlier trials.[325] Accordingly, secondary IPD analysis could save costs of conducting new trials[326] and reduce redundant research.[327] Besides, historical IPD analysis can play an important role in informing

---

Methodology, ATC, structure and principles. https://www.whocc.no/atc/structure_and_principles/. Accessed 26 Mar 2021. The third ATC level, which groups medicines by their therapeutic indications (intended use), is usually used define the relevant drug market. Case COMP/A. 37.507/F3 – AstraZeneca, Commission Decision of 15 June 2005, para 371.

[319] For a detailed analysis of this scenario, see Chap. 8 at Sect. 8.1.4.4, subheading 'The Case of Drug Improvements'.

[320] Such scenarios are analysed in Chap. 8 at Sects. 8.1.4.3 and 8.1.4.4.

[321] Above at Sect. 4.3.4.2, subheading 'The Dominance of the IPD Holder'.

[322] Case C-7/97 *Bronner* [1998] ECLI:EU:C:1998:569, para 41.

[323] ibid para 44.

[324] See Chap. 5 at Sect. 5.4.1.2.

[325] Chalmers and Glasziou (2009), p. 87.

[326] Gøtzsche (2012), p. 237.

[327] On this point, see Chap. 8 at Sect. 8.2.3.

the design of new trials.[328] For instance, when a trial is planned, the effect size[329] is estimated based on the evidence from preceding trials addressing a similar research question. Researchers need to make 'a reasonable assumption'[330] concerning the adequate sample size and statistical power. Without data from earlier exploratory or pilot studies, trial investigators 'are left with no choice but to make a "best guess" at the effect size.[331] In this view, it might be more straightforward to justify the indispensability of IPD sharing based on the public interest in eliminating 'best guesses' and avoiding unnecessary exposure of trial participants to health risks. However, such public interest is unlikely to come within the purview of the indispensability assessment under Article 102 of the TFEU.[332] The industry-wide cost-saving effect might not be relevant either, given that undertakings 'should generally not be required to "help" competitors'[333] from a competition law perspective.

The Exclusion of Effective Competition in a Neighbouring Market

If a neighbouring (drug) market cannot be defined for the reasons explained above, neither can the exclusion of effective competition in that market be plausibly argued at the point when access to IPD is sought. Even if it could be ascertained that secondary exploratory analysis of IPD related to drug X would eventually contribute to the development of its improved version X^1 and, hence, compete with the data holder's product,[334] competitive effects of a refusal could hardly be evaluated as the market conditions cannot be known years before the petitioner for access might succeed in developing and launching drug X^1. In such cases, the exclusion of effective competition would be hard to argue, given that competition by improvement generally tends to be intense already before the first-in-class drug is launched onto the market,[335] even without IPD being shared for exploratory research.

---

[328] Tierney et al. (2015).

[329] An 'effect size' of medical intervention refers to a measure correlating with the outcome of interest, a measurable effect of a health intervention on the clinical manifestations of a disease (e.g. relief of symptoms).

[330] Massaro (2009), p. 46.

[331] ibid.

[332] See Drexl (2017a), para 179 (drawing an analogy with the REACH Regulation and noting that it 'facilitates access to information beyond the remedies available under competition law' and pointing out that, while the obligation to share information under the REACH Regulation is motivated by a public interest in avoiding duplicative testing on animals, 'EU competition law does not exempt the petitioner from making the same investment as the holder of the essential facility').

[333] Crémer et al. (2019), p. 101.

[334] For a detailed analysis of this scenario, see Chap. 8 at Sect. 8.1.4.4, subheading 'The Case of Drug Improvements'.

[335] DiMasi and Faden (2011).

Moreover, it cannot be excluded that a drug eventually resulting from IPD analysis would not compete with any product of the data holder. Such situations would arguably not fall within the ambit of Article 102 of the TFEU at all.[336]

Furthermore, if effects on competition in an existing neighbouring market cannot be adequately assessed, the question arises whether the assessment should consider the potential effects of a refusal on *competition in innovation* as a standalone form of competition preceding the emergence of a market.[337] Innovation is a key parameter of competition in the drug industry; research-based drug companies compete by R&D investment 'to win the race to discover new products'[338] and earn supra-competitive returns on R&D.[339] However, the analysis of what constitutes elimination of effective competition beyond an existing product market, how market dominance should be assessed and how the rules on unilateral conduct can apply in this context is a rather uncharted area of competition law analysis.[340] The analytical complexity of defining the conditions under which intervention by competition law in competition in innovation would be justified, to a large extent, stems from a fundamental uncertainty regarding the relationship between competition and innovation, especially R&D intensity and innovation.[341]

The Appearance of a New Product for Which There Is Potential Consumer Demand

In a typical refusal-to-supply case, the refused input would be needed for manufacturing[342] and entering an existing identifiable product market.[343] Can a refusal to grant access to IPD for hypothesis-generating research be deemed anti-competitive in that it obstructs the development of a new drug and thus limits

---

[336] See Crémer et al. (2019), p. 100 (noting that refusals to grant access to data sought 'for analytics purposes unrelated to the market in which the data holder is active [. . .] are difficult to bring under the scope of Article 102 TFEU').

[337] Drexl (2012), p. 507.

[338] Charles River Associates (2004) Innovation in the pharmaceutical sector, p. 63. https://www.crai.com/insights-events/publications/innovation-pharmaceutical-sector/. Accessed 26 Mar 2021.

[339] On innovation as the key parameter of competition in the pharmaceutical sector, see Chap. 8 at Sect. 8.1.2.4.

[340] Drexl (2012).

[341] See e.g. Ben-Asher (2000), p. 292 (pointing out that the concept of competition in innovation 'lacks a proper theoretical and empirical foundation because no economic theory or empirical basis conclusively links less research and development competition to less research and development, or less research and development to welfare losses' (with further references)); Drexl (2012), p. 507 (noting that 'modern competition law, which strongly focuses on market analysis, may face a major problem in addressing restraints of competition in innovation appropriately').

[342] As emphasised by the Commission, the methodology for assessing refusals to supply set out in its Guidance on enforcement priorities 'deals only with this type of refusal'. Enforcement priorities guidance (n 287), para 76.

[343] As in information-related cases *Magill*, *IMS Health* and *Microsoft*.

'technical development to the prejudice of consumers' under Article 102 lit. b) of the TFEU? Access to IPD for exploratory analysis, by definition, is needed to gain knowledge that can facilitate medical research and contribute to drug R&D. Concerns have been raised that restricted access to IPD can 'slow the rate of scientific discovery and advancement'.[344] However, proving the 'stifling' effect on technical development in a particular case can be challenging for the same reasons as pointed out above—the probabilistic nature of exploratory research and the fact that drug discovery is, in principle, possible without analysing historical IPD. Thus, it is highly doubtful whether the 'appearance' of a new product might be stretched that far to include preclinical research in cases where access to data is petitioned for exploratory research purposes.[345]

The Objective Justification of a Refusal

When imposing a duty to supply as a remedy under competition law, the potential impact on innovation incentives should be carefully considered. A data holding company could argue that it intends to use IPD in exploratory research, and thus the duty to share IPD with competitors would diminish its advantage in competition in innovation. Such likelihood cannot be established in the abstract. Theoretically, the larger the aggregated data pool, the less probable it is that parallel exploratory analyses would overlap.[346] In other words, anti- and pro-competitive effects of a refusal appear equally speculative and therefore challenging to establish, given the probabilistic nature of exploratory IPD analysis and the remoteness in time of its results.

4.3.4.3 Conclusion on the Application of Competition Law

The preceding discussion shows that the competitive effects of a refusal to grant access to IPD for exploratory research can hardly be adequately assessed.[347] In view

[344] Institute of Medicine of the National Academies (2015), p. 141.

[345] Besides, it is uncertain whether a new product criterion is 'the only parameter which determines whether a refusal to license an intellectual property right is capable of causing prejudice to consumers within the meaning of Article 82(b)'. See *Microsoft* (n 298), para 647. Recall that the CJEU proceeded based on a presumption that trade secrets 'must be treated as equivalent to intellectual property rights'. ibid para 289. It remains to be seen whether the 'new-product rule' should be applied to data not protected either under exclusive IP rights or as trade secrets. Drexl (2017a), para 140.

[346] For a detailed analysis, see Chap. 8 at Sect. 8.3.

[347] This is likely to be the case when access to 'big data' is sought for exploratory analysis. See MPI position statement on data ownership and access to data (n 300), para 37 (noting that when 'access concerns data that are much larger in volume and of unknown or unspecified contents [and when] products or services that might be developed on the basis of such data cannot be readily ascertained at the time when access is granted [. . .] it is not (yet) possible to adequately evaluate the dynamic effects on competition of the refusal to grant access in such cases').

of the identified analytical ambiguities and the practical hurdles, such as lengthy enforcement procedures, compulsory access under competition law cannot be viewed as a systematic way of enabling access to IPD held by drug companies for secondary research.[348] The case of access to IPD can illustrate that competition law might play a limited role in regulating access to data as an innovation resource.[349] Nevertheless, an appreciation of the potential effects of IPD disclosure on competition and innovation can inform the policy analysis. When designing a sector-specific regime of access to IPD, it should be considered that IPD is generated and processed in the context of competition in innovation and that there is a fine line between protecting the advantage of data holders in competition in innovation and preventing 'data lock-ins'.

4.4 Conclusion on Chapter 4

Based on the analysis in this chapter, the status quo of clinical trial data under EU law can be summarised as follows. *De lege lata* IPD—either personal or anonymised—cannot be subject to property-type rights. Nevertheless, trial sponsors can exercise de facto exclusive control over IPD due to the regulatory requirement to protect all information and data gathered in trials against *unauthorised* access by third parties.

As far as access to clinical trial data is concerned, the approach under the existing EU framework relies predominantly on the transparency regulation. While fostering transparency in clinical studies and regulatory decision making is highly important, such approach appears limited in several aspects. First, at EU level, the right of access to documents can be exercised only regarding clinical trial data held by the EMA, which leaves out a considerable amount of data, most importantly, IPD[350] and data from unsuccessful trials.[351] Second, access to non-summary trial data held by the EMA is subject to the exception for the protection of commercial interests of trial sponsors, which scope is uncertain. Third, access to data held by the public authorities for *exploratory* research purposes, arguably, lies beyond the ambit of the right of access to documents. As for IPD held by drug companies, the analysis has not

[348] This finding is consistent with the view on competition law as an instrument of access to data in the context of data-driven innovation. See MPI position statement on data ownership and access to data (n 300), paras 31–32.

[349] On the potential limitations of applying competition law to unilateral practices in the context of competition in innovation, see generally Drexl (2012). See also Drexl (2013), pp. 317–318 (observing that 'it may well be that EU competition law lacks any remedy against unilateral restraints that obstruct R&D efforts of competitors prior to the emergence of a new technology or product market or the entry of the firms concerned into existing technology and product markets').

[350] See Chap. 6 at Sect. 6.5.1.2.

[351] Currently, only summary results from such trials are required to be submitted to the EU database. Reg 536/2014/EU, art 37(4).

identified a viable legal basis under the EU framework on which access to such data for research purposes could be claimed.

References

Anson W (2005) Traditional valuation methodologies. In: Anson W, Suchy DP, Ahya C (eds) Intellectual property valuation: a primer for identifying and determining value. American Bar Association, Chicago, pp 30–36

Anson W, Halligan RM, Weyand RF (2005) The economic valuation of trade secret assets. In: Anson W, Suchy DP, Ahya C (eds) Intellectual property valuation: a primer for identifying and determining value. American Bar Association, Chicago, pp 84–94

Ben-Asher D (2000) In need of treatment? Merger control, pharmaceutical innovation, and consumer welfare. J Leg Med 21:271–349

Beunen AC (2007) Protection for databases: the European database directive and its effects in the Netherlands, France and the United Kingdom. Wolf Legal Publishers, Nijmegen

Chalmers I, Glasziou P (2009) Avoidable waste in the production and reporting of research evidence. Lancet 374(9683):86–89. https://doi.org/10.1016/S0140-6736(09)60329-9

CIOMS (2016) International ethical guidelines for health-related research involving humans. CIOMS, Geneva

Coleman JL, Kraus J (1989) Rethinking the theory of legal rights. Yale Law J 95(7):1335–1371

Cottier T, Meitinger B, Meitinger I (2000) The protection of test data submitted to governmental authorities: the impact of the TRIP's Agreement on EC Law. In: Meng W, Stein T (eds) Marketing authorization for pharmaceutical products and the protection of submitted data. Nomos, Baden-Baden, pp 53–72

Crémer J, de Montjoye Y-A, Schweitzer H (2019) Competition policy for the digital era. Publications Office of the European Union, Luxembourg

Cummings SR, Browner WS, Hulley SB (2007) Conceiving the research question. In: Hulley SB et al (eds) Designing clinical research, 3rd edn. Wolters Kluwer Health, Philadelphia, pp 17–26

Curtin D, Mendes J (2014) Article 42 – right of access to documents. In: Peers S et al (eds) The EU Charter of Fundamental Rights. A commentary. Hart Publishing, Oxford, pp 1099–1120

Day S (2007) Dictionary for clinical trials. John Wiley & Sons, Hoboken

de Carvalho NP (2018) The TRIPS regime of patents and test data, 5th edn. Wolters Kluwer, Alphen aan den Rijn

DiMasi JA, Faden LB (2011) Competitiveness in follow-on drug R&D: a race or imitation? Nat Rev Drug Discov 10(1):23–27. https://doi.org/10.1038/nrd3296

Drazen JM (2002) Who owns the data in a clinical trial? Sci Eng Ethics 8(3):407–411. https://doi.org/10.1007/s11948-002-0062-x

Drexl J (2012) Anti-competitive stumbling stones on the way to a cleaner world: protecting competition in innovation without a market. J Compet Law Econ 8(3):507–542. https://doi.org/10.1093/joclec/nhs019

Drexl J (2013) AstraZeneca and the EU sector inquiry. In: Drexl J, Lee N (eds) Pharmaceutical innovation, competition and patent law. A trilateral perspective. Edward Elgar, Cheltenham, pp 290–322

Drexl J (2017a) Designing competitive markets for industrial data – between propertisation and access. JIPITEC 8(4):257–292

Drexl J (2017b) Neue Regeln für die Europäische Datenwirtschaft? Ein Plädoyer für einen wettbewerbspolitischen Ansatz – Teil 1. NZKart 5(7):339–344

Drexl J (2017c) Neue Regeln für die Europäische Datenwirtschaft? Ein Plädoyer für einen wettbewerbspolitischen Ansatz – Teil 2. NZKart 5(8):415–421

Drexl J (2017d) On the future EU legal framework for the digital economy: a competition-based response to the 'ownership and access' debate. In: Lohsse S, Schulze R, Staudenmayer D (eds) Trading data in the digital economy: legal concepts and tools. Nomos, Baden-Baden, pp 223–243

Drexl J (2018) Data access and control in the era of connected devices. BEUC, Brussels

European Commission (2015) 50 years of EU pharmaceutical regulation in 12 stories. Publications Office of the European Union, Luxembourg

Fisher R et al (2018) Study in support of the evaluation of Directive 96/9/EC on the legal protection of databases. Publications Office of the European Union, Luxembourg

Gøtzsche PC (2012) Strengthening and opening up health research by sharing our raw data. Circ Cardiovasc Qual Outcomes 5(2):236–237. https://doi.org/10.1161/CIRCOUTCOMES.112.965277

Graef I, Husovec M, Purtova N (2018) Data portability and data control: lessons for an emerging concept in EU law. German Law J 19(6):1359–1398. https://doi.org/10.1017/S2071832200023075

Hervey TK, McHale JV (2015) European Union health law: themes and implications. CUP, Cambridge

Higgins JPT, Altman DG, Sterne JAC (2017) Assessing risk of bias in included studies. In: Higgins JPT et al (eds) Cochrane handbook for systematic reviews of interventions. Cochrane, pp 8:1–8:73

Hoof K (2010) Property rights in legal history. In: Bouckaert B (ed) Encyclopedia of law and economics, vol 5, 2nd edn. Edward Elgar, Cheltenham, pp 5–30

Institute of Medicine of the National Academies (2015) Sharing clinical trial data: maximizing benefits, minimizing risk. The National Academies Press, Washington DC

Jager MF (2002) The critical role of trade secret law in protecting intellectual property assets. In: Goldscheider R (ed) Licensing best practices: the LESI guide to strategic issues and contemporary realities. John Wiley & Sons, New York, pp 127–138

Kim D (2017) Transparency policies of the European Medicines Agency: has the paradigm shifted? Med Law Rev 25(3):456–483

Krumholz HM (2015) Why data sharing should be the expected norm. BMJ 350:h599. https://doi.org/10.1136/bmj.h599

Leistner M (2017) Big data and the EU Database Directive 96/9/EC. In: Lohsse S, Schulze R, Staudenmayer D (eds) Trading data in the digital economy: legal concepts and tools. Nomos, Baden-Baden, pp 27–58

Lynskey O (2015) The foundations of EU data protection law. OUP, Oxford

Massaro J (2009) Experimental design. In: Robertson D, Williams GH (eds) Clinical and translational science: principles of human research: principles of human research. Elsevier, Amsterdam, pp 41–58

Mattei UA (1998) Property rights in civil law. In: Newman P (ed) The new Palgrave dictionary of economics and the law, vol 3. Macmillan Press, London, pp 157–164

OECD (2015) Data-driven innovation: big data for growth and well-being. OECD Publishing, Paris. https://doi.org/10.1787/9789264229358-en

Osborne Clarke LLP (2016) Legal study on ownership and access to data. Publications Office of the European Union, Luxembourg

Penner JE (1996) The 'bundle of rights' picture of property. UCLA Law Rev 43:711–820

Prins C (2006) Property and privacy: European perspectives and the commodification of our identity. In: Guibault L, Hugenholtz PB (eds) The future of the public domain. Identifying the commons in information law. Kluwer Law International, Alphen aan den Rijn, pp 233–258

Pugatch Consilum (2015) Clinical trial data and disclosure policies. The European Union, Member States, and international best practices. US Chamber of Commerce, Washington DC

Purtova N (2011) Property in personal data: second life of an old idea in the age of cloud computing, chain informatisation, and ambient intelligence. In: Gutwirth S et al (eds) Computers, privacy and data protection: an element of choice. Springer, Dordrecht, pp 39–64

Rathi V (2012) Sharing of clinical trial data among trialists: a cross sectional survey. BMJ 345: e7570. https://doi.org/10.1136/bmj.e7570
Shaikh OH (2016) Access to medicine versus test data exclusivity: safeguarding flexibilities under international law. Springer, Berlin, Heidelberg
Tierney JF et al (2015) How individual participant data meta-analyses have influenced trial design, conduct, and analysis. J Clin Epideiol 68(11):1325–1335. https://doi.org/10.1016/j.jclinepi.2015.05.024
van Erp S, Akkermans B (2010) Property rights: a comparative view. In: Bouckaert B (ed) Encyclopedia of law and economics, vol 5, 2nd edn. Edward Elgar, Cheltenham, pp 31–52
Vickers AJ (2006) Whose data set is it anyway? Sharing raw data from randomized trials. Trials 7:15. https://doi.org/10.1186/1745-6215-7-15
von Bar C, Drobnig U (2004) The interaction of contract law and tort and property law in Europe. A comparative study. Sellier European Law Publishers, Munich

Chapter 5
Implications of IPD Disclosure for Statutory Innovation Incentives

Abstract This chapter examines the intersection between disclosure of non-summary clinical trial data and statutory innovation incentives from a *de lege lata* perspective. In view of the allegations of the research-based pharmaceutical companies that mandatory disclosure would 'impede' their innovation incentives, it is essential to clarify how IPD disclosure might affect the applicable protection afforded under the existing EU framework. In particular, the analysis focuses on the implications of data disclosure for patents and sector-specific exclusivity-based forms of protection. Overall, the analysis shows that the industry's concerns regarding the offsetting effect on incentives can be justified only to a limited extent.

5.1 The Impediment-to-Innovation-Incentives Claim

5.1.1 Arguments Submitted During the EMA Public Consultation

During the public consultation that had preceded the EMA publication policy 0070, research-based pharmaceutical companies argued relentlessly that third-party access to clinical trial data, including for non-commercial scientific research purposes, would impede their incentives to innovate.[1] For instance, the German Association of Research-Based Pharmaceutical Companies claimed that the implementation of the EMA publication policy

> would have a substantial detrimental impact on Germany as a center of research, as well as the research landscape in Europe: the publication of approval information jeopardizes research and development investments in the billions, and therefore threatens to undercut

[1] While this overview focuses on the debate in the EU, analogous arguments regarding the impeding effect of clinical trial data disclosure on innovation incentives were raised by pharmaceutical companies during the public consultation conducted by the USFDA on the availability of masked and de-identified non-summary safety and efficacy data. See Availability of anonymized non-summary safety and efficacy data; request for comments (4 Jun 2013) https://www.regulations.gov/document/FDA-2013-N-0271-0001/comment. Accessed 26 Mar 2021.

D. Kim, *Access to Non-Summary Clinical Trial Data for Research Purposes Under EU Law*, Munich Studies on Innovation and Competition 16,
https://doi.org/10.1007/978-3-030-86778-2_5

> the research and development of new and innovative medicinal products for the promotion of public health.[2]

The Danish Association of the Pharmaceutical Industry and the Romanian Association of International Medicines Manufacturers alleged that the EMA policy is 'a threat to research and innovative medicine development'.[3] According to Novo Nordisk, CSRs contain

> detailed strategic and operational information revealing general company know-how about the efficient and competitive set-up of clinical studies, such as trial site performance [. . .] disclosure of such information could provide competitors with a roadmap to facilitate their own product development programs.[4]

Some submissions alluded to IP protection; for instance, the BioIndustry Association alleged:

> It should not be ignored that allowing third parties access to [clinical trial] data held by the [European Medicines] Agency may negatively impact on the value, competitiveness, ownership of trade secrets and intellectual property rights of undertakings.[5]

The EuropaBio Association asserted that the EMA publication policy

> risks negatively impacting incentives for biomedical innovation [as the p]remature disclosure of [commercially confidential information], misappropriation of research know-how and the lack of a predictable EU intellectual property/trade secret protection framework are factors that can undermine the ability of EuropaBio members to operate and innovate, and ultimately 'their value proposition' for potential investors.[6]

Furthermore, some companies pointed out that data disclosure contradicts the protection obligations under Article 39 of the TRIPS Agreement.[7]

5.1.2 Arguments Raised Before the CJEU

Similar arguments were raised before the CJEU. In *PTC Therapeutics v EMA*, the claimant alleged that the EU Transparency Regulation

---

[2] EMA (2 Oct 2014) Overview of comments received on 'Publication and access to clinical-trial data' (EMA/240810/2013). EMA/349245/2014, p. 23.

[3] ibid p. 74; EMA (2 Oct 2014) Overview of comments received on 'Publication and access to clinical-trial data' (EMA/240810/2013). EMA/344107/2014, p. 19.

[4] EMA (2 Oct 2014) Overview of comments received on 'Publication and access to clinical-trial data' (EMA/240810/2013). EMA/354914/2014, pp. 70–71.

[5] EMA (2 Oct 2014) Overview of comments received on 'Publication and access to clinical-trial data' (EMA/240810/2013). EMA/342115/2014, p. 12.

[6] EMA (2 Oct 2014) Overview of comments received on 'Publication and access to clinical-trial data' (EMA/240810/2013). EMA/344107/2014, pp. 27–28.

[7] ibid p. 86 (citing the submission of Pfizer); EMA (2 Oct 2014) Overview of comments received on 'Publication and access to clinical-trial data' (EMA/240810/2013). EMA/342115/2014, p. 32 (citing the submission of EFPIA).

> must be interpreted in such a way as to preserve [...] the balance that the EU legislature intended to strike between, on the one hand, the requirement for transparency, the protection of public health and the need to avoid the duplication of trials and, on the other hand, the objective of protecting the confidentiality of industrial and commercial information and the need to encourage innovation.[8]

In *AbbVie v EMA*, the pharmaceutical company argued that even if access was granted to one student,[9] there was a risk that 'the confidential information could be disclosed to anybody, including current or potential competitors' of AbbVie.[10] The damage would then result from the competitors' use of the disclosed data in two ways. First, disclosed data could be used for regulatory purposes to support the marketing approval for generic drugs,[11] while the claimant 'would have no effective means of knowing whether their competitors had actually used the clinical study reports in order to obtain approval for a medicinal product which would compete with the medicinal product Humira, in particular in the case of an approval sought outside the European Union'.[12] Second, by analysing data contained in the clinical study reports at issue, competitors would be able 'to develop a medicinal product which would compete with the medicinal product Humira'.[13]

An analogous argument was raised by InterMune, in particular, that

> damage would arise [...] from the future use of the information [contained in the] requested documents by the InterMune companies' competitors – and specifically by Boehringer Ingelheim GmbH, which has already requested that the EMA disclose that information – in order to *develop* a medicinal product which would compete with the medicinal product Esbriet.[14]

PTC Therapeutics submitted that competitors would be able

> to mine the data in the report at issue in order to restructure their own clinical trials and avoid some of the *trial-and-error* development which it had to undertake. The release of the report at issue in its entirety would confer a *competitive advantage* on those of the applicant's competitors that are seeking to produce a *direct rival* to Translarna in the European Union [which would be] *safer, more effective* or otherwise *clinically superior* [...] In particular, such release would provide [competitors] with information on the *development* of the precedent orphan product and the design of clinical trials [...] that is particularly valuable

[8] Case C-513/16 P(R) *EMA v PTC Therapeutics International* [2017] ECLI:EU:C:2017:148, para 55.

[9] In that case, access was requested by a university science student in connection with the preparation of a master's thesis. Case T-44/13R *AbbVie v EMA* [2013] ECLI:EU:T:2013:221, para 20.

[10] ibid para 46.

[11] In particular, AbbVie alleged that it would suffer 'the damage [...] of a financial nature [that] in practice [...] is impossible to assess [...] given the many ways in which the clinical study reports could be used by an indeterminate number of competing undertakings throughout the world, before an indeterminate number of regulatory authorities'. Case C-389/13 *EMA v AbbVie* [2013] ECLI: EU:C:2013:794, para 31.

[12] ibid para 32.

[13] ibid.

[14] C-390/13 P(R) *EMA v InterMune* [2013] ECLI:EU:C:2013:795, para 34 (emphasis added).

in the case of 'ultraorphan' conditions. In addition, the report at issue could be used in the European Union to support [marketing authorisation] applications for similar products or, in the long run, for generic [marketing authorisation] applications once Translarna has lost all exclusivity.[15]

To summarise, competitive concerns raised by the research-based drug companies tend to fall into two categories—the one related to *generic* competition (if the disclosed dossiers could be 'reused' by competitors to obtain the marketing authorisation for a generic product), and the other one related to competition in *innovation* (if the disclosed data could facilitate competitors' R&D and the development of a new product).

5.1.3 Restrictive Provisions Under the Industry Data-Sharing Policies

Data-sharing policies have been implemented by many research-based pharmaceutical companies and overall share the common concept and approach.[16] While companies generally demonstrate their willingness to share anonymised IPD for 'legitimate research' purposes,[17] access is usually subject to a set of conditions and reservations. For instance, research proposals submitted through the Clinical Study Data Request portal[18] are reviewed by an independent review panel concerning the following aspects:

- the scientific rationale and relevance of the proposed research to medical science or patient care;
- the ability of the proposed research plan (design, methods and analysis) to meet the scientific objectives;
- potential identification of individual research participants;
- the publication plan for the research;
- real or potential conflicts of interest that may impact the planning, conduct or interpretation of the research and proposals to manage these conflicts of interest;
- qualifications and experience of the research team to conduct the proposed research.[19]

---

[15]Case T-718/15 R *PTC Therapeutics International v EMA* [2016] ECLI:EU:T:2016:425, para 92 (emphasis added).

[16]Doshi (2014).

[17]PhRMA and EFPIA (18 Jul 2013) Principles for responsible clinical trial data sharing. Our commitment to patients and researchers, pp. 1, 4. https://www.efpia.eu/media/25189/principles-for-responsible-clinical-trial-data-sharing.pdf. Accessed 26 Mar 2021.

[18]https://www.clinicalstudydatarequest.com. Accessed 26 Mar 2021.

[19]Independent review panel charter (2 Mar 2020), p. 2 (emphasis added). https://www.clinicalstudydatarequest.com/Documents/Independent_Review_Panel_Charter.pdf. Accessed 26 Mar 2021.

Similarly, the PhRMA and the EFPIA, in their joint Principles for responsible clinical trial data sharing, put forward that research proposals should include and be evaluated against the following particularities: the description of the data being requested, including the hypothesis to be tested, the rationale for the proposed research, the analysis plan, the publication plan, qualifications and experience of the proposed research team, the description of any potential conflicts of interest, including possible competitive use of the data, etc.[20] The potential 'competitive use' of data refers to using the sponsors' data 'to help gain approval of a potentially competing medicine'.[21] Furthermore, the joint Principles state that

> while companies may enter into agreements to co-develop medical products, these data sharing Principles are not intended to allow *freeriding* or degradation of incentives for companies to invest in biomedical research. Accordingly, it would be *appropriate* [. . .] for companies to refuse to share proprietary information with their competitors.[22]

Individual data sharing policies of drug companies contain similar provisions. For example, Merck Serono declares that it can provide access to clinical data to 'qualified researchers with appropriate competencies, engaged in rigorous, independent, and novel scientific research'.[23] However, the likelihood of a conflict of interest will be examined and 'data will not be released to individuals with significant conflict of interest or individuals requesting data access for competitive, commercial or legal interests'.[24]

Sometimes, the terms and conditions explicitly state that data shall not be shared with competitors. For instance, Merck Serono's policy stipulates:

> Following approval of a new product or a new indication for an approved product in both the European Union and the United States, Merck Serono will share study protocols, anonymized patient level, and study level data and redacted clinical study reports from clinical trials in patients with *qualified scientific and medical researchers*, upon researcher request, as necessary for conducting legitimate research [However,] data will not be shared with Merck Serono *competitors*.[25]

Roche's policy states that 'access to analyzable datasets from clinical trials through a secure system, following an independent assessment of the *scientific*

---

[20] PhRMA and EFPIA (18 Jul 2013) Principles for responsible clinical trial data sharing. Our commitment to patients and researchers, p. 4. https://www.efpia.eu/media/25189/principles-for-responsible-clinical-trial-data-sharing.pdf. Accessed 26 Mar 2021.

[21] ibid.

[22] ibid (emphasis added).

[23] Merck (Nov 2020) Procedure on access to clinical trial data, p. 1. https://www.merckclinicaltrials.com/pdf/ProcedureAccessClinicalTrialData.pdf. Accessed 26 Mar 2021.

[24] ibid.

[25] Merck Serono. Summary of Merck's responsible data sharing policy for researchers, p. 1 (emphasis added). https://www.merckgroup.com/content/dam/web/corporate/non-images/research/healthcare/Summary_of_Mercks_Responsible_Data_Sharing_Policy_EN.pdf. Accessed 26 Mar 2021.

merit of a rigorously defined research question from a third party'.[26] Along the same lines, Pfizer limits the eligibility of data requestors by declaring that clinical trial data 'will not be released to applicants with significant conflicts of interest, including individuals requesting access for commercial/competitive or legal purposes'.[27] Not only actual competitors might be denied access to non-summary clinical trial data at the trial sponsor's 'sole discretion and without further arbitration',[28] but also if a request for access poses a potential competitive risk.[29] According to the statistics on the inquiries submitted via the Clinical Study Data Request portal,[30] only in the case of five studies access was denied because the proposed research 'competed with the sponsor's publication plan'.[31]

Overall, the similarity of data-sharing provisions might suggest that pharmaceutical companies adhere to an 'industry code', which is based on the principle that it is 'appropriate'[32] to reject an application for non-summary data if third-party access and data analysis raise competitive concerns.

5.2 Dissecting the Claim

As a starting point, let us discern the critical elements within the impediment-to-incentives claim and clarify their meaning.

[26] Roche (1 Jun 2013) Roche global policy on sharing of clinical trials data, p. 1 (emphasis added). https://www.roche.com/dam/jcr:1c46aa73-cea0-4b9b-8eaa-e9a788ed021b/en/roche_global_policy_on_sharing_of_clinical_study_information.pdf. Accessed 26 Mar 2021.

[27] Pfizer (2015) A guide to requesting Pfizer patient-level clinical data, p. 2. http://www.pfizer.com/files/research/research_clinical_trials/A_Guide_to_Requesting_Pfizer_Patient-Level_Clinical_Data_April_2015.pdf. Accessed 26 Mar 2021.

[28] Amgen. Clinical trial data sharing request. https://www.amgen.com/science/clinical-trials/clinical-data-transparency-practices/clinical-trial-data-sharing-request. Accessed 26 Mar 2021.

[29] ibid. See also Independent review panel charter (2 Mar 2020), p. 2 (stating that study sponsors 'may, in exceptional circumstances, veto a request to access data where they feel there is a potential conflict of interest or an actual or potential competitive risk'). https://www.clinicalstudydatarequest.com/Documents/Independent_Review_Panel_Charter.pdf. Accessed 26 Mar 2021.

[30] Metrics. https://www.clinicalstudydatarequest.com/Metrics.aspx. Accessed 9 Jun 2021. In total, 611 research proposals were submitted between 1 Jan 2014 and 30 Apr 2021.

[31] Reasons why access cannot be provided (enquiries for Sanofi studies). https://www.clinicalstudydatarequest.com/Metrics/Reasons-Access-Not-Provided-Sanofi.aspx. Accessed 9 Jun 2021.

[32] PhRMA and EFPIA (18 Jul 2013) Principles for responsible clinical trial data sharing. Our commitment to patients and researchers, p. 4. https://www.efpia.eu/media/25189/principles-for-responsible-clinical-trial-data-sharing.pdf. Accessed 26 Mar 2021.

5.2.1 *Innovation*

The term of innovation can be broadly defined to encompass 'all those scientific, technological, organisational, financial and commercial steps which actually, or are intended to, lead to the implementation of innovations includ[ing] R&D that is not directly related to the development of a specific innovation'.[33] R&D refers to 'creative and systematic work undertaken [. . .] to increase the stock of knowledge and to devise new applications of available knowledge'.[34]

Accordingly, pharmaceutical innovation includes research preceding the discovery of a new target or novel pharmacological properties of a molecule and pre-clinical and clinical developmental directed at bringing a new medicinal product to the market.[35] The notion of innovation includes 'entirely new goods and services [and] *significant* improvements to existing products'.[36] In pharmaceutical innovation, 'entirely new' products are known as 'first-in-class' drugs, while improvements over existing drugs and significant improvements are also called 'follow-on' and 'best-in-class' drugs, respectively.[37]

5.2.2 *Innovation Incentives*

In broad terms, innovation incentives can comprise various factors that can bear on the motivation and decision making to undertake the development and commercialisation of new products and services. In the economics of innovation, the term 'incentive' often refers to the means and conditions of earning returns on innovative activity,[38] benefits that 'make the investment worthwhile'.[39] Appropriability is defined as the conditions that determine 'possibilities to earn profits from an innovative activity [particularly, by] protecting innovations from *imitation*'.[40] The conditions of appropriability (or internalisation) of benefits can

---

[33] OECD and Eurostat (2005), p. 91. See also OECD and Eurostat (2018), p. 20 (emphasising that the 'requirement for implementation differentiates innovation from other concepts such as invention, as an innovation must be implemented, i.e. put into use or made available for others to use').

[34] OECD and Eurostat (2018), p. 87. See also Hall et al. (2010), p. 1035.

[35] Nightingale and Mahdi (2006), p. 74.

[36] OECD and Eurostat (2005), p. 17 (emphasis added). See also OECD and Eurostat (2018), p. 20 (defining innovation as 'a new or improved product or process (or combination thereof) that differs significantly from the [the innovator's] previous products or processes and that has been made available to potential users').

[37] Potential effects of IPD disclosure on competition in the respective markets are discussed in detail in Chap. 8 at Sect. 8.1.4.

[38] Breschi and Malerba (2005), p. 135.

[39] Levin et al. (1987), p. 983.

[40] Breschi and Malerba (2005), p. 135 (emphasis added).

take the form of legal (e.g. IP rights) or factual protection (e.g. the first-mover advantage[41] or control over complementary assets).[42]

Industrial sectors are characterised in terms of high or low appropriability conditions. High appropriability means that efficient ways exist to protect innovation against imitation.[43] Low appropriability implies that an environment is conducive to knowledge spillovers,[44] i.e. the third-party use and benefits from knowledge embedded in an innovative product cannot be excluded (or internalised) by the knowledge producer once the product is placed in the market.[45] Industries differ starkly in terms of conditions and mechanisms of appropriability.[46]

5.2.3 The Problem of Appropriability in Drug Innovation

The problem of innovation incentives is also known as the *public-good* problem in innovation, the proposition that a perfectly competitive market fails to provide optimal incentives for innovation. Such market failure is commonly attributed to the non-excludability of R&D results (technological knowledge is a classic example[47]) and unrestricted imitation that can hinder the innovators' ability to earn returns on R&D and thus discourage firms from undertaking an innovative activity.[48]

The problem of appropriability is especially acute in the pharmaceutical sector due to substantial R&D costs, on the one hand, and the relative ease of reverse-engineering of technologies underlying (mainly chemical) drugs and the relatively low costs of producing and commercialisation of their generic copies, on the other hand. Several policy instruments can support innovative activity in the research-based pharmaceutical sector. These instruments are surveyed below before the potential effect of disclosure of non-summary clinical trial data on the protection afforded by them can be analysed.

---

41 Cohen (2010), p. 138.

42 Breschi and Malerba (2005), p. 135 (with further references).

43 ibid.

44 ibid.

45 Knowledge spillovers are discussed in more detail in Chap. 7.

46 Breschi and Malerba (2005), p. 135 (emphasis added).

47 See e.g. den Hertog J (2010) Review of economic theories of regulation. Utrecht School of Economics Discussion Paper Series No 10-18, p. 16. https://ideas.repec.org/p/use/tkiwps/1018.html. Accessed 26 Mar 2021; Stiglitz (1999), pp. 308–309.

48 According to Arrow, indivisibilities, inappropriability and uncertainty are 'three classical reasons for the possible failure of perfect competition to achieve optimality in resource allocation' and the 'discrimination against investment in inventive and research activities'. Arrow (1962), pp. 609, 616.

5.2.4 *Innovation Incentives*

5.2.4.1 Patents and SPCs

Pharmaceutical patents can protect a product (e.g. an active ingredient, its hydrates, salts, esters, metabolites and intermediates, as well as combinations of active ingredients), as well as a process (e.g. a method of manufacturing an active ingredient, methods or uses of medical treatment, formulation methods). Cross-industry empirical studies find that patent protection can play a prominent role in supporting innovation in the pharmaceutical sector.[49] A 2012 review of empirical research on the economics of patents concludes that

> if there is an increase in innovation due to patents, it is likely to be centered in the pharmaceutical, biotechnology, and medical instrument areas [. . .] This conclusion relies mostly on survey evidence from a number of countries that shows rather conclusively that patents are *not* among the important means to appropriate returns to innovation, *except* [for these areas].[50]

In addition, due to the lengthy processes of drug development and marketing authorisation, an extension of protection was introduced to compensate for the reduction of the effective patent term that would be otherwise 'insufficient to cover the investment put into the research'.[51] Supplementary protection certificates ('SPCs') for medicinal products were introduced in the EU in 1992 and can be granted to the holders of national and European patents.[52] The protection conferred by an SPC extends only to 'the product covered by the authorisation to place the

---

[49] See e.g. Orsenigo L, Sterzi V (2010) Comparative study of the use of patents in different industries. Università Commerciale Luigi Bocconi KITeS Knowledge, Internationalization and Tech. Studies Working Paper 33/2010, p. 7 (arguing that 'the role of patents is likely to be higher [. . .] where imitation is easier, i.e. when the ratio between the costs of imitation to the costs of innovation is lower (e.g. chemicals, pharmaceuticals, machinery)'); Encaoua et al. (2006), p. 1425 (noting that '[w]hen imitating is as costly as inventing, or when firms have economic and technical means for protecting their inventions then [. . .] there is no need for further legal protection'); Bessen and Meurer (2005), pp. 26–27 (noting that 'patent premium is greatest in the pharmaceutical industry' and that the patent system 'provides critical incentives for research and development in the pharmaceutical and a few other industries'); Levin et al. (1987), p. 796 (reporting that only in the drug industry product patents were regarded by the majority of respondents 'as strictly more effective than other means of appropriation'); Mansfield (1986), p. 175. But see Qian (2007), p. 450 (finding 'no statistically significant relationship between national pharmaceutical patent protection and [. . .] domestic R&D'); US Senate, Subcommittee On Antitrust and Monopoly (1961), p. 119 (reporting an 'overwhelming' finding that '[d]rugs discovered in foreign countries without product patents outnumber those discovered in countries with such protection in the order of 10 to 1', while 'drugs which are among the most widely used in the world were discovered in countries which have never awarded patents on pharmaceutical products').

[50] Hall and Harhoff (2012), p. 548 (emphasis added).

[51] Reg 469/2009/EC, rec 4. See also Case C-457/10 P *AstraZeneca AB v European Commission* [2012] ECLI:EU:C:2012:770, para 8.

[52] Council Regulation (EEC) No 1768/92 of 18 June 1992 concerning the creation of a supplementary protection certificate for medicinal products (2 Jul 1992) OJ L 182, repealed by Regulation

corresponding medicinal product on the market and for any use of the product as a medicinal product that has been authorised before the expiry of the certificate'.[53] A 'product' is defined as an active ingredient or combination of active ingredients of a medicinal product.[54] A 'medicinal product' refers to

> any substance or combination of substances presented for treating or preventing disease in human beings or animals and any substance or combination of substances which may be administered to human beings or animals with a view to making a medical diagnosis or to restoring, correcting or modifying physiological functions in humans or in animals.[55]

The underlying policy rationale is to stimulate research and the introduction of new medicinal products.[56] An SPC can extend the protection conferred by a basic patent for an additional period maximum of five years. A 'basic patent' is defined as a patent protecting a product as such, a process to obtain a product or an application of a product.[57] Overall, the SPC holder can benefit from a maximum 15-year period of protection from obtaining the first marketing authorisation for a medicinal product in the EU.[58] Furthermore, to encourage research on paediatric applications of medicines, the legislation provides a six-month extension of SPCs for drugs with paediatric indications.[59]

5.2.4.2 Incentive Instruments Under the Sector Legislation

Besides patents, research-based pharmaceutical companies can benefit from the sector-specific forms of protection, namely, test data exclusivity and market exclusivities for drugs with orphan and paediatric indications.

Test data exclusivity, as explained earlier, prevents a drug authority from using safety and efficacy data submitted by the originator company to authorise a generic product.[60] The innovation rationale for test data exclusivity appears straightforward: the delay of generic entry is meant to allow drug originators to recover costs of

---

(EC) No 469/2009 of the European Parliament and of the Council of 6 May 2009 concerning the supplementary protection certificate for medicinal products (16 Jun 2009) OJ L 152.

[53] Reg 469/2009/EC, art 4.

[54] Reg 469/2009/EC, art 1(b).

[55] Reg 469/2009/EC, art 1(a).

[56] Case C-431/04 *Massachusetts Institute of Technology* [2006] ECLI:EU:C:2006:291, para 19 (recollecting that the proposal for the Regulation introducing SPCs concerned 'only new medicinal products' defined as 'an active substance in the strict sense', while 'minor changes', including a new dose or the use of a different salt or ester, 'will not lead to the issue of a new [SPC]'). See also Case C-484/12 *Georgetown University v Octrooicentrum* [2013] ECLI:EU: C:2013:828, para 31.

[57] Reg 469/2009/EC, art 1(c).

[58] Reg 469/2009/EC, rec 9.

[59] Reg 1901/2006/EC, art 36 (specifying the conditions).

[60] On the duration, scope and nature of protection, see Chap. 4 at Sect. 4.2.4.

conducting clinical trials.[61] From a healthcare economics perspective, this can be viewed as an attempt to strike a balance between affordable and innovative drugs. According to the EU legislator, the main reason for exempting 'essentially similar' drugs from conducting full-scale trials is to avoid 'repetitive tests on humans or animals without over-riding cause'.[62] At the same time, it should be ensured that 'innovative firms are not placed at a disadvantage'.[63] Notably, the existing framework does not prohibit either a generic company from conducting duplicative trials or a drug authority from approving a generic drug based on the efficacy and safety data submitted by a generic applicant during the test data exclusivity term of protection.[64]

Test data exclusivity has been criticised for its effect on access to affordable drugs, especially in developing countries.[65] Besides, its role in promoting drug innovation in the presence of patent protection[66] is not straightforward. According to some authors, the claim of the research-based pharmaceutical industry that data exclusivity 'can be a key consideration in the business decision to introduce new innovative drugs into a market'[67] lacks empirical support.[68] Two normative questions are debated: whether data exclusivity can be justified as an incentive to innovate; if so, whether a *sui generis* exclusivity-based protection is an optimal means towards that goal.[69]

---

[61] Such argument, however, contradicts the inducement theory of patents, according to which patents induce not only an invention but also its development and commercialisation. Mazzoleni and Nelson (1998), p. 276 ff.

[62] Dir 2001/83/EC, rec 10.

[63] Dir 2001/83/EC, rec 9.

[64] As discussed below at Sect. 5.4.1.2.

[65] Ragavan (2018); Baird (2013); Correa (2008).

[66] Patent protection usually outlasts test data exclusivity. See European Commission (8 Jul 2009) Pharmaceutical sector inquiry final report, p. 73 [hereinafter Pharmaceutical sector inquiry final report]. https://ec.europa.eu/competition/sectors/pharmaceuticals/inquiry/staff_working_paper_part1.pdf. Accessed 26 Mar 2021. According to the report, the test data protection term outlasted the duration of the corresponding patent protection (including SPC protection) in 51 out of the examined 713 cases.

[67] IFPMA (2011) Data exclusivity: encouraging development of new medicines, p. 5. https://www.ifpma.org/wp-content/uploads/2016/01/IFPMA_2011_Data_Exclusivity__En_Web.pdf. Accessed 26 Mar 2021.

[68] Palmedo (2013). The study examined whether test data exclusivity protection impacted pharmaceutical investment in 45 countries, including the EU member states and found no statistically significant relationship between data exclusivity protection and the level of investment by the pharmaceutical industry in the examined jurisdictions. Apart from data exclusivity, the variables included gross national income, population and ease of doing business. But see Gaessler and Wagner (2020) (showing that data exclusivity can be an effective policy instrument in cases where a drug cannot be protected under patent rights). See also Shaikh (2016), pp. 30 ff (providing an overview of studies on the impact of test data exclusivity).

[69] See Reichman (2009), p. 42 ff. See also Shaikh (2016), p. 25 (pointing out that '[a]lternate incentives such as rewards, prizes, public-private partnerships in R&D, purchase commitments for the final output of the pharmaceutical R&D, tax benefits etc are more direct and targeted as

Apart from test data exclusivity, the EU legislation provides for market exclusivity for drugs with *orphan*[70] and *paediatric*[71] indications to promote the development of medicines 'for otherwise neglected populations with unmet medical needs, namely children and patients with rare diseases'.[72] An 'orphan medicinal product' means that a drug

(a) is intended for the diagnosis, prevention or treatment of a life-threatening or chronically debilitating condition affecting not more than five in 10 thousand persons in the Community when the application is made, or

is intended for the diagnosis, prevention or treatment of a life-threatening, seriously debilitating or serious and chronic condition in the Community and that without incentives it is unlikely that the marketing of the medicinal product in the Community would generate sufficient return to justify the necessary investment; and

(b) no satisfactory method of diagnosis, prevention or treatment of the condition in question [exists] that has been authorised in the Community or, if such method exists, that the medicinal product will be of significant benefit to those affected by that condition.[73]

---

compared to test data exclusivity'). The debate features the opposition between protection by exclusive rights vs. the liability rule-based cost-sharing approach. On the cost-sharing approach to pharmaceutical test data protection, see Weissman (2006); Fellmeth (2004). On the normative underpinnings of test data exclusivity, see Reichman (2009), pp. 36 ff; Shaikh (2016), pp. 21 ff; Kim D (forthcoming) Test data exclusivity: an elusive pursuit to strike a balance between affordable drugs and protection of returns on investment. In: Bonadio E, Goold P (eds), The Cambridge handbook of non-creative intellectual property. CUP, Cambridge.

[70] Regulation (EC) No 141/2000 of the European Parliament and of the Council of 16 December 1999 on orphan medicinal products (22 Jan 2000) OJ L 18. For an overview of incentives implemented to support research, marketing and development of orphan medicinal products at EU and national level, see European Commission (22 Dec 2016) Inventory of Union and member state incentives to support research into, and the development and availability of, orphan medicinal products – state of play. SWD(2015) 13 final. Besides market exclusivity, incentives include tax reduction, direct reimbursement after marketing authorisation, joint procurement, etc.

[71] Regulation 1901/2006/EC of the European Parliament and of the Council of 12 December 2006 on medicinal products for paediatric use and amending Regulation (EEC) No 1768/92, Directive 2001/20/EC, Directive 2001/83/EC and Regulation (EC) No 726/2004 (27 Dec 2006) OJ L 378/1. For an overview of regulatory incentives for R&D for paediatric drugs in the EU, see EMA (9 Sep 2014) Report to the European Commission on companies and products that have benefited from any of the rewards and incentives in the Pediatric Regulation and on the companies that have failed to comply with any of the obligations in this Regulation. EMA/24516/2014 Corr.3. https://ec.europa.eu/health/sites/default/files/files/paediatrics/2013_annual-report.pdf. Accessed 26 Mar 2021.

[72] EMA (27 Oct 2016) 10-year Report to the European Commission. General report on the experience acquired as a result of the application of the Paediatric Regulation. EMA/231225/2015, p. 66. https://ec.europa.eu/health/sites/default/files/files/paediatrics/2016_pc_report_2017/ema_10_year_report_for_consultation.pdf. Accessed 26 Mar 2021. See also Reg 141/2000/EC, rec 8; Reg 1901/2006/EC, rec 19.

[73] Reg 141/2000/EC, art 3(1).

Orphan drugs can enjoy market exclusivity for ten years starting from the date of marketing authorisation.[74] Protection is referred to as 'market exclusivity' since, unlike in the case of data exclusivity, a generic drug cannot be approved irrespective of whether subsequent applicants submit data generated in their trials. Derogation from market exclusivity is envisaged in three cases, namely, (i) upon the consent of the original marketing authorisation holder, (ii) due to the inability of the original marketing authorisation holder to supply sufficient quantities, and (iii) if the second medicinal product is demonstrated to be safer, more effective or otherwise clinically superior.[75]

Several incentive instruments explicitly target the development of drugs with paediatric indications.[76] First, the developers of medicinal products, which are exclusively intended for the paediatric population and are not covered by IP rights, are eligible for the Paediatric Use Marketing Authorisation (PUMA) under the EU Paediatric Regulation. The PUMA holders can benefit from ten years of data and market exclusivity protection.[77] According to the evaluation carried out by the EMA in 2016, the EU Paediatric Regulation 'has had a very positive impact on paediatric drug development, as shown by the data collected over the first nine years since its inception'.[78] Second, an SPC can be extended by six months for medicinal products developed according to an agreed paediatric investigation plan, provided other

---

[74]Reg 141/2000/EC, art 8(1). During the term of protection, a drug authority 'shall not [. . .] accept another application for a marketing authorisation, or grant a marketing authorisation [. . .] for the same therapeutic indication, in respect of a similar medicinal product'. A 'similar medicinal product' means 'a medicinal product containing a similar active substance or substances as contained in a currently authorised orphan medicinal product, and which is intended for the same therapeutic indication'. A 'similar active substance' refers to 'an identical active substance, or an active substance with the same principal molecular structural features (but not necessarily all of the same molecular features) and which acts via the same mechanism'. Reg 847/2000/EC, art 3(3) (b) and (c), respectively.

[75]Reg 141/2000/EC, art 8(3). See also European Commission (23 Sep 2008) Guideline on aspects of the application of article 8(1) and (3) of Regulation (EC) No 141/2000: assessing similarity of medicinal products versus authorised orphan medicinal products benefiting from market exclusivity and applying derogations from that market exclusivity. OJ C 242.

[76]A 'paediatric population' refers to the population aged between birth and 18 years. Reg 1901/2006/EC, art 2(1). A 'paediatric use marketing authorisation' is defined as

> a marketing authorisation granted in respect of a medicinal product for human use which is not protected by a supplementary protection certificate under Regulation (EEC) No 1768/92 or by a patent which qualifies for the granting of the supplementary protection certificate, covering exclusively therapeutic indications which are relevant for use in the paediatric population, or subsets thereof, including the appropriate strength, pharmaceutical form or route of administration for that product.

Reg 1901/2006/EC, art 2(4).

[77]Reg 1901/2006/EC, art 38.

[78]EMA (27 Oct 2016) 10-year Report to the European Commission. General report on the experience acquired as a result of the application of the Paediatric Regulation. EMA/231225/2015, p. 9.

conditions are met.[79] Third, orphan drugs for children can qualify for two-year market exclusivity, in addition to the existing ten years under the EU Orphan Drugs Regulation.[80]

Although the above-reviewed regulatory exclusivities differ in their duration and prerequisites for protection, they underlie the same 'incentive logic' of allowing originator drug companies to recover R&D costs by delaying generic competition.[81] In so doing, they resemble the rationale and mechanism of patent protection. However, given that protection terms often overlap, it might be unfeasible to assess the actual impact of each instrument on the originator's capacity to earn returns on investment and motivation to engage in further innovative activity. Not to mention that the existence of a causal relationship between returns on investment and innovation cannot be unequivocally presumed.

The expiration of all forms of protection is known as the 'loss of exclusivity'.[82] Let us consider next how disclosing non-summary clinical trial data can affect the protection afforded by the policy instruments reviewed above and under what conditions it can facilitate the loss of exclusivity.

5.3 Implications of IPD Disclosure for Patent Protection

5.3.1 Concerns of Drug Companies

During the EMA's public consultation, the German Pharmaceutical Industry Association claimed that

> [i]f pre-clinical and/or clinical data are published by the authorities, subsequent patents can no longer be obtained if the therapeutic effect on animals has been published (e.g. *in vivo* or *in vitro*), or if clinical trials indicate that the medicine would be effective on humans.[83]

Similarly, the European Confederation of Pharmaceutical Entrepreneurs alleged that, due to the data disclosure policies by drug authorities,

> patent protection may be difficult to achieve where a new therapeutic indication for a well-known substance is subject of the marketing authorisation. In these circumstances, the

---

[79] Reg 1901/2006/EC, art 36.

[80] Reg 1901/2006/EC, art 37.

[81] See Reg 141/2000/EC, rec 8 (stating that 'the strongest incentive for the pharmaceutical industry to invest in the development and marketing of orphan medicinal products is where there is a prospect of obtaining market exclusivity for a certain number of years during which part of the investment might be recovered').

[82] Pharmaceutical sector inquiry final report (n 66), p. 23.

[83] EMA (2 Oct 2014) Overview of comments received on 'Publication and access to clinical-trial data' (EMA/240810/2013). EMA/349245/2014, p. 18.

> [EMA] Policy could dis-incentivise companies, from filing applications for marketing authorisations or variations in the EU.[84]

In *EMA v InterMune*, the originator company InterMune argued that, as a result of the disclosure of CSRs by the EMA to a competitor (Boehringer Ingelheim GmbH), 'damage would arise [. . .] from the InterMune companies' loss of the opportunity to obtain *future* patents, since that information would henceforth constitute prior art'.[85]

Notably, the EMA Clinical trial advisory group on legal aspects took note of the argument that the proactive disclosure of trial data under the EMA publication policy can affect the ability of drug companies to obtain patents for future inventions, especially as far as second generation inventions are concerned. In particular, its advice highlights:

> Patents do not only relate to active substances but also to, inter alia, formulations, isomeric and crystal forms, pro-drugs and metabolites, processes, further medical uses, dosing regimes, combination therapies, drug-drug interactions, contra-indications and safety measures, etc. Information *underpinning inventions* relating to any of those can be found in clinical and non-clinical-trial data, and it is possible that marketing authorisation applicants *create these inventions* through analysis of the information provided in the MAA [marketing authorisation application]. Once information in MAA is disclosed, it becomes "prior art" and cannot later serve as the basis for an invention and patent application. Thus, marketing authorization applicants would no longer be able to use the currently confidential information to obtain patents for the *inventions relating* to the information in a MAA if the MAA is disclosed to the public. Hence, the Agency's proactive publication policy could prejudice later patent filing on subsequent inventions made on known products.[86]

As acknowledged by the EMA, disclosure of data on *exploratory* endpoints[87] 'may preclude obtaining patents that would cover biomarkers/diagnostics themselves, as well as method of use patents directed to patient subpopulations'.[88] Moreover, legal scholars argued that clinical trial data disclosure could lead to 'the broad increase to the "prior art" and "common general knowledge" [that] may render many related new drugs unpatentably obvious'.[89]

At the outset, it should be clarified that the above-cited concerns mainly relate to the patentability of the so-called 'second generation' pharmaceutical inventions, follow-on inventions building on the pioneering ('basic') invention.[90] 'Secondary'

---

[84] ibid p. 46.

[85] Case C-390/13 P(R) *EMA v InterMune UK* [2013] ECLI:EU:C:2013:795, para 34 (emphasis added).

[86] EMA (30 Apr 2013) Advice to the European Medicines Agency from the Clinical Trial Advisory Group on Legal Aspects (CTAG5) – final advice, lines 311–321 (emphasis added). https://www.ema.europa.eu/en/documents/other/ctag5-advice-european-medicines-agency-clinical-trial-advisory-group-legal-aspects-final-advice_en.pdf. Accessed 26 Mar 2021.

[87] Exploratory endpoints refer to the variables, such as biomarkers, tested in the hypothesis-generating studies that do not support the label claim of an investigational product.

[88] EMA publication policy 0070, p. 19.

[89] Price and Minssen (2015), p. 686.

[90] The terms 'primary' and 'secondary' patents correspond to the distinction between 'basic' and 'follow-on' inventions. 'Primary' ('basic') patents claim an active ingredient (the lead compound),

pharmaceutical patents can claim an alternative route of administration, different dosage forms, specific pharmaceutical formulations that combine the active ingredient with other substances to promote the therapeutic effect of a drug (e.g. by enhancing absorption), etc.[91] Applications for secondary patents are usually filed during the late stages of developing a candidate compound or after the product launch.[92]

There is little doubt that disclosure of non-summary clinical trial data can considerably enlarge prior art, as CSRs and IPD present far more comprehensive records compared to scientific publications and summary information reported in trial registers. Once the disclosed data constitutes the relevant prior art, it can reduce the chances of obtaining a patent. Thus, if access-to-data policies were to be designed to avoid a prejudicial effect on patentability, one first needs to examine in what particular way data disclosure can impact patentability.

5.3.2 Implications of Non-summary Clinical Trial Data Disclosure for Patentability

In a series of cases, the EPO Boards of Appeal examined the implications of disclosure of trial-related information on the patentability of follow-on pharmaceutical inventions. Notably, all identified decisions concern second-generation inventions.[93] The prior art documents, which gave rise to the objections to patentability, were scientific publications regarding the ongoing or completed clinical trials. In

---

while 'secondary' patents claim modifications and derivatives of an active ingredient (e.g. salts, esters, ethers, polymorphs, metabolites and isomers), combinations with other molecules and new methods for the production of the known active ingredients. See Pharmaceutical sector inquiry final report (n 66), p. 100. Such inventions improve the basic active ingredient and can offer therapeutic benefits, such as improved potency and enhanced absorption. ibid p. 187. On the typology of pharmaceutical inventions and patents, see Ahn (2014), p. 34 ff.

[91] Pharmaceutical sector inquiry final report (n 66), p. 51; Ahn (2014), p. 56.

[92] Applications for basic patents are often filed during the lead identification and optimisation process, i.e. before clinical trials begin. Pharmaceutical sector inquiry final report (n 66), p. 381.

[93] The review presented in this section covers the following cases: T 2506/12 *Pegylated Liposomal Doxorubincin/Pharma Mar S.A.* (4 Oct 2016) ECLI:EP:BA:2016:T250612.20161004; T 1859/08 *Anti-ErbB2 antibodies/GENENTECH, Inc.* (5 Jun 2012) ECLI:EP:BA:2012:T185908.20120605; T 158/96 *Obsessive-compulsive-disorder/PFIZER* (28 Oct 1998) ECLI:EP:BA:1998:T015896.19981028; T 0652/12 *Treatment for systemic onset juvenile idiopathic arthritis/CHUGAI* (8 Dec 2016) ECLI:EP:BA:2016:T065212.20161208; T 1031/00 Amlodipine/SEPRACOR (23 May 2002) ECLI:EP:BA:2002:T103100.20020523; T 0385/07 *Aplidine/PHARMA MAR* (5 Oct 2007) ECLI:EP:BA:2007:T038507.20071005; T 1493/09 *Human papillomavirus vaccines/GSK*) (1 Oct 2014) ECLI:EP:BA:2014:T149309.20141001; T 128/82 *Pyrrolidin-Derivate/Hoffmann-La Roche* (12 Jan 1984) ECLI:EP:BA:1984:T012882.19840112; T 0715/03 *Use of ziprasidone for treating Tourette's syndrome/PFIZER* (16 Jan 2006) ECLI:EP:BA:2006:T071503.20060116; T 1745/12 *Temozolomide/JIANGSU* (4 Jun 2018) ECLI:EP:BA:2018:T174512.20180604.

some cases, patent applicants were the sponsors of those trials. While the EPO Boards of Appeal applied the established EPO methodology for the patentability assessment, several specific aspects were clarified concerning the prior art documents related to clinical trials.

5.3.2.1 Novelty

A prior art document is novelty-destroying if it discloses a combination of all claimed features.[94] The claimed subject matter lacks novelty if it is 'derivable directly and unambiguously' from a prior art document, taking into consideration 'any features *implicit* to a person skilled in the art in what is expressly mentioned in the document'.[95] Only a 'clear and unmistakable teaching' of the prior art can be prejudicial to novelty.[96]

The EPO Boards of Appeal followed this approach when the objections to novelty were based on the prior art documents related to clinical trials. Overall, the reviewed case law suggests that whether the disclosed information about a clinical trial can be prejudicial to novelty depends on the particular events reported from that trial. For instance, in T 158/96,[97] the prior-art document (a scientific publication) disclosed that compound sertraline intended to treat obsessive-compulsive disorder was undergoing a phase II trial. The EPO technical Board of Appeal (TBA) reasoned that, based on that publication, the skilled person 'would not *realistically* have concluded that evidence of sertraline effectiveness had already been produced'.[98] Therefore, the document at issue was not considered to indicate any therapeutic effectiveness of sertraline for the claimed indication.[99] Neither could the information regarding the planned phase III trial serve as an indicator that the claimed therapeutic effect could be derived 'directly and unambiguously'.[100] In that case, the scientific publication reported the results of the clinical trials, which involved the ingredients of the therapeutic combination claimed in a patent application. However, none of the trials tested the combined therapy in humans. The TBA held that the prior art

[94] Separate prior art items cannot be combined when assessing novelty, in contrast to inventive step. EPO (Mar 2021) Guidelines for examination in the European Patent Office, pt G, VI-1. https://www.epo.org/law-practice/legal-texts/html/guidelines/e/g_vi_1.htm. Accessed 26 Mar 2021.

[95] ibid VI-2 (emphasis added). https://www.epo.org/law-practice/legal-texts/html/guidelines/e/g_vi_2.htm. Accessed 26 Mar 2021.

[96] Established case law of the EPO Boards of Appeal. EPO (2016), p. 104.

[97] Case T 158/96 *Obsessive-compulsive-disorder/PFIZER* (28 Oct 1998) ECLI:EP:BA:1998:T015896.19981028.

[98] ibid para 3.6.2 (emphasis added).

[99] ibid.

[100] Case T 1859/08 *Anti-ErbB2 antibodies/GENENTECH, Inc.* (5 Jun 2012) ECLI:EP:BA:2012:T185908.20120605, para 21.

document could not be interpreted as 'encouraging results' that translated into a clinical benefit 'measured by increased time to disease progression'.[101]

Similarly, in T 2506/12,[102] the claimed subject matter involved a combinational therapy for treating ovarian cancer undergoing a phase I trial; yet the results were still unknown. As considered by the TBA, the therapeutic efficacy of the combination at issue was 'implicitly disclosed'[103] in the scientific publications. However, its safety could not be established as the skilled person would not exclude the possibility that the tested combination 'might interact to produce unacceptable adverse effects'.[104]

In contrast, in T 1031/00,[105] the TBA found that the claimed subject matter was anticipated by the disclosure in the prior art document.[106] The contested claim was directed at the first medical use of amlodipine for the treatment of hypertension. The scientific publication disclosed that amlodipine was undergoing a phase III trial and that *in vitro* evaluation confirmed its efficacy for the claimed indication.

In sum, a prior art document related to a clinical trial can be prejudicial to novelty if it allows the skilled person to derive the therapeutic application, which is claimed for patent protection, with a high degree of certainty regarding both efficacy and safety. In other words, the novelty-destroying effect of trial results disclosure cannot be presumed. Instead, it will depend on the case's circumstances, particularly the specific effects of a medical intervention on the health state observed in a trial and reported in a prior art document. As noted by the TBA in T 158/96, 'a pharmacological effect observed in an early investigation may *directly* and *unambiguously* reflect a therapeutic effect [. . .] underlying a therapeutic application. [. . .] Yet this is not a general or absolute rule.'[107]

While the reviewed cases mainly concerned scientific publications, one would assume that disclosure of IPD could pose a greater risk to novelty. However, such novelty-destroying effect cannot be presumed. One should consider that scientific publications reporting the trial outcomes usually present the results of IPD analysis. In principle, the results of an independent *secondary* IPD analysis should not substantially deviate from the conclusions regarding the treatment effect reported by the trial investigators. In other words, secondary IPD analysis of reliable and robust IPD is unlikely to produce new teaching regarding the therapeutic application,

[101] ibid para 20.

[102] Case T 2506/12 *Pegylated Liposomal Doxorubincin/Pharma Mar S.A.* (4 Oct 2016) ECLI:EP:BA:2016:T250612.20161004.

[103] ibid para 2.11 (further holding that '[t]he possibility that each drug might cancel out the other's pharmacological activity is remote and would not have been considered a realistic outcome without actual experimental evidence').

[104] ibid para 2.12.

[105] Case T 1031/00 *Amlodipine/SEPRACOR* (23 May 2002) ECLI:EP:BA:2002:T103100.20020523.

[106] ibid para 2.1.1.

[107] Case T 158/96 *Obsessive-compulsive-disorder/PFIZER* (28 Oct 1998) ECLI:EP:BA:1998:T015896.19981028, para 3.5.2 (emphasis added).

in addition to what was reported by the original investigators. However, this might not always be the case. The lack of inferential reproducibility can explain the divergence between the conclusions of primary and secondary IPD analyses.[108] Besides, IPD might contain information related to the endpoints or unforeseen events[109] that were not subsequently reported in a scientific publication. However, such additional observations are often uncertain and, at best, can suggest a hypothesis for further research.[110] In any case, even though disclosure of data on exploratory endpoints is unlikely to destroy novelty due to the probabilistic nature of the results, such concerns can be mitigated by adjusting the timing of IPD disclosure.

5.3.2.2 Inventive Step

The EPO's methodology of assessing inventive step (the so-called 'problem-solution approach') is a three-prong inquiry:

- first, the 'closest prior art' is determined;
- second, the 'objective technical problem' to be solved is defined;
- third, it is considered whether the claimed invention would have been obvious to the skilled person, given the closest prior art and the objective technical problem at hand.[111]

When answering the third question, the 'could-would' test is applied. In particular, it is considered whether any teaching in the prior art *would*—as opposed to could—have prompted the skilled person to modify or adapt the closest prior art, thereby 'arriving at something falling within the terms of the claims, and thus achieving what the invention achieves'.[112] Overall, the obviousness of the claimed solution can depend on individual facts of a case, including factors such as the technical problem at issue,[113] state of the art in a particular technical field,[114] and earlier disclosed teachings available to the skilled person. The reviewed cases related to the disclosure of information about clinical trials illustrate this premise.

In T 1745/12,[115] the objective technical problem was defined as providing a formulation allowing the prolonged release of temozolomide to the central nervous

---

[108] Inferential reproducibility is defined as the reproducibility of the primary analysis results through independent confirmatory analysis. Chow and Liu (2004), p. 82.

[109] Unforeseen effects can be beneficial, neutral or adverse.

[110] On exploratory endpoints, see Chap. 3 at Sect. 3.2.2.3.

[111] EPO (Mar 2021) Guidelines for examination in the European Patent Office, pt G, VII-5. https://www.epo.org/law-practice/legal-texts/html/guidelines/e/g_vii_5.htm. Accessed 26 Mar 2021.

[112] ibid VII-5.3. https://www.epo.org/law-practice/legal-texts/html/guidelines/e/g_vii_5_3.htm. Accessed 26 Mar 2021.

[113] EPO (2016), p. 185.

[114] ibid p. 186.

[115] Case T 1745/12 *Temozolomide/JIANGSU* (4 June 2018) ECLI:EP:BA:2018:T174512.20180604.

system. The TBA considered whether the claimed solution was obvious to the skilled person in light of several scientific publications discussing various aspects of temozolomide administration. The publications reported the ongoing clinical trials investigating the efficacy and safety of temozolomide in treating different types of systemic tumours. The Board concluded that the prior art documents at issue would lead the skilled person *directly* to the claimed subject matter.[116] Consequently, the inventive step could not be established.[117]

Neither could the inventive step be confirmed in T 1493/09[118] concerned with a vaccine against cervical cancer.[119] The prior art documents reported results from monkey studies involving the same vaccine. The TBA held that such results—alone or in combination with the disclosed information that the vaccine at issue was undergoing a clinical trial directed at the same purpose as the invention—would allow the skilled person to arrive at the claimed solution without making an inventive step.[120]

The inventive step was also denied in light of state of the art in T 0652/12.[121] In that case, the TBA found that a prior art document reporting successful clinical trials of etanercept for treating a subtype of arthritis 'would have had a significant positive impact on the skilled person's expectations regarding the likely success of such therapies'.[122]

In T 2506/12,[123] the claimed subject matter involved a combinational therapy for treating ovarian cancer and was undergoing a phase I trial, as reported by two scientific publications, while the results were unknown. The TBA acknowledged that it was not known whether and to what extent the claimed combination potentiated the risk of increased toxicity.[124] The interactions between the drugs in the combined treatment typically increase toxicities, which can nevertheless be balanced by reducing the dosages of each drug.[125] However, it was considered that the skilled person would have reasonable expectations of success in finding the 'adequately safe combinations of dosages'.[126] Therefore, irrespective of whether 'the outcome of the reported clinical trial could be success or failure, no particular reason was known that

[116] ibid para 3.5.3.

[117] ibid para 3.5.4.

[118] T 1493/09 *Human papillomavirus vaccines/GSK*) (1 October 2014) ECLI:EP:BA:2014:T149309.20141001.

[119] ibid para 10.

[120] ibid para 19.

[121] Case T 0652/12 *Treatment for systemic onset juvenile idiopathic arthritis/CHUGAI* (8 Dec 2016) ECLI:EP:BA:2016:T065212.20161208.

[122] ibid para 19.

[123] Case T 2506/12 *Pegylated Liposomal Doxorubincin/Pharma Mar S.A.* (4 Oct 2016) ECLI:EP:BA:2016:T250612.20161004.

[124] ibid para 3.14.

[125] ibid.

[126] ibid.

would have *discouraged* the person skilled in the art from carrying out an experimental evaluation to confirm the usefulness of the combination treatment'.[127] Accordingly, the Board concluded that 'finding out in this straightforward manner that useful dosage combinations providing both efficacy and safety indeed existed cannot be regarded as an invention'.[128]

In contrast, in T 0385/07,[129] the TBA held that, given the prior art document disclosing a phase I clinical trial, the skilled person would not have a reasonable expectation of success when using the drug tested in that trial to treat a different type of cancer.[130] As noted by the Board, the types of cancer affecting different organs differ in terms of aetiology,[131] the underlying molecular alterations and the way of producing metastases.[132] Knowing that different types of cancer—and even patients having the same type of tumour—cannot be treated uniformly with the same treatment, the skilled person would not be able 'to predict whether or not a drug shown to be effective in the treatment of one type of cancer would also be effective against a different type of cancer'.[133]

In T 0715/03,[134] the TBA concluded that none of the cited prior art documents would 'give any hint' to the skilled person searching for the compounds suitable for treating Tourette's syndrome to the use ziprasidone.[135]

Overall, the reviewed case law suggests that the assessment of the inventive step is highly context-dependent. Whether clinical trial data disclosure can have a prejudicial effect on inventive step would depend on the particular effects reported from a trial, the specific characteristics of a disease, the problem intended to be solved, and the state of the relevant prior art.[136] Accordingly, the reviewed cases cannot be generalised to allege that clinical trial data disclosure is detrimental to the patentability of inventions that might be developed based on knowledge gained in clinical trials.

---

[127] ibid para 3.15 (emphasis added).

[128] ibid.

[129] Case T 0385/07 *Aplidine/PHARMA MAR* (5 Oct 2007) ECLI:EP:BA:2007:T038507.20071005.

[130] ibid para 17.

[131] Aetiology is a discipline that studies the causes that set a disease process 'in motion'. Porth (2011), p. xix.

[132] Case T 0385/07 *Aplidine/PHARMA MAR* (5 Oct 2007) ECLI:EP:BA:2007:T038507.20071005, para 16.

[133] ibid.

[134] Case T 0715/03 *Use of ziprasidone for treating Tourette's syndrome/PFIZER* (16 Jan 2006) ECLI:EP:BA:2006:T071503.20060116.

[135] ibid para 2.4.3.

[136] In contrast to novelty, the assessment of inventive step considers separate prior art references in combination, whereby '[e]ven an implicit prompting or implicitly recognisable incentive [could be] sufficient to show that the skilled person would have combined the elements from the prior art'. EPO (Mar 2021) Guidelines for examination in the European Patent Office, pt G, VII-5.3. https://www.epo.org/law-practice/legal-texts/html/guidelines/e/g_vii_5_3.htm. Accessed 26 Mar 2021.

Compared to scientific publications constituting prior art in the above-discussed cases, CSRs and IPD provide a far more comprehensive account of clinical effects observed in a trial. Based on such data, it can be assumed that the skilled person would be able to gain a more detailed understanding of pharmacological properties of the tested substance(s) and, thus, have a stronger (or weaker) motivation to pursue further experimentations directed at solving a problem at hand. However, disclosure of IPD might not necessarily pose a greater challenge for inventive step relative to the disclosure of trial results through a scientific publication. As noted above, the reason is that the conclusions of primary and secondary IPD analyses, in principle, should not differ to the extent that secondary data analysis can produce *new* technical teaching.[137] In contrast to novelty, however, the assessment of inventive step can accommodate the probabilistic nature of effects observed in a trial, given that 'absolute certainty is not required for finding that the skilled person would have a reasonable expectation of success'.[138] Nevertheless, the concept of 'reasonable expectation of success' cannot be stretched to embrace the 'hope to succeed'.[139]

5.3.2.3 Summary of Implications of Data Disclosure for Patentability

In light of the reviewed case law of the EPO Boards of Appeal, concerns expressed by drug companies regarding the detrimental effect of IPD disclosure on the patentability of pharmaceutical inventions might be justified only in limited situations. In general, data disclosure would not affect the patentability of first-generation inventions, for which patent applications are usually filed before conducting trials.[140] As for second-generation inventions, implications are not straightforward. On the one hand, disclosure of non-summary data—CSRs and IPD—can undoubtedly enlarge the scope of the prior art. On the other hand, the prejudicial effect on patentability—relative to what can be *already disclosed* in scientific publications, i.e. summary-level data—cannot be presumed as the outcome would depend on the contextual assessment. Disclosure of trial-related information and data is unlikely to be novelty-destroying, given that a relatively high threshold of certainty of the effect of a therapeutic application is required for it to be considered derivable from the prior art. As for the inventive step assessment, the effect of IPD disclosure on the skilled person's expectation of success would depend on various factors such as the technical problem at issue, the state of the relevant prior art and specific pharmacological characteristics of a product tested in a trial.

---

[137] See above (n 108) and the accompanying text.

[138] EPO (2016), p. 185. To prejudice novelty, the claimed therapeutic effect would need to be 'derivable directly and unambiguously'. See above (nn 95–96) and the accompanying text.

[139] While the 'hope to succeed' implies 'merely the expression of a wish', a reasonable expectation of success means a 'scientific appraisal of available facts'. EPO (2016), p. 186.

[140] Pharmaceutical sector inquiry final report (n 66), p. 51; Turner (2007), p. 6 ff.

The situation might be more complex if multiple effects on the health state were observed in a trial, while some might not be subsequently disclosed in the scientific publications. Exploratory endpoints, for instance, can lay a foundation for a potentially patentable therapeutic application, distinct from the investigational product. In this regard, it would be pertinent to consider how prejudicial effects on patentability can be mitigated. In particular, access-to-data measures could foresee a reservation to allow the trial sponsors to file patent applications related to the additional treatment effects that emerged from the trial observations.

5.3.2.4 Implications of Disclosure for SPC Protection

Disclosure of clinical trial data may impact SPC protection insofar as it can affect either the grant or the validity of the basic patent.[141] As noted earlier, where the disclosed trial data constitutes relevant prior art, it can affect the patentability of the claimed therapeutic effect. However, the effect on patentability would ultimately depend on the particular events observed during a trial and subsequently reported. Besides, the possibility that secondary analysis of disclosed data might give rise to post-grant validity challenges cannot be excluded. If the basic patent underlying SPC protection gets invalidated in light of the non-summary test data,[142] the corresponding SPC will lapse.[143]

In addition, the protection under an SPC will cease if the product covered by it is no longer placed on the market following the withdrawal of the corresponding drug marketing authorisation.[144] In some exceptional cases, one could speculate that this provision may apply if secondary analysis of non-summary data reveals inaccurate conclusions regarding the benefit-risk balance of the medicinal product and eventually leads to the withdrawal of the marketing authorisation at issue. Beyond such situations, data disclosure is unlikely to affect protection under SPCs.

[141] Reg 469/2009/EC, art 13. For a definition of a 'basic patent', see above (n 57).

[142] Patent invalidation proceedings can be based on the prior art already considered during the examination or opposition proceedings and the newly introduced prior art. See Hess et al. (2014) (finding that the newly introduced prior art frequently leads to patent invalidation).

[143] On the grounds of SPC invalidation, see Reg 469/2009/EC, art 15.

[144] Reg 469/2009/EC, art 14(d).

5.4 Implications of IPD Disclosure for Sector-Specific Incentives

5.4.1 Implications of Data Disclosure for Data Exclusivity Protection in the EU

5.4.1.1 Concerns Regarding the Circumvention of Data Exclusivity Protection

As mentioned earlier, the EMA transparency policies permit access to CSRs and IPD for non-commercial scientific research and explicitly precludes using the released data for regulatory purposes.[145] Nevertheless, research-based drug companies put forward the argument that

> [p]roactive disclosure would have the effect of undermining data exclusivity and would support [marketing authorisation application] by *innovators or generic* companies [. . .] *either in the EU or elsewhere*, by allowing third parties to *circumvent* existing regulatory data protection rules or by taking advantage of the *absence* of such rules.[146]

The raised concerns should be examined in the jurisdictions with test data exclusivity protection, such as in the EU, and without it.

5.4.1.2 Implications of Data Disclosure for Test Data Exclusivity Protection in the EU

The above-cited concerns underlie a premise that, during the term of data exclusivity, the EMA or other drug authorities can accept and review an application for the authorisation of a generic product submitted not through the abbreviated pathway[147] but based on a full dossier comprising efficacy and safety data. Even though such scenario is highly hypothetical,[148] it is important to clarify the legal basis of the

---

[145] EMA publication policy 0070, annex 1, para 3; annex 2, para 3.

[146] EMA (30 Apr 2013) Advice to the European Medicines Agency from the Clinical Trial Advisory Group on Legal Aspects (CTAG5) – final advice, lines 231–237 (emphasis added). https://www.ema.europa.eu/en/documents/other/ctag5-advice-european-medicines-agency-clinical-trial-advisory-group-legal-aspects-final-advice_en.pdf. Accessed 26 Mar 2021. See also EMA (2 Oct 2014) Overview of comments received on 'Publication and access to clinical-trial data' (EMA/240810/2013). EMA/344107/2014, p. 85 (citing the argument by Pfizer that competitors 'may circumvent Regulatory Data Protection rules or take advantage of their non-existence, especially outside the EU, when detailed, non-public domain data are disclosed in the EU'). See also Case T-33/17 *Amicus Therapeutics v EMA* [2018] ECLI:EU:T:2018:595, para 84.

[147] That is, according to the requirements under Article 8(3) of Directive 2001/83/EC.

[148] It is well-known that the costs of conducting full-scale clinical trials are prohibitive for generic companies. See e.g. Pharmaceutical sector inquiry final report (n 66), p. 35. See also Gaessler and Wagner (2020), p. 10 (reporting that 'interviews with executives from pharmaceutical companies did not reveal any cases where firms duplicated clinical trials').

originator companies' concerns that competitors can reuse data disclosed by the EMA for regulatory purposes. Directive 2001/83/EC that lays down the rules on data exclusivity protection in the EU does not explicitly outlaw this option. According to the EMA,

> no applicant is restricted to making only generic applications under Art. 10(1) of Directive 2001/83/EC, and they are free to make applications also under other legal bases. [...] The regulatory data exclusivity provisions [...] prevent for a certain time from relying on information submitted in an application for another product, but they do not prevent, however, providing *own* data on the same or related substance.[149]

Thus, the EMA does not exclude that a generic drug can be approved during the term of data exclusivity protection based on 'own' data. Some legal commentators share this view.[150] However, if we assume that a generic drug can, theoretically, be approved based on their 'own' data, we first need to clarify what 'own' means. It is logical to assume that 'own' data implies data generated in *different* trials than those conducted by the sponsor of the originator drug. In other words, one cannot simply download CSRs from the EMA's portal and resubmit them as if one's 'own'. First of all, such conduct would violate the terms of the agreement stipulated by the EMA that explicitly preclude using data for drug marketing authorisation.[151] Second, it would, in effect, amount to the violation of data exclusivity protection since the drug authority would approve a generic drug by directly relying on the originator's data. Detecting such instances appears straightforward because CSRs include the trial identification information (including the unique trial identification number), making it possible to establish whether the applicant attempts to 'reuse' data submitted elsewhere. Accordingly, concerns that disclosed originator's data could be misused— resubmitted by generic companies as if 'their own'—should be viewed as a matter of enforcement of the requirement to provide 'own' data (which is not excluded under the current regime).[152] Such measures should not be problematic in a

---

[149] E-mail of 26 Sep 2017, inquiry number ASK-32704 (on file with the author) (emphasis added).

[150] See e.g. de Carvalho (2018), p. 286 (arguing that the term 'market protection' applies only to protection of the products 'whose marketing has been obtained with the support of protected test data [but not] to bio-equivalent products whose registration has been eventually obtained by a generics manufacturer using its own data')); Shaikh (2016), p. 9 (noting that test data exclusivity 'does not necessarily result in monopoly prices in the relevant market', and that other drug producers, 'both originators and generics, may target the same market and may get their product approved by generating their own test data').

[151] EMA publication policy 0070, annex 1, para 3; annex 2, para 3. Clinical trial data released by the EMA bears the watermark that states that the use of the documents for commercial purposes is prohibited. ibid p. 6.

[152] As noted by the CJEU in *PTC Therapeutics International v EMA*, even if competitors could gain access to CSRs under the EMA transparency policies, they 'would still have to carry out its own relevant studies and trials'. Case T-718/15 *PTC Therapeutics International v EMA* [2018] ECLI: EU:T:2018:66, para 91. Thus, the Court held that 'the claim that the report at issue must be considered confidential in its entirety on the ground that its disclosure might enable competitors to apply for [marketing authorisation] is unfounded in law'. ibid.

system where trials are subject to the authorisation and registration that assigns a unique trial identifier.

5.4.1.3 The Risk of Misusing Disclosed Data Outside of the EU

Next, let us consider the argument that generic competitors might be able to take undue advantage of IPD and CSRs disclosed by the EMA outside the EU.[153] The claim that test data can be reused for marketing authorisation needs to be considered in two regulatory settings—with and without test data exclusivity.

Regarding jurisdictions that provide test data protection,[154] the reasoning is, in principle, the same as in the preceding section. Even if a national drug authority might accept and review an application for a generic product during the term of test data exclusivity protection (provided that the applicant submits 'own' efficacy and safety data), 'own' should be interpreted as referring to data generated in trials *other* than those conducted by the originator companies. For instance, regulations in China and Russia explicitly state that the subsequent applicants can be approved based on 'own' data during the term of test data exclusivity. In particular, according to the Chinese Drug Administration Law, the generic applicant for the drug marketing authorisation can be approved during the term of protection of the originator's test

---

[153] See above (nn 12, 146) and the accompanying text. During the public consultation conducted by the USFDA concerning the availability of masked and de-identified non-summary safety and efficacy data, the PhRMA submitted that, 'to the extent that competitors could unmask the data, they could obtain approval of competing products with little effort or investment'. PhRMA (24 Jul 2013) Comments to Docket No. FDA-2013-N-0271: availability of masked and de-identified non-summary safety and efficacy data. Request for comments, p. 4. https://www.regulations.gov/comment/FDA-2013-N-0271-0003. Accessed 26 Mar 2021. See also Institute of Medicine of the National Academies (2015), p. 143 (pointing out the risk that disclosed clinical trial data 'may be used for "unfair commercial purposes", such as wholesale copying of originator data sets for purposes of receiving regulatory approval in jurisdictions with limited regulatory data protection laws').

[154] For an overview of data exclusivity regimes, see IFPMA (2011) Data exclusivity: encouraging development of new medicines. https://www.ifpma.org/wp-content/uploads/2016/01/IFPMA_2011_Data_Exclusivity__En_Web.pdf. Accessed 26 Mar 2021.

data only if *independently acquired* data is submitted.[155] The same is possible under the Russian legislation.[156]

The US Federal Food, Drug, and Cosmetic Act does not state explicitly whether a generic drug can be approved based on the submission of a new drug application (NDA)[157] during the term of data exclusivity protection.[158] However, as clarified by the USFDA, such protection does not preclude 'the marketing of a duplicate version of the same drug product if the duplicate version is the subject of a full new drug application submitted under 505(b)(1) [of the US Federal Food, Drug, and Cosmetic Act]'.[159] Accordingly, where a generic product can be authorised based on the full dossier, including efficacy and safety data generated in new trials, the argument that competitors might take undue advantage of data disclosed elsewhere appears unfounded.

As for jurisdictions without test data exclusivity, concerns that data disclosed by the EMA might be reused to obtain marketing authorisation for a generic product[160] might be relevant if generic applicants in such jurisdictions are required to submit full efficacy and safety data. Whether this might be the case in any country nowadays

---

[155] Regulations for Implementation of the Drug Administration Law of the People's Republic of China, Decree of the State Council of the People's Republic of China No. 360 of 4 August 2002. See also WTO (1 Oct 2001) Report of the Working party on the accession of China. WT/ACC/CHN/49 (01-4679), para 284 (reporting the confirmation from the representative of China that

> China would, in compliance with Article 39.3 of the TRIPS Agreement, provide effective protection against unfair commercial use of undisclosed test or other data submitted to authorities in China as required in support of applications. During this period, any second applicant for market authorization would only be granted market authorization if he submits *his own data* (emphasis added).

[156] Federal Law of the Russian Federation No 61-FZ of April 4, 2010 'On Marketing of Medicinal Products', art 18. See also WTO (17 Nov 2011) Report of the Working party on the accession of the Russian Federation to the World Trade Organization' WT/ACC/RUS/70, WT/MIN(11)/2, para 1295 (reporting that the representative of the Russian Federation confirmed that Russia had enacted legislation to comply with the requirement under Article 39(3) of the TRIPS Agreement and that, accordingly, during the six-year protection term, 'any subsequent application for marketing approval or registration would not be granted, unless the subsequent applicant submitted his own data').

[157] An NDA refers to a new drug application, a complete set of documents filed under section 505(b)(1) of the US Federal Food, Drug, and Cosmetic Act.

[158] Sections 505(c)(3)(E) and 505(j)(5)(F) of the US Federal Food, Drug, and Cosmetic Act.

[159] USFDA (2 Nov 2016) Frequently asked questions for new drug product exclusivity. https://www.fda.gov/Drugs/DevelopmentApprovalProcess/SmallBusinessAssistance/ucm069962.htm. Accessed 26 Mar 2021.

[160] For instance, according to the industry association EuropaBio, competitors might take advantage of data disclosed by the EMA to obtain marketing authorisation for generic products, in particular, 'in regions where the originator company does not have a marketing authorisation, or where no stewardship or adequate protection for [test data] exists'. EMA (2 Oct 2014) Overview of comments received on 'Publication and access to clinical-trial data' (EMA/240810/2013). EMA/344107/2014, p. 27.

is highly doubtful. The EMA's Clinical trial advisory group on legal aspects notes in this regard:

> It has been widely argued that generic manufacturers will use clinical-trial data to obtain marketing authorisations in jurisdictions without patent protection. It has not, however, been shown that the regulatory authorities in any such jurisdiction even require detailed clinical-trial data for their granting. If they demanded such data, this would surely imply that generic manufacturers would not be able to obtain a marketing authorisation in those jurisdictions today.[161]

In some jurisdictions, drug authorities do not carry out a scientific assessment of efficacy and safety data. Instead, a drug can be authorised based on the approval issued by other regulators, such as the USFDA or the EMA.[162] In such countries, competitors would not be able to take undue advantage by resubmitting the originator's safety and efficacy data disclosed elsewhere since such data is not required in the first place.[163]

Overall, there is hardly a legal basis that could support the claim that generic companies might misuse non-summary clinical trial data released by the EMA for regulatory purposes. Not surprisingly, such arguments did not succeed at the CJEU.[164]

5.4.2 Implications of Data Disclosure for Orphan Drug Exclusivity

In general, concerns that the released IPD can affect market exclusivity for drugs with orphan indications appear to be unwarranted, given that protection precludes a drug authority from accepting another application for marketing authorisation for the

[161] EMA (30 Apr 2013) Advice to the European Medicines Agency from the Clinical Trial Advisory Group on Legal Aspects (CTAG5) – final advice, lines 93–98 (emphasis added). https://www.ema.europa.eu/en/documents/other/ctag5-advice-european-medicines-agency-clinical-trial-advisory-group-legal-aspects-final-advice_en.pdf. Accessed 26 Mar 2021.

[162] For instance, in India, drugs are authorised predominantly based on their earlier approval in other jurisdictions, such as Australia, Canada, the European Union, Japan, the UK and the US. Irrespective of whether the originator drug has been authorised for marketing in India, a generic company is only required to 'prove that the drug has been approved and marketed in another country and submit confirmatory test and other data from clinical studies on a very few (in some cases as few as 16) Indian patients'. Institute of Medicine of the National Academies (2015), p. 262 (with further references).

[163] ibid p. 261 (noting that, in the jurisdictions where 'competitors can rely for their marketing applications on approval of the molecule by the FDA or the EMA, data release may confer little marginal advantage to the competitor').

[164] The CJEU held that the allegation concerning the potential misuse of CSRs is 'vague' and no specific argument was put forward 'to show that the alleged danger in certain third countries is real'. Case T-33/17 Amicus Therapeutics v EMA [2018] ECLI:EU:T:2018:595, para 84.

same therapeutic indication in respect of a similar medicinal product.[165] An exception to such immunity would be a situation where secondary IPD analysis can contribute to developing a *clinically superior* drug for the same therapeutic indication, which can subsequently be approved by derogating market exclusivity of the first marketing authorisation holder.[166] For instance, *PTC Therapeutics v EMA*, the drug company PTC Therapeutics—the sponsor and holder of the marketing authorisation for drug Translarna—alleged that

> [t]he release of the report at issue in its entirety would confer a *competitive* advantage on those of the applicant's competitors that are seeking to produce a *direct rival* to Translarna in the European Union in relation to the nonsense mutation DMD indication by claiming that their new product is *safer, more effective* or otherwise *clinically superior* [because] such release would provide them with *information* on the *development* of the precedent orphan product and the design of clinical trials.[167]

The case facts do not mention for which purposes the 'unknown' pharmaceutical company requested access to the CSRs at issue.[168] Hence, one could only speculate to what extent concerns expressed by the claimant were justified. What is known at the time of writing is that no other treatment has been approved in the EU for the same indication as Translarna (Duchenne muscular dystrophy),[169] while the Community Register of orphan medicinal products contains 33 entries[170] for the orphan medicine designation for the indication of the treatment of Duchenne muscular dystrophy filed by more than 20 companies between 2007 and 2021.[171] This fact is remarkable as it shows that pre-market competition among drug companies can be highly intense, even in orphan diseases.

One could hypothesise whether and to what extent IPD related to Translarna—if accessed by other drug developers—could facilitate the development of the clinically superior treatment. As evident from the reviewed entries in the Community Register of orphan medicinal products,[172] products that have to date received the orphan designation for Duchenne muscular dystrophy indication differ in terms of

[165] For a definition, see above (n 74).

[166] See above (n 75) and the accompanying text.

[167] Case T-718/15R *Therapeutics International v EMA* [2016] ECLI:EU:T:2016:425, para 92 (emphasis added).

[168] ibid paras 12, 38.

[169] According to the Community register of orphan medicinal products, the orphan market exclusivity for Translarna is expected to expire on 5 Aug 2024. https://ec.europa.eu/health/documents/community-register/html/h902.htm. Accessed 15 Jun 2021.

[170] An application for an orphan designation can be filed at any stage of developing a medicinal product before applying for marketing authorisation of that product. Reg 141/2000/EC, art 5(1).

[171] European Commission. Community register of orphan medicinal products. https://ec.europa.eu/health/documents/community-register/html/reg_od_act.htm?sort=a. Accessed 15 Jun 2021. In the Community register of orphan medicinal products, an 'active' application corresponds to a granted valid application for the orphan designation, while a 'non-active' application means that it was withdrawn, refused or expired.

[172] Above (n 171).

their active ingredients and the types of treatment (e.g. pharmaceutical drugs, gene and stem cell therapies). It is unclear whether clinical trial data related to Translarna after being submitted to the EMA could have contributed to R&D directed at those products. At the same time, it cannot be denied that aggregated IPD from all R&D projects related to Duchenne muscular dystrophy could improve the understanding of the pathophysiology and mechanisms of the disease.

The argument raised by PTC Therapeutics[173] suggests that data holders would have an inherent incentive not to share IPD if its secondary analysis might facilitate the development of new treatments with a more favourable benefit-risk balance relative to the existing drugs. However, the questions of whether such outcome is socially optimal and how the situation can be improved in this regard lie beyond the scope of analysis *de lege lata* and call for the analysis *de lege ferenda*.

5.5 Conclusion on Chapter 5

This chapter examined whether the argument that public disclosure of non-summary clinical trial data impedes innovation incentives has a legal basis. Overall, the analysis shows that concerns regarding the detrimental impact of data disclosure on statutory incentives—patents, test data exclusivity and sector-specific market exclusivities—can be justified only in limited situations. Table 5.1 summarises specific conditions under which the offsetting effect on innovation incentives might take place.

These findings can inform the design of the regulatory measures enabling access to IPD that would not interfere with the policy instruments protecting innovation incentives in the drug sector. However, before designing and implementing such measures, one needs to define the relevant justification for regulatory intervention enabling access to IPD for secondary analysis, beyond improving transparency in regulatory decision making related to drug marketing authorisation.

[173] Above (n 167).

Table 5.1 Summary of potential effects of IPD disclosure on the statutory innovation incentives

Innovation incentives	Implications of non-summary clinical trial data disclosure for protection
Patents	The potential impact of non-summary data disclosure on patentability cannot be determined in the abstract but would depend inter alia on the specific events reported from a clinical trial, the technical problem at issue and other relevant prior art.
SPCs	Protection can be affected insofar as IPD disclosure can impact the validity of the basic patent.
Test data exclusivity in the EU	The disclosed test data cannot be misused by circumventing the originator's data exclusivity, given that a generic applicant might be approved during the test data exclusivity term only based on a full dossier, including 'own' efficacy and safety data.
Orphan drug exclusivity in the EU	Protection might be affected to the extent to which third-party IPD analysis might contribute to developing a clinically superior drug.
Test data protection outside the EU	Competitors cannot resubmit data disclosed by the EMA in the jurisdictions with test data exclusivity where a generic drug might be approved during the term of protection based on 'own' data. Neither can data be 'reused' for regulatory purposes in jurisdictions without test data exclusivity where drug authorities do not carry out the scientific assessment of data but rely on marketing authorisation granted in other jurisdictions.

References

Ahn H (2014) Second generation patents in pharmaceutical innovation. Nomos, Baden-Baden

Arrow KJ (1962) Economic welfare and the allocation of resources for invention. In: Nelson RR (ed) The rate and direction of inventive activity: economic and social factors. Princeton University Press, Princeton

Baird S (2013) Magic and hope: relaxing Trips-plus provisions to promote access to affordable pharmaceuticals. Boston Coll J Law Soc Just 33(1):107–145

Bessen J, Meurer MJ (2005) Lessons for patent policy from empirical research on patent litigation. Lewis Clark Law Rev 9(1):1–27

Breschi S, Malerba F (2005) Sectoral innovation systems: technological regimes, Schumpeterian dynamics, and spatial boundaries. In: Edquist C (ed) Systems of innovation. Technologies, institutions and organizations. Routledge, London, New York, pp 130–156

Chow SC, Liu JP (2004) Design and analysis of clinical trials: concepts and methodologies, 2nd edn. Wiley, Hoboken

Cohen WM (2010) Fifty years of empirical studies of innovative activity and performance. In: Hall BH, Rosenberg N (eds) Handbook of the economics of innovation, vol 1. Elsevier, Amsterdam, pp 129–213

Correa C (2008) Designing patent policies suited to developing countries' needs. Econômica Rio de Janeiro 10:82–105

de Carvalho NP (2018) The TRIPS regime of patents and test data, 5th edn. Wolters Kluwer, Alphen aan den Rijn

Doshi P (2014) From promises to policies: is big pharma delivering on transparency? BMJ 348: g1615. https://doi.org/10.1136/bmj.g1615

Encaoua D, Guellec D, Martínez C (2006) Patent systems for encouraging innovation: lessons from economic analysis. Res Policy 35(9):1423–1440

EPO (2016) Case law of the Boards of Appeal of the European Patent Office, 8th edn. EPO, Nördlingen
Fellmeth AX (2004) Secrecy, monopoly, and access to pharmaceuticals in international trade law: protection of marketing approval data under the TRIPS Agreement. Harv Int Law J 45 (2):443–502
Gaessler F, Wagner S (2020) Patents, data exclusivity, and the development of new drugs. Rev Econ Stat:1–49. https://doi.org/10.1162/rest_a_00987
Hall BH, Harhoff D (2012) Recent research on the economics of patents. Annu Rev Econ 4:541–565
Hall BH, Mairesse J, Mohnen P (2010) Measuring the returns to R&D. In: Hall BH, Rosenberg N (eds) Handbook of the economics of innovation, vol 2. Elsevier, Amsterdam, pp 1034–1082
Hess P, Müller-Stoy T, Wintermeier M (2014) Sind Patente nur "Papiertiger"? Mitteilungen der deutschen Patentanwälte 105(10):439–452
Institute of Medicine of the National Academies (2015) Sharing clinical trial data: maximizing benefits, minimizing risk. The National Academies Press, Washington DC
Levin RC et al (1987) Appropriating the returns from industrial research and development. Brook Pap Econ Act 3:784–831
Mansfield E (1986) Patents and innovation: an empirical study. Manag Sci 32(2):173–181
Mazzoleni R, Nelson RR (1998) The benefits and costs of strong patent protection: a contribution to the current debate. Res Policy 27(3):273–284. https://doi.org/10.1016/S0048-7333(98)00048-1
Nightingale P, Mahdi S (2006) The evolution of pharmaceutical innovation. In: Mazzucato M, Dosi G (eds) Knowledge accumulation and industry evolution: the case of pharma-biotech. CUP, Cambridge, pp 73–111
OECD, Eurostat (2005) Oslo Manual. Guidelines for collecting and interpreting innovation data, 3rd edn. OECD Publishing, Paris
OECD, Eurostat (2018) Oslo Manual. Guidelines for collecting, reporting and using data on innovation, 4th edn. OECD Publishing, Paris
Palmedo M (2013) Do pharmaceutical firms invest more heavily in countries with data exclusivity? Currents: Int Trade Law J 21:38–47
Porth CM (2011) Preface. In: Porth CM (ed) Essentials of pathophysiology: concepts of altered health states, 3rd edn. Wolters Kluwer, Philadelphia, pp ix–xii
Price WN II, Minssen T (2015) Will clinical trial data disclosure reduce incentives to develop new uses of drugs? Nat Biotechnol 33(7):685–686. https://doi.org/10.1038/nbt.3243
Qian Y (2007) Do national patent laws stimulate domestic innovation in a global patenting environment? A cross-country analysis of pharmaceutical patent protection 1978–2002. Rev Econ Stat 89(3):436–453
Ragavan S (2018) The (re)newed barrier to access to medication: data exclusivity. Akron Law Rev 51:1163–1196
Reichman JH (2009) Rethinking the role of clinical trial data in international intellectual property law: the case for a public goods approach. Marq Intell Prop Law Rev 13(1):1–68
Shaikh OH (2016) Access to medicine versus test data exclusivity: safeguarding flexibilities under international law. Springer, Berlin, Heidelberg
Stiglitz JE (1999) Knowledge as a global public good. In: Kaul I, Grunberg I, Stern MA (eds) Global public goods: international cooperation in the 21st century. OUP, Oxford, pp 308–325
The US Senate, Subcommittee On Antitrust and Monopoly (1961) The study of administered prices in the drug industry 87th Congress, 1st session. US Govt. Print. Off., Washington DC
Turner JR (2007) New drug development: design, methodology, and analysis. Wiley, Hoboken
Weissman R (2006) Data protection: options for implementation. In: Roffe P, Tansey G, Vivas-Eugui D (eds) Negotiating health: intellectual property and access to medicines. Earthscan, London, pp 151–178

Part III
Analysis *De Lege Ferenda:* Exclusively Controlled or Readily Accessible?

Part III presents the normative analysis. The key questions are whether a policy intervention enabling access to IPD can be justified on the grounds of promoting drug innovation and, if so, how it should be designed to protect diverse interests at stake and achieve multiple policy objectives in a balanced way. The analysis proceeds as follows. Chapter 6 provides a detailed problem analysis that could inform the 'intervention logic' of access measures. Chapter 7 outlines a pertinent theoretical framework for analysing an innovation policy dilemma over access to IPD as a knowledge resource. Chapter 8 explores how theoretical propositions apply in the specific case of clinical trial data. Finally, Chap. 9 evaluates the legislative options for implementing the findings of the legal-theoretical analysis.

Chapter 6
Defining the Intervention Logic of Access-To-Data Measures: A Problem Analysis

Abstract Given the overall objective of proposing the rules on access to non-summary clinical trial data, this chapter examines why the status quo of access to trial data is problematic and how a policy analysis could be conducted. It starts with a brief overview of the basic principles of designing a regulatory intervention and methodology for the problem analysis developed and applied by the European Commission. After taking a closer look at the available evidence on the industry's data-sharing practice, concerns regarding the ability of trial sponsors to exercise de facto exclusive control over non-summary data are discussed in detail. Finally, given these concerns, the problem drivers, policy objectives and the overall intervention logic of access measures are outlined.

6.1 General Principles of Regulatory Intervention

6.1.1 Regulatory Intervention as an Exception

The basic principle of policymaking is that intervention into the private sector and market is an exception that needs to be justified by 'an important public objective that an unregulated marketplace cannot provide'.[1] It reflects the idea that 'the freedom of individual action and minimization of governmental coercion [are] basic societal values'.[2] In the EU, this principle corresponds to one of the fundamental freedoms—the freedom of business undertakings to conduct a business and exercise economic activities guaranteed under Article 16 of the Charter of Fundamental Rights.[3] Thus, any government intervention needs to be justified by proving that, absent such intervention, an important societal objective cannot be achieved or

[1]Breyer (1979), p. 552.

[2]Ibid.

[3]Such freedom is subject to limitations that can be introduced in line with the principle of proportionality and 'only if they are necessary and genuinely meet objectives of general interest recognised by the Union or the need to protect the rights and freedoms of others'. CFR, art 52(1).

D. Kim, *Access to Non-Summary Clinical Trial Data for Research Purposes Under EU Law*, Munich Studies on Innovation and Competition 16,
https://doi.org/10.1007/978-3-030-86778-2_6

that it can be achieved with a more favourable cost-benefit ratio relative to the non-intervention scenario.[4]

6.1.2 *The Grounds for Policy Intervention*

Justifications for a policy intervention usually fall into one of the following categories:

- a market failure (where an unregulated market fails to deliver an efficient outcome);[5]
- a regulation failure (where an implemented public policy action that appeared justified failed to solve the problem adequately or caused new problems);[6]
- equity considerations in situations where an efficient market outcome 'may not be the most desirable one for the policy in question';[7]
- the inability of individuals to act or decide based on their own best interests due to behavioural biases.[8]

Public policies that intervene in market operations are often analysed from a market failure perspective. A market failure refers to 'the failure of a more or less idealized system of price-market institutions to sustain "desirable" activities or to estop "undesirable" activities'.[9] The desirability of activities is typically evaluated relative to an 'explicit or implied *maximum-welfare* problem'.[10]

Among the potential sources of a market failure are *externalities*—external effects that cannot be internalised via a market transaction or do not bear on the decision making of their generator.[11] Externalities that cannot be feasibly overcome through voluntary negotiations[12] provide grounds for regulatory intervention. The most common examples are found in the areas of public health, safety and environmental protection.[13]

[4] Below (nn 18–19).

[5] European Commission (19 May 2015) Better Regulation 'Toolbox' supplementing SWD(2015) 111 [hereinafter Better Regulation 'Toolbox'], p. 67 (further defining efficiency as 'a situation where no one can be made better off without someone else being made worse off').

[6] Ibid (emphasis added).

[7] Ibid.

[8] Ibid.

[9] Bator (1958), p. 351.

[10] Ibid (emphasis added).

[11] Better Regulation 'Toolbox' (n 5), p. 68.

[12] Todorova T (2014) Archive the transaction-cost roots of market failure. MPRA Paper No 66757, p. 7 (arguing that 'all forms of market failure could be explained by transaction cost').

[13] See e.g. Better Regulation 'Toolbox' (n 5), p. 140.

The concepts of market failure and externalities are of particular relevance for the present analysis. As it will be shown later, the problem of access to non-summary clinical trial data arises at the intersection of several market failures and involves different types of externalities.

6.1.3 Social Welfare as a Normative Benchmark

Social welfare is a widely accepted normative criterion applied in planning, designing, and evaluating public policies.[14] It is a multifaceted concept and, in broad terms, can be defined as well-being in all aspects of societal life.[15] Welfare economics is a branch of economics that studies how welfare can be defined and measured and which social institutions and arrangements can promote it most optimally.[16]

Economic efficiency is one of the central concepts in economic policies and welfare economics.[17] The latter seeks to define the conditions under which efficiencies can be maximised and how trade-offs between different efficiencies can be resolved. Welfare analysis is applied in different areas of public policy and law-making. It involves weighing up social costs and benefits when planning a policy intervention and assessing alternative regulatory measures vis-à-vis the pursued objective(s).[18] The choice among the policy options shall favour the regime, which is superior in terms of the *net* impact on social welfare (effectiveness) and the *cost-benefit ratio* (efficiency).[19]

At the same time, one should acknowledge the limitations of the cost-benefit analysis. In the words of Just et al.:

> Although a social welfare function is a convenient and powerful concept in theory, its practical usefulness has been illusory. Many attempts have been made to specify a social welfare function sufficiently to facilitate empirical usefulness but none have been widely accepted.[20]

Regulation of the pharmaceutical sector presents an apt example, where welfare analysis is highly complex and where multiple competing interests can lead to policy trade-offs.[21]

---

[14] See e.g. Bator (1957), pp. 57–58; Breyer (1982), pp. 15–34; Ogus (1994), pp. 29–46.

[15] See e.g. Scott (2012), pp. 36–37; Baldock (2007), p. 21.

[16] Feldman (2008), p. 721.

[17] Reiter (2018), p. 765 (noting that efficiency has been at the centre of economic theory 'since ancient times, and is an essential element of modern microeconomic theory').

[18] Better Regulation 'Toolbox' (n 5), p. 338.

[19] OECD (2009), p. 97.

[20] Just et al. (2004), p. 41.

[21] As discussed in Chap. 8.

6.1.4 Necessity and Proportionality as the Guiding Principles

The overarching principles of necessity, proportionality, subsidiary, transparency, accountability, accessibility and simplicity should govern all stages of policymaking to ensure the high quality of regulation.[22] The general principles of necessity and proportionality mean that a policy measure—either legislative or non-legislative—should be designed, implemented and evaluated vis-à-vis the pursued objective and should not exceed what is necessary and appropriate for its accomplishment.[23] Under EU law, this idea is also reflected in the principle of subsidiary, which means that the Union should act only, first, '[i]f, and in so far as, the objective of the action cannot be achieved sufficiently by the Member States (at national, regional and local levels); second, where the objective can be better achieved at Union level by reason of the scale or effects of the proposed action'.[24]

6.2 The European Commission's Methodology for Problem Analysis

In 2016, the European Commission adopted the Better Regulation Guidelines, a set of methodological tools providing guidance for planning, designing and implementing EU policies.[25] To determine whether a prospective legislative or non-legislative action is justified, the Guidelines recommend conducting an impact assessment to verify the existence of a problem, identify its underlying causes, assess whether the action at EU level is needed and evaluate relative costs and benefits of the policy alternatives.[26]

[22] Mandelkern Group on Better Regulation (13 Nor 2001) Final report, p. i [hereinafter Mandelkern report].

[23] Bergkamp (2003), pp. 677 ff.

[24] Better Regulation 'Toolbox' (n 5), p. 22. Article 5 of the EC Treaty stipulates that '[a]ny action by the Community shall not go beyond what is necessary to achieve the objectives of th[e] Treaty'.

[25] European Commission (19 May 2015) Better Regulation Guidelines. SWD(2015) 111 final [hereinafter Better Regulation Guidelines].

[26] Ibid pp. 16–17. See also Better Regulation 'Toolbox' (n 5), p. 59; Mandelkern report (n 22), p. ii (defining regulatory impact assessment as 'an effective tool for modern, evidence-based policy making, providing a structured framework for handling policy problems [that] should be an integral part of the policy making process at EU and national levels [as] it allows that decision to be taken with clear knowledge of the evidence').

6.2.1 *The 'Intervention Logic'*

6.2.1.1 The Components of the Intervention Logic

The 'intervention logic' is defined as the logical links connecting the problem, the policy objective, the underlying problem drivers and the available policy options.[27] Intervention logic should be defined *ex ante* when assessing the prospective impact and evaluated *ex post*.[28] The starting point of the policy design is analysing a problem, which is likely to persist in the absence of policy intervention.[29]

6.2.1.2 Causality

Regulatory measures are designed 'to address the identified problem by causing direct and indirect changes to the behaviour of those influencing it (i.e. the problem drivers)'.[30] The sound intervention logic implies that *verifiable causal* links—'a chain of impacts'[31]—between the problem drivers and the policy outcomes can be established. Two elements of the 'chain' are distinguished: first, a policy measure should trigger a change in the behaviour of certain actors; second, the change in the behaviour should produce the anticipated policy outcome.[32] Nevertheless, the limitations of the causality analysis need to be realised. According to the European Commission's Better Regulation Guidelines, the causal relationships are 'challenging to prove, particularly when evaluating EU policies which operate in a complex environment influenced by a wide range of factors falling outside the scope of the EU intervention'.[33]

6.2.1.3 The Choice of a Policy Measure

Once the problem and problem drivers are identified, the policy objectives can be formulated, and the different approaches can be assessed in terms of their effectiveness and efficiency.[34] The cost-benefit analysis presents 'a structured framework for informing the consideration of the range of options available'[35] and forms the basis of the regulatory impact assessment. A policy measure is effective if its benefits

---

[27] Better Regulation Guidelines (n 25), pp. 90–91.
[28] Better Regulation 'Toolbox' (n 5), p. 67.
[29] Ibid p. 66.
[30] Ibid p. 97.
[31] Ibid.
[32] Ibid p. 269.
[33] Ibid pp. 269–270.
[34] Better Regulation 'Toolbox' (n 5), pp. 65 ff.
[35] Mandelkern report (n 22), p. 82.

exceed costs (in discounted terms) and efficient if the benefit-cost ratio is higher in comparison to the benefit-cost ratio of an alternative policy instrument[36] (including the non-intervention scenario[37]).

6.2.2 *The Framework for the Problem Analysis*

For identifying a problem and its causes, the European Commission recommends addressing the following set of questions.

- What is the status quo?
- What and whose behaviour needs to be changed and why?
- To which broader policy objectives does a prospective regulatory measure relate?
- What are the causes ('drivers') of the problem?[38]

Let us apply this framework to consider how the problem analysis could have been conducted regarding the clinical trial data access policy.[39]

6.3 Defining the Status Quo of Access to Non-Summary Clinical Trial Data

As the starting point of the problem analysis, the Better Regulation 'Toolbox' of the European Commission recommends to 'succinctly describe the current situation (the 'status quo')'.[40] For the subsequent analysis, two aspects of the status quo of access to clinical trial data need to be distinguished: the currently applicable legal regime and the existing data-sharing practice.

---

[36] Ibid p. 97 (emphasising the importance of the benefit-cost ratio, as it allows to rank the alternatives based on efficiency). See also Better Regulation Guidelines (n 25), p. 29; Patton, Sawiski and Clark (2016), p. 263 (noting that where ratios of discounted benefits to discounted costs of alternative instruments are compared, the 'alternative with the highest benefit-cost ratio does not necessarily have the highest net present value').

[37] Better Regulation Guidelines (n 25), p. 23.

[38] Ibid pp. 65–67.

[39] The adoption of the EU Clinical Trials Regulation was preceded by the impact assessment. While the Regulation introduced the mandatory publication of summaries of CSRs and the optional sharing of IPD among the novelties, the Impact Assessment Report of the European Commission does not contain specific problem analysis underlying these requirements. European Commission (17 Jul 2012) Impact assessment report on the revision of the 'Clinical Trials Directive' 2001/20/EC accompanying the document Proposal for a Regulation of the European Parliament and of the Council on clinical trials on medicinal products for human use, and repealing Directive 2001/20/EC, SWD(2012) 200 final, vol. I and II.

[40] Better Regulation 'Toolbox' (n 5), p. 65.

6.3.1 Summarising the Legal Status Quo of Access to Clinical Trial Data

As concluded in Chap. 4, while neither personal nor anonymised patient-level ('raw') data is subject to exclusive rights under EU law, trial sponsors can exercise de facto exclusive control over non-summary clinical trial data. Such control stems from the duty imposed on the trial sponsors under the EU Clinical Trials Regulation to store and protect all data and information from trials against unauthorised access[41] and can be reinforced through contractual arrangements.

6.3.2 Evidence on Industry Data-Sharing Practice

The ability of trial sponsors to control third-party access to data does not render IPD completely inaccessible. Some commentators viewed the data-sharing initiatives implemented by drug companies as a 'change of paradigm'[42] and a 'sea change'.[43] Others took a more critical stance. For instance, the 2014 review by Doshi[44] identified the following common provisions under the companies' data-sharing policies that can restrict access to data.

- As a rule, access is granted to qualified researchers for 'legitimate research' purposes based on evaluating a research proposal.
- Access to data is often provided only on screen, without the possibility to download the datasets; a condition viewed as 'a convenient way to monitor requestors and prevent truly independent research'.[45]
- Usually, IPD is available for the drugs approved for a particular indication, which leaves out the data from trials investigating off-label uses and, most importantly, failed trials.[46]
- Companies redact data and information deemed as commercially sensitive.[47]

[41] Reg 536/2014/EU, art 56(1).

[42] Institute of Medicine of the National Academies (2015), p. 64.

[43] Krumholz et al. (2014), p. 499 (concluding, based on the review of clinical data sharing practices of 12 leading research-based pharmaceutical companies, that '[i]t is clear that a sea change in concept and action has occurred, at least in industry').

[44] Doshi (2014).

[45] Ibid.

[46] On the importance of such data, see Chap. 8, Sect. 8.2.3.2. Notably, the statistics on access to data through the ClinicalStudyDataRequest.com portal evidence a considerable interest in non-listed data, which the trial sponsors did not declare as accessible. CSDR. Metrics. https://www.clinicalstudydatarequest.com/Metrics.aspx. Accessed 26 Mar 2021.

[47] Doshi (2014).

Due to these limitations, the industry practice of data-sharing was characterised as a 'controlled access' mode.[48]

A 2017 review by Goldacre et al.[49] examined transparency policies of 42 pharmaceutical companies and found that

- 22 companies have policies to make IPD from phases I-III of clinical trials available upon request;
- 14 companies can share data from phase IV studies, while some companies explicitly exclude such data;
- Only one company (GSK) declared its commitment to share IPD from trials on the unlicensed products and off-label uses of the licensed products;
- Companies tend to provide access to IPD from relatively recent trials.[50]

The study concludes that, while transparency commitments can vary considerably among the companies in terms of the scope of accessible data, many policies are 'poorly worded and internally inconsistent'.[51] No policy was found to integrate all possible elements of best transparency practices in clinical research. The 2019 study by Miller et al.[52] identified only 25% of large pharmaceutical companies fully meeting data sharing 'best practices'.[53]

Apart from these surveys, several studies have reflected on individual experiences of researchers in obtaining access to individual patient-level data through direct negotiations with the trial sponsors. For instance, the 2017 study by Nevitt et al. assessed the success rate of retrieving data by researchers for conducting IPD meta-analyses.[54] Remarkably, only 25% of the sample was based on 100% of IPD eligible for meta-analysis, and 43%—on 80% of eligible IPD. Overall, the study did not establish a statistically significant correlation between the IPD retrieval rate and the timing of publication of IPD meta-analyses. As concluded by the authors, despite the changes in the culture of clinical trial data sharing, the IPD retrieval rate has not improved significantly over the past decades. Neither did the study find a statistically significant relationship between the IPD retrieval rate and the source of funding. Notably, the success rate among the studies sponsored by the industry was higher than that among the academic studies. Although the question of how incomplete IPD could have affected the conclusions of the meta-analyses was not addressed, it was

[48] Ibid. But see Sydes et al. (2015) (pointing out the risk of identification, self-identification, data distortion and data dredging that might justify the 'controlled access approach' to clinical trial data sharing).

[49] Goldacre et al. (2017).

[50] The median start date of the policy commitments was 2011; the median start date of adopting data sharing policies among the surveyed companies was 2012.

[51] Ibid.

[52] Miller et al. (2019).

[53] Ibid.

[54] Nevitt et al. (2017). The sample included 760 IPD meta-analyses published between 1987 and 2015.

emphasised that researchers 'must consider [. . .] the potential biases introduced by missing such data'.[55]

The 2016 study by Murugiah et al. analysed the availability of IPD from large cardiovascular trials conducted by several major pharmaceutical companies.[56] Out of 60 analysed trials, IPD was available for external researchers from 15 trials and equally from 15 trials unavailable. For the remaining 30 trials, data availability could not be established with certainty either because of no response from the companies or additional requirements.

Furthermore, research points out difficulties in negotiating with the data holders access terms and using the accessed data in research. For instance, Mayo-Wilson, Doshi and Dickersin mention the slow review of data requests and 'unusability' of the accessed data. The authors conclude that 'despite manufacturers' publicized support for research transparency, processes for sharing data remain opaque [and] might have a chilling effect on efforts to obtain information.[57] Geifman et al. characterise their experience in retrieving data via the ClinicalStudyDataRequst.com portal as 'time consuming and demotivating'.[58] Other factors restricting research have been reported, such as the inability to download[59] or combine it with other datasets.[60] Another notable constraint is the contractually imposed limitation on publishing by trial investigators the results of industry-sponsored trials.[61]

On the contrary, positive user experience was reported[62] with the Immunology Database[63] administered by the National Institute of Allergy and the Project Data Sphere database.[64]

Overall, the above-reviewed evidence on industry data-sharing policies and practice appears fragmented and ambivalent on the extent to which non-summary clinical trial data is accessible. Despite the overall tendency towards broader access, commentators have argued that clinical trial data, by and large, remains unavailable for secondary analysis.[65]

---

[55] Ibid.

[56] Murugiah et al. (2016). The sample comprised 60 trials sponsored by 20 leading pharmaceutical companies and involved over 5000 participants.

[57] Mayo-Wilson et al. (2015).

[58] Greifman et al. (2015).

[59] Nisen and Rockhold (2013), p. 477. One of the investigators, whose research project was approved but discontinued, characterised the experience of using the ClinicalStudyDataRequst.com portal as 'not user-friendly' and 'highly inconvenient' due to the password system, further mentioning that 'every click is tracked' and that it is 'difficult to export data'. E-mail of 27 Aug 2017 (on file with the author).

[60] Nisen and Rockhold (2013), p. 477.

[61] See generally Gøtzsche et al. (2006).

[62] Greifman et al. (2015).

[63] www.immport.org. Accessed 26 Mar 2021.

[64] www.projectdatasphere.org. Accessed 26 Mar 2021.

[65] Institute of Medicine of the National Academies (2015), p. 1; Doshi et al. (2013).

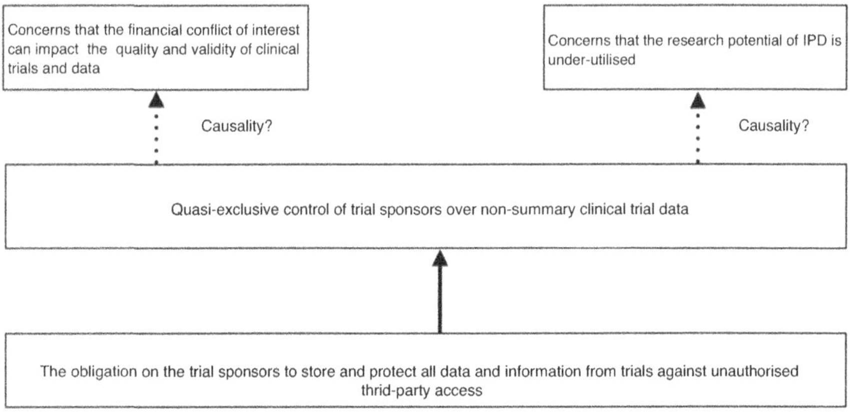

Fig. 6.1 The 'problem tree'

6.4 Dissecting the Problem of Access

6.4.1 The 'Problem Tree'

When designing policy measures, it is recommended to depict a 'problem tree' showing the causal links between the problem, problem drivers and effects.[66] The effectiveness of a policy intervention crucially depends on whether the relevant problem drivers can be identified and verified, whether the causal links can be established between the allegedly problem-driving conduct and the socially undesirable effects.[67] For clinical trial data, the 'problem tree' can be represented in Fig. 6.1.

As shown, at least three levels within the problem of access to non-summary clinical trial data can be identified. Which level should a policy measure address? One would argue that the ultimate target should not be the drug sponsors' control over data but the concerns at the 'crown of the tree' regarding research quality and 'foregone' research opportunities.[68] Two issues need to be addressed: to what extent such concerns can be justified, and whether causality between such concerns and the ability of trial sponsors to control access to data can be established. Let us next take a look at the research on this subject and available evidence.

[66] Better Regulation 'Toolbox' (n 5), p. 66.

[67] Ibid p. 67. See also Better Regulation Guidelines (n 25), p. 20.

[68] Institute of Medicine of the National Academies (2015), p. 18 (with further references).

6.4.2 *The Issue of Reproducibility of Clinical Trials*

6.4.2.1 The Concept of Research Reproducibility

Concerns regarding the quality of trials and data are often discussed under the heading of 'irreproducible' research. As the fundamental principle of scientific research, reproducibility is viewed as 'the demarcation between science and nonscience'.[69] It is defined as the ability 'to duplicate the results of a prior study using the same materials and procedures as were used by the original investigator'.[70] In this respect, the accessibility of primary research data can be viewed as a 'scientific safeguard', which can expose data quality and the robustness of study findings.[71]

Reproducibility is one of the basic statistical concepts used in the trial design, conduct, analysis, and reporting of results. In general terms, it refers to the probability that the same results can be observed in a new trial if the original study protocol is followed.[72] However, in the context of clinical trials—and biomedical research in general—reproducibility cannot be guaranteed due to the inherent variability of biological characteristics,[73] even where the study subjects are selected according to the same criteria as defined in the original protocol.[74]

The reproducibility of trial *results* should be distinguished from the reproducibility of *conclusions* of the primary data analysis. The latter means that similar conclusions are derived by re-analysing the source data from the original trial (the so-called 'inferential reproducibility').[75] Irreproducibility of conclusions can be attributed to the errors in the data analysis methodology (e.g. due to the exclusion of individual trial results from the analysis).[76]

As an aspect of the trial methodological quality, reproducibility relates to the concept of *trial validity*, the methodological quality of a trial, which means that a trial is designed to answer the research question in an unbiased manner. Trial

[69] Mullane et al. (2018), p. 2.

[70] Bollen et al. (2015), p. 4.

[71] As discussed in Chap. 3, Sect. 3.2.2.2.

[72] Chow and Liu (2004), p. 82.

[73] Due to the biological variability of the study subjects, both reproducibility and replicability are generally difficult to achieve in biomedical research. See Begley and Ioannidis (2015), p. 117. Perfect replicability in the field of biomedical research is practically unfeasible. See Mullane et al. (2018), p. 4; Porter (2016), p. 447 (observing that 'few results can be expected to replicate with high precision'). In this regard, it is worth emphasising that research replicability can be problematic not only in the case of industry-sponsored trials but academic research as well. See Institute of Medicine of the National Academies (2012), p. 69 (with further references).

[74] While two pivotal well-controlled trials are usually required to fulfil the regulatory requirement for the substantial evidence on efficacy, it cannot be guaranteed that the results can be reproduced even if the same trial protocol is followed. Chow and Liu (2004), p. 82.

[75] Goodman et al. (2016).

[76] See generally Nüesch et al. (2009); Tierney and Stewart (2005).

validity, in turn, determines the quality of evidence and data robustness—the degree of accuracy of the estimation of the treatment effect.[77]

6.4.2.2 Systematic Errors (Research 'Biases')

Random errors in clinical trials should be distinguished from those of a *systematic* nature. In the former case, multiple replications of the same study can produce different results but, on average, may give the right answer.[78] In the latter case, multiple replications of the original study would, on average, produce the wrong answer.[79] Systematic errors (also known as 'research biases') refer to the errors *intentionally* introduced into the methodology of trial design, conduct, or data analysis, leading to overestimating or underestimating the actual effects of medical intervention.[80]

The risk of bias is one of the factors evaluated when the quality of research and the body of evidence are assessed.[81] For instance, the estimates of treatment effects can be distorted due to the inappropriate allocation concealment, exclusions after randomization, or lack of double-blinding.[82] According to the Cochrane Handbook for Systematic Reviews of Interventions, currently used as a standard methodology for the risk of bias assessment, the main types of biases in clinical trials include selection bias, performance bias, attrition bias, comparator bias, reporting bias (or selective reporting) and detection bias.[83]

6.4.2.3 The 'Industry Bias'

Reproducibility in biomedical research is generally difficult to achieve. At the same time, concerns have been particularly vocal regarding clinical trials sponsored by the pharmaceutical industry. The contention is that commercial sponsorship inherently implies the risk that financial interests of drug sponsors 'may unduly influence professional judgments [and] threaten the integrity of scientific investigations'.[84]

---

[77] Atkins et al. (2004). In the context of this study, two types of reproducibility—reproducibility of the trial results and conclusions of the primary analysis—are used as defined in this section.

[78] Higgins et al. (2017), p. 8:3.

[79] Ibid.

[80] Ibid p. 8:1. See also the International Council for Harmonisation of Technical Requirements for Pharmaceuticals for Human Use (ICH) (1998) ICH harmonised tripartite guideline. Statistical principles for clinical trials. E9, p. 32 (defining operational and statistical biases as '[t]he systematic tendency of any factors associated with the design, conduct, analysis and evaluation of the results of a clinical trial to make the estimate of a treatment effect deviate from its true value').

[81] Higgins et al. (2017), p. 8:4.

[82] Schulz et al. (1995).

[83] For definitions, see Higgins et al. (2017), pp. 8:7–8:9.

[84] Lo and Field (2009). See generally Montaner et al. (2001).

Besides, it can be speculated that *systematic errors* can be introduced intentionally into research to yield more favourable results for the sponsored treatment and secure private returns on drug R&D.[85]

Some authors have referred to the prevalence of the results favouring the treatment of the trial sponsor in the industry-sponsored research compared to research with other sources of funding as the 'industry bias' and proposed to single it out as a specific category of biases. For instance, Lundh et al. argued that 'industry sponsorship should be treated as *bias-inducing* and industry bias should be treated as a separate domain'[86] because industry sponsorship might entail risks of bias that are not covered by the standard methodology of assessing the risk of bias.[87]

Concerns regarding the undue influence of the pharmaceutical industry on biomedical research and healthcare provision were articulated, at least, as early as in the 1960s.[88] More recent publications provide a disturbing account of malpractice in the pharmaceutical industry, including in clinical trials.[89] One can hardly ascertain the scope of the problem. As noted by Williams, Mullane and Curtis, 'the perception of the conflict of interest may far exceed[] the actuality [while] isolated examples are taken as the norm'.[90] Notwithstanding 'many examples of the appearance of conflict having *no impact* on an ethically derived outcome, examples of serious conflict of interest however few add fuel to the conflict of interest debate'.[91]

6.4.2.4 Empirical Studies on the 'Industry Bias'

Empirical studies show a statistically significant, consistent association between the industry sponsorship and positive trial outcomes reported in the publications and systematic reviews.[92] The key question is whether more favourable trial results indicate 'false positives', unreliable data, and lower quality of research. Numerous

---

[85] Lexchin (2012). See also Lundh et al. (2017) (observing that commercial sponsors can influence the study results in several potential ways including 'the framing of the question, the design of the study, the conduct of the study, how data are analyzed, selective reporting of favorable results, and spin in reporting conclusions' (with further references)).

[86] Lundh et al. (2017) (emphasis added).

[87] Ibid (arguing that '[t]here are many subtle mechanisms through which sponsorship may influence outcomes, and an assessment of sponsorship should therefore be used as a proxy for these mechanisms').

[88] See e.g. The pharmaceutical persuaders – the industry, the doctor, and the clinical trial (1961). See also Shapiro (1978), pp. 166 ff.

[89] See e.g. Gøtzsche (2013); Goldacre (2014); Dukes et al. (2015); Fisher (2009).

[90] Williams et al. (2018), p. 180 (further noting that '[t]here is also a tendency by the community at large to assume that researchers in industry are automatically guilty of a conflict of interest until proven otherwise' (with further references)).

[91] Williams et al. (2018), p. 179 (with further references).

[92] See e.g. Sismondo (2008); Bekelman (2003); Lexchin et al. (2003); Schott et al. (2010); Golder and Loke (2008).

studies have attempted to evaluate the association between industry funding and the quality of research to understand the extent to which concerns over the 'industry bias' are justified.[93] Literature reviews on this topic can provide helpful insights.

The 2003 review by Lexchin et al. analysed whether industry-funded clinical trials that had reported the more favourable outcomes to the funder differed in terms of the trial methodology compared with the trials with other funding sources.[94] None of the reviewed studies had found that industry-funded studies were of poorer methodological quality.[95] The authors conclude that, even though the methods used in the industry-sponsored trials were found to be at least as good as those used in non-industry funded research, it could not guarantee the absence of bias as the outcome could be influenced by the factors *beyond* the applied quality scores, such as the selection of an inappropriate comparator to the product being investigated, or the publication bias.

The 2017 review by Lundh et al. examined 75 studies investigating drugs, medical devices, and mixed types of interventions published between 1986 and 2015[96] to define whether the studies sponsored by the industry differed in terms of the risk of bias compared to the research with other sources of funding. The following set of factors was applied:

- whether explicit and well-defined criteria that others could replicate were used to select studies for inclusion/exclusion of trial participants;
- whether there was an adequate study inclusion method;
- whether the search for studies was comprehensive;
- whether methodological differences and other characteristics that could introduce bias were controlled for or explored.[97]

Overall, the study did not identify a significant difference between the trials with industry and non-industry funding concerning the standard methodological factors of the risk-of-bias assessment, which led the authors to conclude that more favourable results of the studies sponsored by the industry can be mediated by factors *other* than those classified in the Cochrane methodology.[98]

In sum, the evidence on the alleged 'industry bias' appears inconclusive. Several reasons can explain the divergence in conclusions. First, different samples of trials were examined. Second, various methods were used for assessing research quality.

[93] Bekelman et al. (2003); Lexchin et al. (2003); Friedberg et al. (1999); Cho and Bero (1996); Als-Nielsen et al. (2003); Kjaergard and Als-Nielsen (2002).

[94] Lexchin et al. (2003).

[95] The sample included pharmacoeconomic reports, meta-analyses and systematic reviews identified through searching the Medline database records from January 1966 to December 2002 and the Embase database records from January 1980 to December 2002.

[96] Lundh et al. (2017). The publication is an update of the earlier systematic review published by the authors that examined the association between the industry sponsorship with the publication of the study outcomes favourable to the sponsor. See Lundh et al. (2017).

[97] Ibid.

[98] Ibid.

Third, the standard methodologies might not account for all potential ways of introducing biases into research. Such methods include the choice of comparator, non-publication of negative outcomes, misrepresentation of data submitted to regulatory agencies, 'ghost-writing' of articled favouring an investigational product[99] and the use of 'seeding' trials.[100] Thus, it remains uncertain how the industry sponsorship might be affecting the quality of research and reliability of findings and to what extent the prevalence of more favourable findings reported by commercial sponsors might be mediated by selective reporting or other systematic errors.[101]

6.4.2.5 The Role of Access to Non-summary Data in Improving Research Quality

Views differ as to what level of transparency is sufficient to enable a confirmatory secondary analysis of clinical trial data. Some commentators argue that access to all clinical trial data, including IPD and meta-data, is essential for validating the trial results.[102] Others find that transparency can be improved by 'simpler and more cost-efficient measures',[103] such as prospective trial registration, reporting summary results, making available the trial protocols together with other trial materials that can allow researchers to interpret and replicate trials.[104]

Analytically, it is helpful to distinguish between three levels of trial validity that directly and cumulatively bear on data reliability and robustness of the trial results:

1. the validity of the trial design;
2. robustness of the trial data;

---

[99] Lexchin (2012), p. 254. See also Stern and Lemmens (2011) (proposing to impose fraud liability to prevent the practice of ghost-writing in the medical literature).

[100] Trial 'seeding' refers to the practice of conducting post-marketing studies for 'the sole purpose of getting doctors to start to use a product with the aim of establishing the drug as a regular part of the doctor's prescribing'. Lexchin (2012), p. 255.

[101] Clifford et al. (2002) (pointing out that 'the absence of significant associations between funding source, trial outcome and reporting quality reflects a true absence of an association or [such absence] is an artefact of inadequate statistical power, reliance on voluntary disclosure of funding information, a focus on trials recently published in the top medical journals, or some combination thereof'). See also Lundh et al. (2012) (concluding that 'sponsors are usually involved in the analysis and reporting of results in industry-sponsored trials, but their exact role is not always clear from the published papers' and proposing that '[j]ournals should require more transparent reporting of the sponsors' role in crucial elements such as data processing, statistical analysis and writing of the manuscript and should consider requiring access to trial protocols, independent data analysis and submission of the raw data').

[102] See e.g. Jackson (2019), p. 499; Kaur and Choy (2014), p. 25; Naci et al. (2015); Institute of Medicine of the National Academies (2015), p. 141; AllTrials. What does all trials registered and reported mean? http://www.alltrials.net/find-out-more/all-trials/. Accessed 26 Mar 2021.

[103] Hoffmann et al. (2017).

[104] Ibid. Such materials can include consent forms, statistical analysis plans, blank case report forms and descriptions of measurements and interventions.

3. reproducibility of the conclusions of the primary analysis (or 'inferential' reproducibility).[105]

Let us consider the role of access to non-summary data at each level. Perhaps most obviously, access to complete data—including meta-data such as the trial protocol and specifications—can be required for validating the *conclusions* reported by the trial sponsors drawn from the primary analysis. The problem of incomplete data, however, appears to be pervasive.[106]

Further, IPD re-analysis can play an important role in validating the trial results. Yet, due to the difficulties in obtaining data, confirmatory analysis is usually based on summary-level data. Notably, none of the 75 studies included in the 2017 review by Lundh et al.[107] used IPD re-analysis to identify research biases. As commented by Professor Lexchin, one of the study co-authors,

> it is probably due to the lack of access to this type of data and the resources necessary to re-analyze it [as] using IPD is much more time consuming than using aggregate data and also getting access to IPD is also more difficult as this type of data is either in the possession of the researchers or the companies that financed the research and is therefore not available for re-analysis. This is the case for the trial data about the effects of statins on cardiac disease which the Cholesterol Treatment Trialists Collaborators has refused to release. [. . .] However, just to be clear, having access to IPD would allow researchers to detect bias much better than using aggregate data.[108]

Ultimately, it should be emphasised that at the basis of research quality lies the validity of the trial design. There would be no value in reproducing trial outcomes if the trial was designed so that it could not yield accurate knowledge about the treatment effect in the first place.[109] Thus, the reliability and robustness of data crucially depend on the quality of the trial methodology. Valid methodology means that the trial design allows to answer the research question reliably and provides accurate knowledge regarding the risks and benefits of the investigated medical intervention. The stronger the methodological quality, the more accurately the study results will reflect the true intervention effect.

Thus, a systematic policy approach should, ideally, combine *ex ante* validation of the trial protocol and overall methodology with *ex post* reassessment of the study findings and conclusions.

---

[105] See above (nn 75–76) and the accompanying text.

[106] Hoffmann et al. (2017).

[107] Lundh et al. (2017).

[108] E-mail from Joel Lexchin of 28 Aug 2017 (reproduced with the permission of Professor Lexchin; on file with author). On difficulties in accessing primary research data, see e.g. Doshi et al. (2013).

[109] See e.g. Skovlund (2009), p. 260 (pointing out that 'unless the design and statistical analysis of a clinical trial are appropriate, results cannot be considered reliable and no confidence can be placed in the subsequent clinical interpretation').

6.4.2.6 Implications for Clinical Practice

Reproducibility of the study results and conclusions plays a crucial role in clinical practice and medical research. It affects a broad range of stakeholders, including patients, health professionals, researchers, health authorities, industry sponsors, and others, whose decision making relies on the reported research outcomes.[110] Clinical recommendations should be based on high-quality evidence, demonstrating that the claimed therapeutic effects *undoubtedly* and *consistently* outweigh potential risks.[111] However, flaws in the trial design, conduct, analysis and reporting (not to mention non-reporting of trial outcomes) can distort clinical recommendations[112] at the level of individual patient care as well as the totality of the evidence in a particular therapeutic area.[113]

Studies based on the secondary (confirmatory) IPD analysis provide fragmented and, thus, inconclusive evidence. For instance, the 2004 review by Chan et al.[114] compared published articles with the corresponding trial protocols and found that the majority of the trial outcomes within the analysed sample were not only reported incompletely but also inconsistent with the trial protocols.[115] The authors conclude that the published articles and reviews incorporating individual reports 'may therefore be unreliable and overestimate the benefits of an intervention'.[116] The 2014 study by Ebrahim et al. identified all articles based on the confirmatory analysis of IPD from previously published RCTs available in the MEDLINE database to verify whether IPD re-analyses could have led to different clinical recommendations compared to those drawn based on the primary analyses. Altogether, 36 studies published between 1985 and 2012 were identified.[117] Notably, the majority of the re-analyses were carried out by the investigators, who had conducted the original trials and later re-assessed the intervention effects by applying different statistical methods.[118] Overall, the study found that about two-thirds of the results were successfully reproduced. At the same time, the remaining part diverged in the conclusions and clinical recommendations (e.g. some re-analyses found that different types of patients should be treated[119]). The implications of those variances for

---

[110] CIOMS (2016), p. 1.

[111] Guyatt et al. (2008), p. 925 (noting that quality of evidence is 'a continuum; any discrete categorisation involves some degree of arbitrariness').

[112] Atkins et al. (2004) (explaining that the strength of a recommendation refers to 'the extent to which we can be confident that adherence to the recommendation will do more good than harm').

[113] Robinson and Goodman (2011); McGauran et al. (2010).

[114] Chan et al. (2004).

[115] The sample comprised 101 randomised trials approved in 1994–1995 and corresponding to 122 published journal articles.

[116] Chan et al. (2004), p. 2457.

[117] Ebrahim et al. (2014). The number of the examined studies indicates that publications based on IPD re-analysis are rather rare.

[118] Ibid p. 1030.

[119] Ibid p. 1027.

clinical practice remain unclear. As noted by the authors, it appears 'difficult to assess whether these changes in trial conclusions led eventually to major changes in clinical practice and, if so, how large these changes were'.[120] The authors further acknowledge that the study was underpowered to identify a significant difference between the re-analyses that suggested changes in the clinical recommendations conducted by original trial investigators vis-à-vis those performed by independent authors. In general, the variations in the conclusions between the initial and secondary analyses of clinical trial data can be too subtle to qualify as an adverse event[121] or a serious adverse event.[122] Since such subtle effects might not be immediately observed, it is uncertain how they might be impacting the clinical practice.

Further, secondary IPD analysis might bear important clinical implications not only concerning the indication for which the original drug has been approved but also for its off-label use(s).[123] First, if a drug has not been known or used for a different indication, such new indications can be revealed through IPD re-analysis.[124] Second, if a drug has already been used off-label, secondary data analysis can further refine clinical recommendations and better inform clinical practice. In the opinion of the European Ombudsman, where data analysis can allow for a better understanding of off-label uses, 'there is an overriding public interest in the disclosure of [clinical trial data]'.[125]

---

[120]Ibid.

[121]An adverse event is defined as 'any untoward medical occurrence in a subject to whom a medicinal product is administered and which does not necessarily have a causal relationship with this treatment'. Reg 536/2014/EU, art 2(1)(32).

[122]A serious adverse event is defined as 'any untoward medical occurrence that at any dose requires inpatient hospitalisation or prolongation of existing hospitalisation, results in persistent or significant disability or incapacity, results in a congenital anomaly or birth defect, is life-threatening, or results in death'. Reg 536/2014/EU, art 2(1)(33).

[123]Off-label use refers to the situations where an approved drug is used for treating a different condition or a different age group. Off-label uses can also involve clinically significant variations such as a different dosage regime or route of administration.

[124]On the discovery of off-label uses, see e.g. DeMonaco et al. (2006). The use of the authorised medicinal products in clinical practice, including off-label use, is not regulated at EU level. See Weda et al. (2017), p. 13.

[125]European Ombudsman (8 Jun 2016) Decision on own initiative inquiry OI/3/2014/FOR concerning the partial refusal of the European Medicines Agency to give public access to studies related to the approval of a medicinal product, para 44 (emphasis added).

6.4.3 *The Issue of the Under-Realised Research Potential of Clinical Trial Data*

6.4.3.1 Concerns

Distinct from concerns regarding transparency, accuracy, and trial validity, are arguments that exclusive control over non-summery data can preclude or slow down medical research and drug innovation. In particular, it was submitted that greater data sharing

- 'could open up opportunities for exploratory research that might lead to *new* hypotheses about the mechanisms of disease, more effective therapies, or alternative uses of existing or abandoned therapies that could then be tested in additional research';[126]
- 'could potentially increase the long-term return on grants by catalyzing secondary data analyses and helping to avoid future research that is redundant or based on an unpromising approach';[127]
- 'could make future clinical trials more efficient in the long run since new research could build on secondary analyses of the shared data'.[128]

The 'could language' reflects the idea that the research potential of data might be underutilised under the exclusive control of trial sponsors, and that such unrealised research opportunities represent a 'deadweight loss' and suboptimal return on risks borne by trial participants and resources invested into conducting trials.[129]

6.4.3.2 The Scope of the Problem

As such, trial sponsors' control over data does not exclude secondary data analysis. The problem of 'lost' research opportunities arises only if new research paths remain unexplored because neither the data holders (trial sponsors) nor other drug developers or researchers undertake exploratory data analysis. While data holders might lack the motivation or capacity, academic researchers might be discouraged by the transaction cost of negotiating access, especially where an intended research project requires aggregating multiple datasets.[130]

In the absence of a realistic counterfactual, it appears unfeasible to define to what extent the ability of drug companies to control third-party access to IPD prevents the realisation of data research potential. Any attempt to define what new knowledge

[126] Institute of Medicine of the National Academies (2015), p. 32 (with further references).

[127] Ibid p. 74.

[128] Ibid.

[129] Hoffmann et al. (2017) (pointing out that, where study results are unusable and non-replicable by others, 'the entire trial investment becom[es] a sunk-cost').

[130] As discussed in Chap. 9, Sect. 9.3.2.3.

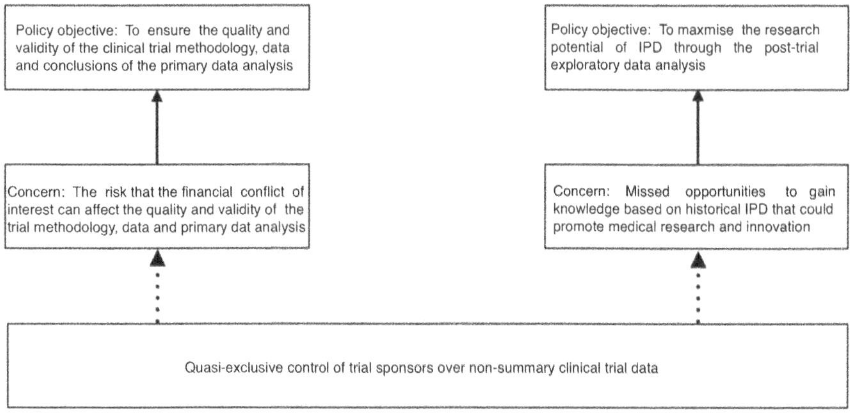

Fig. 6.2 The 'objectives tree'

and, eventually, medicines *might* have been developed if IPD from past trials could have been *readily accessible* for secondary analysis would be highly probabilistic and speculative, especially given the uncertain nature of drug R&D.

6.4.4 *The 'Objectives Tree'*

Based on the problem analysis, specific, measurable, achievable, relevant and time-bound ('S.M.A.R.T.') policy objectives should be formulated.[131] Such objectives 'set the level of policy ambition, fix the yardsticks for comparing policy options and determine the criteria for monitoring and evaluating the achievements of implemented policy'.[132] When multiple interrelated policy objectives are involved, it is recommended to use the 'objectives tree' to align policy goals with the distinct problem issues. The above-discussed concerns can be transformed in Fig. 6.2.

Irrespective of whether or not the stated concerns can be attributed to the control of trial sponsors over data, it is important to consider how and to what extent they are addressed under the existing regulatory framework.

[131] Better Regulation 'Toolbox' (n 5), p. 80.

[132] Ibid.

6.5 The Regulatory Status Quo

6.5.1 The Issue of Research Quality Under the Current Framework

6.5.1.1 Relevant Regulatory Provisions

The current regulatory approach to ensuring the quality of clinical trials and data can be examined in line with the distinction between three aspects of trial validity, namely, trial methodology, data and conclusions of the primary data analysis. The assessment of the trial methodology takes place during the trial authorisation and considers inter alia aspects related to data reliability and robustness such as statistical approaches, the study design and methodology, the sample size, the choice of a comparator, randomisation and endpoints.[133] The EU Clinical Trials Regulation vests the responsibility to ensure that a trial is designed and conducted in accordance with the principles of good clinical practice (GCP) in trial sponsors and investigators.[134] GCP refers to a set of ethical and scientific quality requirements for designing, conducting, performing, monitoring, auditing, recording, analysing and reporting clinical trials that can safeguard the rights, safety and well-being of study subjects and reliability and robustness of trial data.[135] While trial sponsors must 'adequately monitor the conduct of a clinical trial',[136] they shall also determine the extent and nature of monitoring.[137] Furthermore, trial sponsors have to record, process, handle and store clinical trial data so that 'it can be accurately reported, interpreted and verified'.[138]

As for the trial conclusions, the EU Clinical Trials Regulation requires trial sponsors to submit within a year upon the trial completion summary results to the EU database, irrespective of the trial outcome.[139] The summary report has to include the statistical analysis of the investigated endpoints.[140] Besides, the data analysis is mentioned in the context of reporting the suspected unexpected serious adverse

[133] Reg 536/2014/EU, rec 17, art 6(1)(b)(i). The EU Clinical Trials Regulation does not harmonise the division of the tasks and responsibilities between the national competent authorities and ethics committees in respect to the scientific and ethical parts of the assessment. Reg 536/2014/EU, art 4.

[134] Reg 536/2014/EU, art 47.

[135] Reg 536/2014/EU, art 2(2)(30).

[136] Reg 536/2014/EU, art 48, rec 44.

[137] Reg 536/2014/EU, art 48.

[138] Reg 536/2014/EU, art 56(1). Besides, Recital 51 of the Regulation states that 'information generated in a clinical trial should be recorded, handled and stored adequately for the purpose of *ensuring* subject rights and safety, the *robustness* and *reliability* of the data generated in the clinical trial, accurate reporting and interpretation, effective monitoring by the sponsor and effective inspection by Member States' (emphasis added).

[139] Chapter 4, Sect. 4.3.1.1.

[140] Reg 536/2014/EU, annex IV(D).

reactions (SUSARS) events unforeseen in the trial protocol.[141] Based on the evidence, the analysis should consider whether there is 'a reasonable possibility of establishing a causal relationship between the event and the investigational medicinal product'.[142]

While the EU Clinical Trials Regulation does not stipulate the post-trial regulatory assessment of data robustness and reliability, such assessment is usually performed in the course of the drug marketing authorisation. Applicants for marketing authorisation are required to prepare and submit technical dossiers according to the requirements under Directive 2001/83/EC,[143] which follows the common format developed by the ICH.[144] For the EU authorisation, the EMA's Committee for Medicinal Products for Human Use (CHMP) has the exclusive responsibility[145] to draw up the EMA's opinion 'on any matter concerning the admissibility of the files submitted under the centralised procedure, the granting, variation, suspension or revocation of an authorisation to place a medicinal product for human use on the market [. . .] and pharmacovigilance'.[146] In so doing, the CHMP relies on the scientific evaluation and resources that are available to national competent authorities (NCAs), while each NCA is responsible 'to monitor the scientific level and independence of the evaluation'[147] and draw up an assessment report concerning the results of the pharmaceutical and pre-clinical tests and the clinical trials of the medicinal product concerned.[148] Neither Directive 2001/83/EC nor the EU Drug Authorisation Regulation explicitly states whether IPD shall be a part of the dossier.

Furthermore, neither the EU Clinical Trials Regulation nor the EU Drug Authorisation Regulation provides for the post-trial re-analysis of IPD by third parties. The EU Drug Authorisation Regulation mentions that the EMA's scientific committees 'should be able to delegate some of their evaluation duties to standing working parties *open to experts from the scientific world* appointed for this purpose, whilst

[141] Reg 536/2014/EU, art 42. A SUSAR refers to 'a serious adverse reaction, the nature, severity or outcome of which is not consistent with the reference safety information'; a 'serious adverse event' is defined as 'any untoward medical occurrence that at any dose requires inpatient hospitalisation or prolongation of existing hospitalisation, results in persistent or significant disability or incapacity, results in a congenital anomaly or birth defect, is life-threatening, or results in death'. Reg 536/2014/EU, art 2(2)(34) and (33).

[142] Reg 536/2014/EU, annex III, para 2.1(3).

[143] Dir 2001/83/EC, arts 8(3), 10, 10a, 10b and 11 and annex I. See also Reg 726/2004/EC, art 6(1).

[144] Dir 2001/83/EC, annex I, Introduction and general principles, para 2.

[145] Reg 726/2004/EC, rec 23.

[146] Reg 726/2004/EC, art 5(2).

[147] Reg 726/2004/EC, art 61(6).

[148] Dir 2001/83/EC, art 21(4). The results of pharmaceutical and pre-clinical tests and clinical trials 'must enable a sufficiently well-founded and scientifically valid opinion to be formed as to whether the medicinal product satisfies the criteria governing the granting of a marketing authorisation. Consequently, an essential requirement is that the results of all clinical trials should be communicated, both favourable and unfavourable'. Dir 2001/83/EC, annex I, para 5.2(a).

retaining total responsibility for the scientific opinions issued'.[149] However, such possibility does not imply that external experts can readily access trial data upon own initiative for purposes other than the evaluation duties delegated by the EMA. The dispute between the EMA and the Nordic Cochrane Centre illustrates the difficulties that 'experts from the scientific world' were confronted with when seeking access to data held by the EMA.[150]

6.5.1.2 Does the EMA Collect and Hold IPD?

In practice, the EMA has not been routinely collecting IPD, at least not in recent years. According to the EMA's Advisory group on clinical trial data formats, 'raw' patient-level datasets are not regularly requested by the EMA.[151] In this regard, it appears paradoxical that, while the EMA aspires to promote transparency in its decision making by providing access to non-summary clinical trial data, the EMA itself does not re-analyse IPD in the course of marketing authorisation. Significantly limited human resources—particularly as far as biomedical statisticians[152] and clinical assessors validators[153] are concerned—might indicate that IPD might not be routinely analysed in the EU at the level of NCAs either. In this regard, the difference between the practice in the EU and the US is worth highlighting. The

---

[149]Reg 726/2004/EC, rec 25 (emphasis added).

[150]European Ombudsman (19 May 2010) Draft recommendation of the European Ombudsman in his inquiry into complaint 2560/2007/BEH against the European Medicines Agency.

[151]Advice to the European Medicines Agency from the Clinical Trial Advisory Group on Clinical Trial Data Formats (CTAG2)—Final advice to EMA (30 Apr 2013), p. 4. http://www.ema.europa.eu/docs/en_GB/document_library/Other/2013/04/WC500142850.pdf. Accessed 26 Mar 2021. See Koenig et al. (2015), p. 16 (explaining that '[a]t the level of EMA/CHMP, decision making in relation to licensure of new drugs (positive/negative opinion) is currently based on rapporteurs' assessment work without access to clinical trials data on patient level in electronic format'). Such assessment is 'based on thorough review of protocols, analysis plans and clinical trial reports, and usually does not involve processing of patient raw data to replicate analyses carried out by the Sponsor/Applicant' Ibid. See EFSPI (25 Apr 2013) European Federation of Statisticians in the Pharmaceutical Industry (EFSPI) Position on European Medicines Agency (EMA) access to clinical trial data initiative, p. 6 (stating that if 'the new EMA policy will allow re-analysis of patient level data, EFSPI would be interested to know whether it would be possible for EMA to increase their capabilities including expertise and resources to be able to re-analyse patient level data they receive in a regulatory submission, similar to how some other regulatory authorities review regulatory dossiers'). https://www.efspi.org/documents/publications/efspipositiononema250413.pdf. Accessed 26 Mar 2021. See also Senn (2007), p. 463 (referring to the EMA as 'a statistician-free zone').

[152]Skovlund (2009).

[153]European Commission (17 Jul 2012) Impact assessment report on the revision of the 'Clinical Trials Directive' 2001/20/EC accompanying the document Proposal for a Regulation of the European Parliament and of the Council on clinical trials on medicinal products for human use, and repealing Directive 2001/20/EC, SWD(2012) 200 final, vol. II, pp. 22–23 (indicating that the number of clinical and validation assessors in the EU NCAs involved in the assessment of clinical trials ranges between zero and two in the majority of the EU Member States).

USFDA has historically approached the assessment differently and its 'large in-house statistics department [. . .] routinely re-analyses the data of pivotal trials'.[154]

6.5.2 *Provisions Related to Exploratory Data Analysis*

Whereas access to non-summary clinical trial data for exploratory research is not specifically regulated at EU level, certain provisions under the applicable framework can impact IPD accessibility for secondary research purposes.

6.5.2.1 Derogations from Personal Data Protection for Scientific Research

Health-related data constitutes a special category of personal data, which processing is, as a general rule, prohibited.[155] However, the GDPR envisaged several derogations from the rights of data subjects, including for personal data processing for scientific research purposes.[156] Such derogations can be implemented 'in so far as [personal] rights are likely to render impossible or seriously impair the achievement of the specific purposes, and such derogations are necessary for the fulfilment of those purposes'.[157] The implementation of derogations is subject to the conditions and safeguards for the rights and freedoms of data subjects.[158] The GDPR also envisages a certain margin of national discretion as far as the conditions that determine the lawfulness of personal data processing are concerned.[159]

While the GDPR does not define the term scientific research, it states that 'processing of personal data for scientific research purposes should be interpreted in a *broad manner* including for example technological development and demonstration, *fundamental research*, *applied research* and *privately funded research*'.[160] Accordingly, secondary IPD analyses—confirmatory and exploratory—can fall

[154] Koenig et al. (2015), pp. 10–11.

[155] GDPR, art 9(2)(a).

[156] GDPR, art 89(2). Derogations can be applied to the right of access, the right to rectification, the right to restriction of processing and the right to object to processing of personal data.

[157] Ibid.

[158] GDPR, art 89. Such safeguards can be implemented through the organisational and technical measures in line with the principle of data minimisation. Ibid.

[159] In Germany, for instance, the derogations were implemented under Bundesdatenschutzgesetz that, apart from stipulating the appropriate safeguards ('angemessene Maßnahmen') for protecting the interests of data subjects, introduced the criterion 'erheblich überwiegen' (considerably outweigh). This additional requirement means that the data controller's interests in data processing should considerably outweigh the interests of the data subject in restricting data processing. Bundesdatenschutzgesetz vom 30. Juni 2017 (BGB1. I S. 2097) § 27(1).

[160] GDPR, rec 159 (emphasis added).

under personal data processing for scientific research, irrespective of whether it might be conducted by academic institutions, industrial R&D, industry-academia collaboration or partnerships with contract research organisations.

Even though the derogations from personal rights of data subjects do not provide a legal basis to claim access to data for research purposes, their importance should not be underestimated. The very existence of derogations for scientific research indicates the legislator's intent to balance public interests at stake.[161] Besides, the protection of personal data of trial subjects in and of itself might not be a sufficient reason to reject a request for access to IPD for its secondary analysis.

6.5.2.2 Reservations for the Protection of Economic Interests of Trial Sponsors

As concluded in Chap. 4, the current EU framework does not provide a viable legal basis on which access to IPD held by drug companies could be claimed. On the contrary, some provisions can restrict access, such as the exceptions for protecting commercially confidential information under the EU Transparency Regulation and the sector legislation.[162]

6.6 The Summary and Conclusion

Table 6.1 relates the interests of stakeholders, policy objectives concerning access to clinical trial data and the relevant regulatory provisions under the existing EU regulatory framework.

As evident from the table, the stakeholders' interests and concerns related to IPD accessibility are, to some extent, addressed by the existing EU regulatory framework. As regards secondary IPD analysis, the purpose of the independent validation of trial results can, under certain conditions, be achieved through the right of access to documents. However, the current system does not provide a specific[163] legal basis for exploratory IPD analysis.[164]

[161] GDPR, rec 26. Besides, Recital 4 of the GDPR reinforces the idea of balancing personal data protection against other fundamental rights, freedoms and principles recognised under the Charter of Fundamental Rights, including the freedom of the arts and sciences, in accordance with the principle of proportionality.

[162] See Chap. 4, Table 4.1.

[163] Hypothetically, IPD accessed under the EU Transparency Regulation could be used for exploratory research purposes. However, such use would exceed the scope of the right of access to documents, as argued in Chap. 4, Sect. 4.3.3.4.

[164] Exploratory IPD analysis, arguably, falls outside the scope of the substantive and procedural matters covered by the EU Clinical Trials Regulation and the EU Drug Authorisation Regulation.

Table 6.1 Access to clinical trial data in a broader regulatory context

Stakeholder groups	Stakeholders' interests concerning the accessibility of clinical trial data	Policy objectives	Relevant provisions under the existing regulatory framework
Trial participants	Personal data protection	To protect the personal data of trial participants	• The GDPR • The general principle under the EU Clinical Trials Regulation of protecting rights of trial participants, including in relation to personal data
General public	• Reliability of the trial outcomes and data submitted for drug authorisation • Safety and efficacy of the marketed medicinal products • Transparency in the decision making of trial authorities	• To ensure the validity of the trial results • To protect against information asymmetries in the drug market • To ensure transparency in the regulatory decision making regarding drug marketing authorisation	• The general principle of data robustness and reliability proclaimed under the EU Clinical Trials Regulation • The requirements under the EU Drug Authorisation Regulation regarding the evidence from clinical trials • The requirements under the EU Drug Authorisation Regulation concerning the benefit-risk assessment based on the evidence from trials • Transparency provisions under the EU Clinical Trials Regulation and the EU Drug Authorisation Regulation • Provisions under the EU Transparency Regulation enabling access to data held by the EMA
Healthcare professionals/ medical practitioners	• Reliability of trial outcomes and data submitted for drug marketing authorisation • The rapid dissemination of the trial results	• To ensure the reliability and robustness of data submitted for drug marketing authorisation and the accuracy of the benefit-risk assessment • To ensure the timely dissemination of clinical trial results	• The general principle of data robustness and reliability proclaimed under the EU Clinical Trials Regulation • Transparency provisions under the EU Clinical Trials Regulation and the EU Drug Authorisation Regulation

(continued)

Table 6.1 (continued)

Stakeholder groups	Stakeholders' interests concerning the accessibility of clinical trial data	Policy objectives	Relevant provisions under the existing regulatory framework
Researchers and drug developers	Access to source data (IPD) for secondary analysis	**Not addressed under the existing EU framework**	
Drug sponsors (research-based drug companies)	• Protection of the competitive advantage in the existing drug market • Protection of the competitive advantage in drug R&D	To provide economic incentives for drug innovation	• Regulatory incentives, including patents, SPCs, data and market exclusivities • The exceptions for CCI under the EU Transparency Regulation, the EU Clinical Trials Regulation and the EU Drug Authorisation Regulation

Unregulated issues should not necessarily be viewed and treated as regulatory 'gaps'. For that, one would first need to prove that trial sponsors' control over non-summary clinical trial data is a 'problem driver' leading to the diminished validity and reproducibility of trial findings and missed research opportunities. This chapter shows that proving such causal linkage is challenging; the reviewed evidence from empirical studies on this subject is fragmented and inconclusive. The claim regarding the 'underutilised' research potential of IPD is speculative and can hardly be objectively assessed.

Does de facto control over IPD slow scientific discovery and advancement[165] and drug innovation?[166] Alternatively, can it promote innovation by protecting the competitive advantage of research-based drug companies in competition in innovation? This ambiguity necessitates a further inquiry into how the balance of interests in secondary IPD analysis should be understood and implemented.

References

Als-Nielsen B et al (2003) Association of funding and conclusions in randomized drug trials: a reflection of treatment effect or adverse events? JAMA 290(7):921–928. https://doi.org/10.1001/jama.290.7.921

Atkins D et al (2004) Grading quality of evidence and strength of recommendations. BMJ 328 (7454):1490. https://doi.org/10.1136/bmj.328.7454.1490

Baldock J (2007) Social policy, social welfare, and the welfare state. In: Baldock J, Manning N, Vickerstaff S (eds) Social policy. OUP, Oxford

[165] Institute of Medicine of the National Academies (2015), p. 141 (with further references).

[166] Ibid p. 32.

Bator FM (1957) The simple analytics of welfare maximization. Am Econ Rev 47(1):22–69
Bator FM (1958) The anatomy of market failure. Q J Econ 72(3):351–379
Begley GC, Ioannidis JPA (2015) Reproducibility in science: improving the standard for basic and preclinical research. Circ Res 116(1):116–126. https://doi.org/10.1161/CIRCRESAHA.114.303819
Bekelman JE, Li Y, Gross CP (2003) Scope and impact of financial conflicts of interest in biomedical research: a systematic review. JAMA 289(4):454–465. https://doi.org/10.1001/jama.289.4.454
Bergkamp L (2003) European Community law for the new economy. Intersentia, Antwerp
Bollen K et al (2015) Social, behavioral, and economic sciences perspectives on robust and reliable science. Report of the Subcommittee on Replicability in Science Advisory Committee to the National Science Foundation Directorate for Social, Behavioral, and Economic Sciences. National Science Foundation, Washington DC
Breyer S (1979) Analyzing regulatory failure: mismatches, less restrictive alternatives, and reform. Harv Law Rev 92(3):547–609
Breyer SG (1982) Regulation and its reform. Harvard University Press, Harvard
Chan AW et al (2004) Empirical evidence for selective reporting of outcomes in randomized trials: comparison of protocols to published articles. JAMA 291(20):2457–2465. https://doi.org/10.1001/jama.291.20.2457
Cho MK, Bero LA (1996) The quality of drug studies published in symposium proceedings. Ann Intern Med 124(5):485–489. https://doi.org/10.7326/0003-4819-124-5-199603010-00004
Chow SC, Liu JP (2004) Design and analysis of clinical trials: concepts and methodologies, 2nd edn. Wiley, Hoboken
CIOMS (2016) International ethical guidelines for health-related research involving humans. CIOMS, Geneva
Clifford TJ, Barrowman NJ, Moher D (2002) Funding source, trial outcome and reporting quality: are they related? Results of a pilot study. BMC Health Serv Res 2(1):18. https://doi.org/10.1186/1472-6963-2-18
DeMonaco HJ, Ali A, von Hippel E (2006) The major role of clinicians in the discovery of off-label drug therapies. Pharmacotherapy 26(3):323–332. https://doi.org/10.1592/phco.26.3.323
Doshi P (2014) From promises to policies: is big pharma delivering on transparency? BMJ 348: g1615. https://doi.org/10.1136/bmj.g1615
Doshi P et al (2013) Restoring invisible and abandoned trials: a call for people to publish the findings. BMJ 346:f2865. https://doi.org/10.1136/bmj.f2865
Dukes G, Braithwaite J, Moloney JP (2015) Pharmaceuticals, corporate crime and public health. Edward Elgar, Cheltenham
Ebrahim S et al (2014) Reanalyses of randomized clinical trial data. JAMA 312(10):1024–1032. https://doi.org/10.1001/jama.2014.9646
Feldman AM (2008) Welfare economics. In: Durlauf SN, Blume LE (eds) The new Palgrave dictionary of economics, 3rd edn, vol 8. Palgrave Macmillan, Basingstoke, pp 721–731
Fisher JA (2009) Medical research for hire: the political economy of pharmaceutical clinical trials. Rutgers University Press, New Brunswick
Friedberg M et al (1999) Evaluation of conflict of interest in economic analyses of new drugs used in oncology. JAMA 282(15):1453–1457. https://doi.org/10.1001/jama.282.15.1453
Goldacre B (2014) Bad pharma: how drug companies mislead doctors and harm patients. Signal, McClelland & Stewart, Toronto
Goldacre B et al (2017) Pharmaceutical companies' policies on access to trial data, results, and methods: audit study. BMJ 358:j3334. https://doi.org/10.1136/bmj.j3334
Golder S, Loke YK (2008) Is there evidence for biased reporting of published adverse effects data in pharmaceutical industry-funded studies? Br J Clin Pharmacol 66(6):767–773. https://doi.org/10.1111/j.1365-2125.2008.03272.x
Goodman SN, Fanelli D, Ioannidis JPA (2016) What does research reproducibility mean? Sci Transl Med 8(341):341ps12. https://doi.org/10.1126/scitranslmed.aaf5027

Gøtzsche PC (2013) Deadly medicines and organised crime: how big pharma has corrupted healthcare. CRC Press, Boca Raton

Gøtzsche PC et al (2006) Constraints on publication rights in industry-initiated clinical trials. JAMA 295(14):1645–1646. https://doi.org/10.1001/jama.295.14.1645

Greifman N et al (2015) Opening clinical trial data: are the voluntary data-sharing portals enough? BMC Med 13:280. https://doi.org/10.1186/s12916-015-0525-y

Guyatt GH et al (2008) GRADE: an emerging consensus on rating quality of evidence and strength of recommendations. BMJ 336(7650):924–926. https://doi.org/10.1136/bmj.39489.470347

Higgins JPT, Altman DG, Sterne JAC (2017) Assessing risk of bias in included studies. In: Higgins JPT et al (eds) Cochrane handbook for systematic reviews of interventions. Cochrane, pp 8:1-8:73

Hoffmann T et al (2017) Focus on sharing individual patient data distracts from other ways of improving trial transparency. BMJ 357:j2782. https://doi.org/10.1136/bmj.j2782

Institute of Medicine of the National Academies (2012) Informatics needs and challenges in cancer research. The National Academic Press, Washington DC

Institute of Medicine of the National Academies (2015) Sharing clinical trial data: maximizing benefits, minimizing risk. The National Academies Press, Washington DC

Jackson E (2019) Medical law: text, cases, and materials. OUP, Oxford

Just RE, Hueth DL, Schmitz A (2004) The welfare economics of public policy: a practical approach to project and policy evaluation. Edward Elgar, Cheltenham

Kaur S, Choy CY (2014) Ethical considerations in clinical trials: a critique of the ICH-GCP guideline. Dev World Bioeth 14(1):20–28. https://doi.org/10.1111/dewb.12004

Kjaergard LL, Als-Nielsen B (2002) Association between competing interests and authors' conclusions: epidemiological study of randomised clinical trials published in the BMJ. BMJ 325 (7358):249. https://doi.org/10.1136/bmj.325.7358.249

Koenig F et al (2015) Sharing clinical trial data on patient level: opportunities and challenges. Biom J 57(1):8–26. https://doi.org/10.1002/bimj.201300283

Krumholz HM et al (2014) Sea change in open science and data sharing: leadership by industry. Circ Cardiovasc Qual Outcomes 7(4):499–504. https://doi.org/10.1161/CIRCOUTCOMES.114.001166

Lexchin J (2012) Those who have the gold make the evidence: how the pharmaceutical industry biases the outcomes of clinical trials of medications. Sci Eng Ethics 18(2):247–261. https://doi.org/10.1007/s11948-011-9265-3

Lexchin J et al (2003) Pharmaceutical industry sponsorship and research outcome and quality: systematic review. BMJ 326(7400):1167–1170. https://doi.org/10.1136/bmj.326.7400.1167

Lo B, Field MJ (2009) Conflict of interest in medical research, education, and practice. The National Academies Press, Washington DC

Lundh A, Krogsboll LT, Gotzsche PC (2012) Sponsors' participation in conduct and reporting of industry trials: a descriptive study. Trials 13:146. https://doi.org/10.1186/1745-6215-13-146

Lundh A et al (2017) Industry sponsorship and research outcome. Cochrane Database Syst Rev 2 (2):MR000033. https://doi.org/10.1002/14651858.MR000033.pub3

Mayo-Wilson E, Doshi P, Dickersin K (2015) Are manufacturers sharing data as promised? BMJ 351:h4169. https://doi.org/10.1136/bmj.h4169

McGauran N et al (2010) Reporting bias in medical research – a narrative review. Trials 11:37. https://doi.org/10.1186/1745-6215-11-37

Miller J et al (2019) Sharing of clinical trial data and results reporting practices among large pharmaceutical companies: cross sectional descriptive study and pilot of a tool to improve company practices. BMJ 366:l4217. https://doi.org/10.1136/bmj.l4217

Montaner JSG, O'Shaughnessy MV, Schechter MT (2001) Industry-sponsored clinical research: a double-edged sword. Lancet 358(9296):1893–1895. https://doi.org/10.1016/S0140-6736(01)06891-X

Mullane K, Curtis MJ, Williams M (2018) Reproducibility in biomedical research. In: Williams M, Curtis MJ, Mullane K (eds) Research in the biomedical sciences: transparent and reproducible. Academic, London, pp 1–67

Murugiah K et al (2016) Availability of clinical trial data from industry-sponsored cardiovascular trials. J Am Heart Assoc 5(4):e003307. https://doi.org/10.1161/JAHA.116.003307
Naci H, Cooper J, Mossialos E (2015) Timely publication and sharing of trial data: opportunities and challenges for comparative effectiveness research in cardiovascular disease. Eur Heart J Qual Care Clin Outcomes 1(2):58–65. https://doi.org/10.1093/ehjqcco/qcv012
Nevitt SJ et al (2017) Exploring changes over time and characteristics associated with data retrieval across individual participant data meta-analyses: systematic review. BMJ 357:j1390. https://doi.org/10.1136/bmj.j1390
Nisen P, Rockhold F (2013) Access to patient-level data from GlaxoSmithKline clinical trials. N Engl J Med 369(5):475–478. https://doi.org/10.1056/NEJMsr1302541
Nüesch E et al (2009) The effects of excluding patients from the analysis in randomised controlled trials: meta-epidemiological study. BMJ 339:b3244. https://doi.org/10.1136/bmj.b3244
OECD (2009) Regulatory impact analysis. A tool for policy coherence. OECD, Paris
Ogus AI (1994) Regulation: legal form and economic theory. Clarendon Press, Oxford
Patton CV, Sawiski DS, Clark JJ (2016) Basic methods of policy analysis & planning, 3rd edn. Routledge, Abingdon
Porter TM (2016) Depending on numbers. In: Atmanspacher H, Maasen S (eds) Reproducibility: principles, problems, practices, and prospects. Wiley, Hoboken, pp 447–466
Reiter S (2018) Efficient allocation. In: Macmillan Publishers (ed) The new Palgrave dictionary of economics, 3rd edn. Palgrave Macmillan, London, pp 3522–3543
Robinson KA, Goodman SN (2011) A systematic examination of the citation of prior research in reports of randomized, controlled trials. Ann Intern Med 154(1):50–55. https://doi.org/10.7326/0003-4819-154-1-201101040-00007
Schott G et al (2010) The financing of drug trials by pharmaceutical companies and its consequences. A qualitative, systematic review of the literature on possible influences on the findings, protocols, and quality of drug trials. Detsches Ärzteblat Int 107(16):279–285. https://doi.org/10.3238/arztebl.2010.0279
Schulz KF et al (1995) Empirical evidence of bias. Dimensions of methodological quality associated with estimates of treatment effects in controlled trials. JAMA 273(5):408–412. https://doi.org/10.1001/jama.273.5.408
Scott K (2012) Measuring wellbeing: towards sustainability? Routledge, Abingdon
Senn S (2007) Statistical issues in drug development, 2nd edn. Wiley, Hoboken
Shapiro SA (1978) Divorcing profit motivation from new drug research: a consideration of proposals to provide the FDA with reliable test data. Duke Law J:155–183. https://doi.org/10.2307/1372287
Sismondo S (2008) Pharmaceutical company funding and its consequences: a qualitative systematic review. Contemp Clin Trials 29(2):109–113. https://doi.org/10.1016/j.cct.2007.08.001
Skovlund E (2009) Statisticians in European regulatory agencies. Pharm Stat 8(4):259–263. https://doi.org/10.1002/pst.367
Stern S, Lemmens T (2011) Legal remedies for medical ghostwriting: imposing fraud liability on guest authors of ghostwritten articles. PLoS Med 8(8):e1001070. https://doi.org/10.1371/journal.pmed.1001070
Sydes MR et al (2015) Sharing data from clinical trials: the rationale for a controlled access approach. Trials 16:104. https://doi.org/10.1186/s13063-015-0604-6
The pharmaceutical persuaders – the industry, the doctor, and the clinical trial (1961) Lancet 278 (7199):421–423
Tierney JF, Stewart LA (2005) Investigating patient exclusion bias in meta-analysis. Int J Epidemiol 34(1):79–87. https://doi.org/10.1093/ije/dyh300
Weda M et al (2017) Study on off-label use of medicinal products in the European Union. European Union
Williams M, Mullane K, Curtis MJ (2018) Reporting results. In: Williams M, Curtis MJ, Mullane K (eds) Research in the biomedical sciences: transparent and reproducible. Academic, London, pp 146–197

Chapter 7
Access to Clinical Trial Data as a Case on R&D Externalities: A Theoretical Framework

Abstract This chapter outlines an analytical framework that can inform the design of the legal rules on access to IPD as a knowledge resource for research and innovation. It starts by characterising data as an economic good and frames the dual policy objective of regulating access to IPD as an 'access-innovation' dilemma. The analysis finds that such dilemmas typically arise with the private provisioning of public goods and stem from the dual implications of non-excludability of R&D results and knowledge externalities for innovation. Accordingly, an overview of law-and-economics-of-innovation research on R&D externalities is provided. The 'access-innovation' policy dilemma is then restated as a potential trade-off between greater knowledge diffusion and diminished innovation incentives that might arise if data generated in industry-sponsored trials is treated as a non-excludable knowledge resource.

7.1 Framing the Dilemma

7.1.1 Clinical Trial Data as a Non-rivalrous Research Tool

7.1.1.1 Clinical Trial Data as an Inherently Public Good

In broad terms, clinical trials and data can be viewed as *inherently* public goods because the benefits of data analysis in terms of knowledge and innovative medicines can extend to society at large.[1] From an economic perspective, digitalised research data can be characterised as a good non-rivalrous in use[2] and partially

[1] Razzolini (2004), p. 457 (defining public goods as 'goods with benefits that extend to a group of individuals').

[2] Non-rivalry in use means that data can be analysed in parallel research projects without depreciating its inherent value or imposing additional costs on the data producer. On rivalry of benefits of data analysis, see Chap. 8 at Sect. 8.3.4.

D. Kim, *Access to Non-Summary Clinical Trial Data for Research Purposes Under EU Law*, Munich Studies on Innovation and Competition 16,
https://doi.org/10.1007/978-3-030-86778-2_7

excludable by factual, technological and legal means.[3] Thus, it can be defined as an impure[4] or quasi-public good.[5]

The main implication of characterising knowledge gained through clinical trial data analysis as a public good is that, absent the means of appropriability, a decentralised market economy would likely under-invest into its provision.[6] At the same time, any restriction on the consumption of public goods is inefficient.[7] Conventionally, public policies have been predominantly concerned with the supply aspect of R&D,[8] including the design of innovation incentives. The demand aspect of knowledge resources is increasingly emphasised in countering the exclusivity-based incentives.[9]

7.1.1.2 Digital Data as an Intermediate Good and a Research Infrastructure

Knowledge and information constitute both an input and output of the inventive activity.[10] Research data features the same duality: on the one hand, it represents trial results; on the other hand, it can be analysed in the subsequent research directed at

[3] See e.g. Blumenthal (2010), p. 142 (concluding that 'large clinical databases have an aspect of a *quasi*-public good [as they] are definitely excludable'). On the legal determinants of control over and access to clinical trial data under EU framework, see Chap. 4.

[4] A 'pure public good' is characterised by both non-excludability and non-rivalry in consumption; an 'impure public goods' can be partially rival and/or partially excludable. See Cornes and Sandler (1999), p. 4; Stiglitz (1999), pp. 309–310 (referring to knowledge that can be made excludable, e.g. through trade secrets, as an impure public good).

[5] See Kaul I (2013) Public goods: a positive analysis. Discussion draft, UNDP Office of Development Studies, p. 17 (pointing out that excludability of an economic good is 'a social construct' in a sense that it is determined not only by the inherent properties of a good but also by man-made rules).

[6] In the case of clinical trials, several interrelated market failures can be distinguished, namely, the market failure of public goods (suboptimal investment into drug R&D), the market failure of information asymmetry (the risk of inaccurate representation of information regarding the drug effects) and the market failure of insufficient testing. On market failures in the pharmaceutical industry, see Orsenigo et al. (2006), pp. 406–409.

[7] See e.g. Nelson (1969), p. 306; Oakland (1987), p. 485 (noting that, since public goods 'are not used up in the act of consumption [. . .], the marginal cost of extending service to additional users is zero'; therefore, a fee charged to recover the costs of providing a public good 'will usually lead some potential users to forgo consumption, creating a deadweight efficiency loss'); Razzolini (2004), p. 458 (explaining that 'society's welfare would be maximized when the good is available for consumption at no cost to everyone who places a positive value on it: that is, no individual should be excluded from consumption').

[8] Heal (1999), p. 223 (observing that public goods pose the policy questions of how much should be provided and how their provision should be financed).

[9] See generally Frischmann (2012); Frischmann and Lemley (2007); Frischmann (2005).

[10] Arrow (1962), p. 618; Dasgupta and David (1987), p. 535; Antonelli (2017), p. 18.

the questions and hypotheses other than those examined in the original trial.[11] Thus, it can be characterised as an 'intermediate' good, serving as R&D input.[12] One particular type of intermediate non-rivalrous goods are *infrastructures*, 'large-scale indivisible capital goods producing products and services, which become inputs in most or all economic activities on a multi-user basis'.[13] Two kinds of infrastructures are distinguished: traditional or physical (e.g. roads, electricity generating systems, telecommunication systems, etc.) and non-traditional (e.g. institutional and knowledge infrastructures).[14] The latter can encompass various intellectual resources that enable research and innovative activity[15] and can be viewed as a 'research space',[16] 'information space'[17] and 'intellectual space'.[18] Digital data,[19] in general, and scientific data,[20] in particular, fall within this category. Non-summary clinical trial data presents a rich resource for secondary research that can be used for exploring questions beyond the original hypothesis tested in the trial, in which data was gathered.[21]

The view that research tools constitute an infrastructural resource prompted the proposals for creating 'research commons'.[22] In life sciences, the 'commons' model was contemplated, for instance, for governing genomic data,[23] microbial organisms,[24] 'abandoned' molecules[25] and clinical data.[26]

---

[11] As shown in Chap. 3, aggregated IPD has a considerable potential to generate knowledge beyond the benefit-risk assessment of the investigational products.

[12] Intermediate public goods are also known as collective intermediate goods or collective factors of production. See e.g. Oakland (1987), pp. 492–493; Sandmo (1972), p. 149.

[13] Smith (2005), p. 86.

[14] ibid pp. 87–88.

[15] Such resources, for instance, include scientific equipment, archives, collections, e-infrastructures, computing systems, and communication networks. See European Commission. About the research infrastructures. http://ec.europa.eu/research/infrastructures/index.cfm?pg=about. Accessed 26 Mar 2021.

[16] Reichman et al. (2016), p. 475.

[17] Cameron (2001), p. 32.

[18] Rose (2003), p. 90.

[19] OECD (2015), p. 197.

[20] European Commission. About the research infrastructures. http://ec.europa.eu/research/infrastructures/index.cfm?pg=about. Accessed 26 Mar 2021.

[21] Lauer (2010), p. 91.

[22] Dedeurwaerdere (2009), p. 366 (referring to 'research commons' as comprising scientific data, information, materials and research tools).

[23] See e.g. van Overwalle (2014), p. 137; Dedeurwaerdere (2009), pp. 376 ff.

[24] See e.g. Reichman et al. (2016), p. 475 (arguing that 'the disaggregated knowledge assets of microbiological research should be combined and strengthened within a contractually constructed research commons to be organised and managed by the public science community itself').

[25] Reichman (2009), p. 57.

[26] See e.g. Crown (2010), p. 143.

7.1.1.3 Data as a Dual-Purpose Research Tool

A peculiar feature of trial data is that, while it is gathered in the course of the downstream drug development, it provides valuable input for both academic research and industrial R&D, including pre-clinical (upstream) phases. Secondary data analysis can be directed at the questions of fundamental nature (e.g. the underlying factors of disease processes) or the development of the commercial applications of basic knowledge (e.g. when designing new trials). In this view, clinical trial data can be defined as a 'dual-purpose' research tool.[27] Such duality is characteristic of other types of research tools, particularly those used in biomedical research.[28] It can also explain why the debate regarding exclusive control over such tools has been especially intense and polarised in the context of life sciences and drug innovation.[29]

7.1.2 *The 'Access-Incentives Paradox'*

The debate over access to clinical trial data features two ostensibly conflicting propositions. Unrestricted access to non-summary data is posited, on the one hand, to promote biomedical research and drug development[30] and, on the other hand, to affect the innovation incentives of research-based pharmaceutical companies negatively.[31] Thus, the term 'access-incentives dilemma' is used in the context of the present analysis to refer to these dual implications of IPD disclosure.

A policy dilemma of whether to intervene by access measures or not is especially controversial in the case of data gathered in industry-sponsored trials. Such trials represent a situation when inherently public goods—clinical studies and drug innovation—are supplied by the private sector.[32] Private investment inevitably causes tension between the appropriability of R&D benefits and knowledge diffusion.[33]

The fundamental question is how private and social returns on the innovative activity should be balanced to promote innovation optimally. In this regard, the debate over access to clinical trial data illustrates the dilemma that has been in the

[27] The potential to contribute to academic and commercial research is characteristic of other types of research tools used in molecular biology. See Eisenberg (1997), p. 13; Reichman et al. (2016), p. 255; Murray and Stern (2007), p. 648; Lee (2008), p. 73.

[28] Reichman et al. (2016), p. 255.

[29] On this issue, Chap. 8 at Sect. 8.2.2.3.

[30] See generally Chap. 3.

[31] On the role of control over data as a means to protect competitive advantage in competition in drug innovation, see Chap. 8 at Sect. 8.1.4.4, subheading 'Protection of Exploratory Endpoints as Intermediate Research Results'.

[32] Gruber (2016), pp. 197–203.

[33] On the inherent tension between appropriability and diffusion, see below (nn 92–95) in this chapter and the accompanying text.

focus of law and economics of innovation, namely, how to reconcile two policy objectives: to provide adequate incentives to the private sector to undertake R&D and create knowledge, on the one hand, and to ensure rapid diffusion and robust utilisation of once created knowledge, on the other hand.[34]

From an economic perspective, the issue of whether access to and use of data should be conditional on the authorisation of the data holders translates into a question of whether research data should be treated as an *excludable* or a *non-excludable* good. Economic policies generally aim to improve efficiency[35] in allocating resources to the production and consumption of goods. The main implication of characterising research and knowledge as inherently public goods is that, on the one hand, they are prone to the problem of suboptimal investment in R&D; on the other hand, any restriction on their dissemination and use is inefficient.[36] The dilemma suggests that where the output of privately funded research is treated as a *non-excludable* (public) good, a trade-off can arise between the social benefits of unrestricted access and use of such goods, on the one hand, and the social benefits of protecting innovation incentives, on the other hand. At the outset, it does not appear clear whether such trade-off can arise in the case of mandatory disclosure of IPD.

From a law-making perspective, the 'access-incentives dilemma' suggests the choice between two regimes: exclusive control of drug sponsors over IPD, whereby access—including for scientific research purposes—is subject to negotiations with trial sponsors (the non-intervention scenario). Alternatively, regulation could intervene by access measures enabling third-party IPD analysis. The decisive distinction is whether access to data for research purposes should be subject to the authorisation of the initial trial sponsor. Once this question is answered negatively, different ways to calibrate the access regime can be considered.

7.1.3 Limitations of the Welfare Cost-Benefit Analysis

Economic literature suggests that whether certain resources should be privately controlled or commonly accessible should be determined by weighing up social costs and benefits, while the outcome is 'a priori undetermined'.[37] Since this study seeks to define which regime of access to data can promote drug innovation more

[34] Foray (2004), pp. 113–118.

[35] Efficiency is defined as rationing scarce resources for the 'maximum possible satisfaction'. Wadley (2011), p. 97.

[36] Oakland (1987), p. 485 (stating that a fee 'charged by the private firm for the consumption of a public good in order to recover the costs of producing a public good will lead some potential users to forgo consumption [and] would result in a deadweight efficiency loss').

[37] Platteau (2008), p. 22.

optimally, the analysis needs to identify and evaluate the *innovation*-related costs and benefits associated with exclusive control and unrestricted access.[38]

The complexity subsists in that both broad access and exclusive control over data can entail dual implications for innovation, in some way promoting research and innovation activity and, in some way, potentially hindering it. Benefits of access to data can be defined as more knowledge produced through data analysis, while the benefits of control over data can, arguably, be related to the protection of innovation incentives. The choice of an optimal regime requires comparing welfare costs and benefits. However, it would be incorrect to contrast welfare benefits of control with welfare benefits of access or welfare costs of control with welfare benefits of access. Instead, relative costs and benefits under both the status quo and the intervention should be weighed up and compared.

At the outset, conducting a full-fledged welfare cost-benefit analysis in the case at issue appears unfeasible. First, there is a general methodological problem of measuring the impact on innovation. As acknowledged by economists, the effects on innovation (dynamic efficiency) can hardly be predicted and adequately assessed.[39] In the case of IPD, an attempt to define which new knowledge and eventually medicines might be developed under the counterfactual (i.e. were IPD not exclusively controlled by trial sponsors) could, at best, be highly speculative, especially given the uncertain nature of drug R&D. Second, the two conflicting propositions are not directly comparable in terms of their implications for efficiencies in resource allocation in R&D. On the one hand, the claim that data disclosure impedes innovation incentives points to the suboptimal allocation of resources to conducting clinical trials and generating data (the knowledge production activity). On the other hand, the proposition that broad access to data can contribute to subsequent research and new drug development points to efficiency in utilising data as a knowledge resource (the knowledge consumption activity). What could be the common denominator between the two competing propositions, which underlie the 'access-incentives dilemma', allowing them to be compared 'on equal terms'?

---

[38] As shown in Chap. 6, transparency and innovation objectives concerning access to IPD require distinct policy approaches, while *erga omnes* disclosure of IPD is unlikely to be a proportionate measure in either case.

[39] Audretsch et al. (2004), p. 21 (noting that '[c]oncepts such as knowledge and innovative activity do not lend themselves to obvious quantification'). See also de la Mano M (2002) For the customer's sake: the competitive effects of efficiencies on the European merger control. European Commission's Enterprise Directorate-General, Enterprise Papers no. 11, p. 52 (stating that dynamic efficiency is 'the least quantifiable form of efficiency but is almost always the most economically significant component of global efficiency gains'); Kerber (2008), p. 98; Kerber and Schwalbe (2008), para 1-8-097 (defining that the economic process is 'dynamically efficient if [the development and introduction of new goods and production technologies] take place over time at the rate which is socially optimal'); Eagles and Longdin (2011), p. 54 (observing that '[i]nnovative efficiency is loved by all but measured by few.'). On measuring innovative activity, see Audretsch et al. (2004), pp. 21–22.

7.1.4 The Notion of R&D Externalities as the Common Denominator

The argument of drug companies that data disclosure can impede innovation incentives reflects the concern that competitors can 'free-ride' on the investment by analysing IPD in their research projects that may eventually facilitate competing products.[40] This argument can evoke the conventional justification of exclusive rights in knowledge as a means to internalise R&D benefits. The counter-argument that broad access could promote medical research and innovation emphasises the role of secondary analysis of non-summary data from past trials in making subsequent drug R&D more informed, targeted, and efficient. Such effect is associated with the cumulative nature of innovation often captured by the 'standing-on-shoulders-of-giants' metaphor.[41]

In economic terms, the notions of 'free-riding' and 'standing-on-shoulders effect' correspond to the concept of *R&D externalities*, external benefits extending from one firm's R&D to third parties that cannot be excluded or (fully) internalised by the firm generating such benefits.[42] Such effects are attributed to the imperfect excludability of R&D results, such as technological knowledge. Due to the incomplete appropriability of returns to R&D investment, R&D externalities pose dual implications for innovation. The positive aspect is associated with greater knowledge diffusion, while the negative aspect implies the risks of diminished innovation incentives. The 'access-incentives dilemma' in the case of IPD illustrates such duality. Public disclosure of non-summary data can be a catalyst of external benefits. The argument that disclosure impedes innovation incentives reflects the view of R&D externalities as 'a genuine disincentive for firms to undertake R&D'.[43] The proposition that access to IPD can facilitate new drug development highlights the positive aspect of R&D externalities, their potential to optimise innovation efforts at the sector level.[44]

Against this background, the normative question of whether a policy intervention by access measures can be justified on the grounds of promoting drug innovation can be restated as follows. Can de facto exclusive control over non-summary data be justified as a means of internalising R&D benefits and, if so, to what extent? Alternatively, can access measures be justified as promoting the positive effect of R&D externalities? Before these questions can be addressed, the next section outlines the understanding of R&D externalities in the economics of innovation that can further inform the analysis of the case of IPD.

[40] See e.g. PhRMA/EFPIA, 'Principles' 4 (pointing out 'the risks to innovation [by] disclosure to competitors of companies' trade secrets and proprietary information that could allow others to "free ride" off of the substantial investments of innovators') (emphasis added).

[41] Below (nn 75–80) and the accompanying text.

[42] Below at Sect. 7.2.1 in this chapter.

[43] Davidson and Spong (2010), p. 366. See also below at Sect. 7.2.3.2 in this chapter.

[44] Below at Sect. 7.2.3.1 in this chapter.

7.2 R&D Externalities in Innovation Law: A Theoretical Framework

7.2.1 The Concept of R&D Externalities

Externalities are defined as external effects—costs and benefits borne by third parties—that do not bear on the decision making of generators of such effects or are not internalised via a market transaction.[45] The presence of externalities often justifies public policy interventions (environmental and health policies are the common examples).

R&D externalities[46] (also known as 'knowledge spillovers'[47]) occur when R&D results, such as research findings and technological knowledge, created by one firm are disseminated outside the price system.[48] An important characteristic of R&D externalities is that they change the technological relationship between the recipient firm's output and the inputs at its disposal.[49] Such effects are known as 'technological' externalities and are contrasted with 'pecuniary' externalities.[50] Technological externalities arise where information produced by one firm improves the R&D productivity of other firms, for instance, if it can 'trigger new avenues of research, inspire new research projects or find new applications in other firms, sectors or countries,[51] and where the knowledge producer is not compensated. Such 'unintended transfer' is attributed to the imperfect excludability of knowledge, also viewed as the root of the innovation incentives problem.[52]

---

[45] Laffont (2018), p. 4318; Bernstein and Nadiri (1989), p. 249; Schall (1971), p. 983; Hall et al. (2010), p. 1065; Nelson (2009), p. 10.

[46] For an account of R&D externalities in economic literature, see Davidson and Spong (2010); Antonelli (2017), pp. 3–21.

[47] While the term 'knowledge spillover' is widely used in the literature, benefits from others' R&D do not simply 'spill over' unless the knowledge recipient has developed adequate absorptive capacity. See Cohen (2010), p. 186 (pointing out that 'R&D spillovers are not as much of a public good', and that 'the cost of utilising public domain knowledge fruitfully is minimal only for firms which have accumulated sufficient technological capability to absorb external knowledge'). See generally Cohen and Levinthal (1989).

[48] Above (n 45).

[49] Bohm (2018), p. 4314; Laffont (2018), p. 4318.

[50] See Laffont (2018), p. 4318; Griliches (1998), p. 258; Bohanon (1985), p. 306 (pointing out that 'the salient distinction' between 'genuine' and pecuniary externalities is that the latter 'never enter third parties' utility (or production) functions, whereas by definition, technical externalities always enter the utility (or production) functions of third parties').

[51] Hall et al. (2010), p. 1065 (emphasis added).

[52] ibid p. 1065 (observing that 'the more knowledge is codified and the higher is the absorptive capacity of other firms, the more knowledge spillover will take place').

R&D externalities represent one way of knowledge diffusion.[53] They are ubiquitous[54] and can affect the productivity of firms within a sector[55] and in other industries.[56] Imperfect excludability of knowledge, its cumulativeness and non-rivalry in use are the factors that cause the divergence between private and social returns on R&D.[57] External effects can hardly be adequately assessed[58] due to the time lags and the difficulty in identifying the channels through which they are transmitted.[59]

7.2.2 *Imitation Externalities v Research Externalities*

There are different ways to characterise and classify R&D externalities.[60] In the context of this study, it is useful to distinguish between two ways in which externally generated knowledge can be absorbed: through imitation and R&D. While the terminology is not consistent,[61] the present analysis uses the terms 'imitation externalities' and 'research externalities' corresponding to these two ways of knowledge diffusion.

---

[53] Knowledge externalities should be distinguished from technology transfer. While both are the mechanisms of knowledge diffusion, technology transfer implies that externally produced knowledge is acquired under contractual terms, thus, allowing the producer of knowledge to (partially) internalise its value.

[54] Nelson (2009), p. 10 (noting that 'virtually all research and development [. . .] yields externalities, in the sense that some parties not involved in R&D decision-making will be able to learn something useful from its results').

[55] Davidson and Spong (2010), p. 364.

[56] Griliches (1998), p. 258.

[57] Private returns to R&D refer to profits earned by the firm undertaking research and generating knowledge, while social returns comprise private returns as well as benefits received by customers and other firms. The rate of social returns is considerably higher than the rate of private returns. See e.g. Griliches (1998), p. 264; Hall et al. (2010), pp. 1034, 1065; Jaffe (1998), p. 12.

[58] Griliches (1998), pp. 251–252.

[59] ibid p. 262.

[60] Antonelli (2017), pp. 83–84.

[61] For instance, according to Hall, Mairesse, Mohnen, 'rent spillovers' occurs 'when a firm or consumer purchases R&D *incorporated* goods or services at prices that do not reflect their user value, because of [inter alia] imperfect appropriability and imitation', while 'knowledge spillovers' happen 'when an R&D project produces knowledge that can be useful to *another* firm in doing its *own research*'. Hall et al. (2010), p. 1065 (emphasis added). Jaffe uses the terms 'market spillovers' and 'knowledge spillovers' to refer to the same phenomena. Jaffe (1998), pp. 11–12. Antonelli defines imitation externalities as 'the opportunity for imitators to replicate the innovation introduced by the 'inventor'' and knowledge externalities as 'the opportunity to use the knowledge embodied in an innovation to generate new knowledge'. Antonelli (2017), pp. 16, 18. See also Martin (2002), pp. 1–2 (referring to the effect the costs of research for the firms-recipients of external knowledge is reduced as 'input spillovers' and the effect when diminished appropriability of R&D efforts of the firms generating knowledge as 'output spillovers').

Imitation externalities occur when firms can derive economic benefits by copying the products that embody knowledge (technology or know-how).[62] Research externalities refer to the situations where research findings produced by one firms are used as information inputs by other firms and can 'inspire new research projects',[63] facilitate the creation of new knowledge[64] or positively impact their R&D productivity.[65] Knowledge spillovers are viewed as the drivers of economic growth by enabling knowledge diffusion and the creation of new knowledge.[66] The pharmaceutical sector can serve as an example.[67]

From a policy and law-making perspective, the distinction between imitation and research externalities matters because each type involves different implications for balancing private and social returns on R&D.

7.2.3 Multiple Implications of Knowledge Externalities for Innovation

Even though literature often refers to externalities as 'negative' or 'positive', the distinction is arbitrary and depends on the baseline against which they are assessed.[68] External effects are often symmetric and reciprocal.[69] What constitutes a benefit for the externality recipient can correspond to the cost borne by the externality generator and vice versa. Knowledge externalities are not an exception in this regard. For instance, unrestrained imitation—due to the absence of legal, technological, or other means of exclusion—of innovative products embodying technological knowledge, benefits derived by competitors would correspond to a loss in profits incurred by the innovating firm (the so-called 'rent-stealing effect' of knowledge 'spillovers'[70]).[71]

To understand how R&D externalities can interact and cause policy trade-offs, let us take a closer look at their implications for efficiencies.

---

[62] Griliches (1998), pp. 251–252; Jaffe (1998), p. 11.

[63] Hall et al. (2010), p. 1065 (emphasis added).

[64] Antonelli (2017), p. 21.

[65] Jaffe (1998), pp. 11–12.

[66] Hall et al. (2010), p. 1065. See also Audretsch et al. (2004), p. 23 (noting that knowledge re-use accounts for total factor productivity); Lucas (2002), p. 6. For an overview of growth models that incorporate knowledge externalities, see Braunerhjelm (2011), pp. 180–182; Antonelli (2017), p. 4.

[67] Henderson and Cockburn (1996), p. 56.

[68] Duffy (2005), p. 1086.

[69] ibid.

[70] Jaffe et al. (2005), p. 167.

[71] In other words, the knowledge-producing activity would generate external benefits for other firms within the same or different industries, while the knowledge consumption by an imitating firm can diminish returns on R&D of the knowledge-producing firm. Jaffe (1986).

7.2.3.1 The Efficiency-Enhancing Effect of R&D Externalities

The beneficial aspect of R&D externalities can be defined as an efficiency-enhancing or a cost-reducing effect.[72] The use of externally generated 'spilling over' knowledge can improve the productivity of other firms,[73] including by reducing R&D costs.[74] Benefits derived by third parties are external if the knowledge producer cannot internalise them.[75]

Economic literature especially emphasises the importance of intertemporal externalities in the context of cumulative innovation[76] and the role of research tools in enabling such an effect.[77] The metaphor 'standing on the shoulders of giants'[78] captures the idea that each generation of innovators benefits from the predecessors' achievements, both successful and unsuccessful.[79] In other words, the efficiency-enhancing effect of knowledge externalities is associated with making subsequent research and innovative activity more informed, targeted, and efficient. Such effect reflects the idea that openness of knowledge allows validating the accuracy of research findings, increases chances of making further discoveries and reduces duplicative research.[80] Furthermore, R&D externalities can cause a complementary

---

[72] Spence (1984), p. 116; Hall et al. (2010), p. 1065 (noting that knowledge 'spillovers' can reduce the production costs of rival firms).

[73] Hall et al. (2010), p. 1065.

[74] ibid. See also Antonelli (2017), p. 4; Antonelli C, Colombelli A (2017) The locus of knowledge externalities and the cost of knowledge. LEI&BRICK Working Paper 11/2017, pp. 1–2.

[75] See e.g. Jones and Williams (2000), p. 70 (referring to the 'standing on shoulders' effect as intertemporal knowledge spillovers that occur when firms that create knowledge today cannot appropriate its value for future research); Jaffe (1998), pp. 11–12 (defining 'knowledge spillovers' as 'knowledge created by one agent [that] can be used by another without compensation, or with compensation less than the value of the knowledge').

[76] Sena (2004), pp. 324–325; Greenstein (2010), pp. 500–501.

[77] Cockburn IM, Henderson R, Stern S (2018) The impact of artificial intelligence on innovation. NBER Working paper 24449, p. 8 (noting that 'an increasing body of evidence suggests that research tools and the institutions that support their development and diffusion play an important role in generating intertemporal spillovers' (with further references)).

[78] Newton's epigram is often evoked in the literature on cumulative innovation and economics of knowledge. See e.g. David (2003), p. 30; Scotchmer (1991), p. 5; Eisenberg (1989), pp. 1055–1056; Merton (1974), p. 275 (observing that Newton's aphorism expresses 'a sense of indebtedness to the common heritage and a recognition of the essentially cooperative and selectively cumulative quality of scientific achievement').

[79] Thus, earlier innovation 'creates the seeds' for the later innovation and 'a positive externality' allows the later innovator to build on the past advances. Rockett (2010), p. 339. On the importance of the dissemination of unsuccessful research results, see e.g. Cohen (2010), p. 192; Jaffe (1998), p. 11 (noting that 'one firm's abandonment of a particular research line signals to others that the line is unproductive and hence saves them the expense of learning this themselves'); Callon (1998), p. 245 (observing that knowledge generated and disclosed by one firm 'may inspire [other firms] to rethink the direction of their own research').

[80] Foray (2004), p. 166.

effect on incentives of other firms to invest in the absorptive capacity to assimilate external knowledge,[81] thereby raising the level of R&D activity at the sector level.[82]

7.2.3.2 The Efficiency-Reducing Effect of R&D Externalities

The benefits of knowledge non-excludability can be offset by two paradoxically opposing effects on innovation incentives: insufficient R&D (underinvestment) and excessive R&D (a 'racing' behaviour). Each is briefly explained in turn.

The Disincentive Effect of R&D Externalities

The disincentive effect of R&D externalities is attributed to the public-good characteristics of knowledge. If competitors can, absent the legal or factual protection, freely imitate a product embodying new technical teaching, the innovator may not recover R&D costs due to the enhanced price competition. This issue is known as a failure of a perfectly competitive market to provide sufficient incentives to the private sector to supply public goods and invest in R&D[83] (often referred to as the 'free-riding' problem in innovation[84]).

While the disincentive effect of knowledge externalities is mainly associated with competition by imitation, it can occur in competition by improvement if, due to knowledge spillovers, competitors might improve their products[85] and reduce the pioneer company's market share. In this view, knowledge non-excludability serves as a mainstream justification of exclusive rights such as patents as a means of internalising R&D benefits.[86]

A Trade-off Between Knowledge Diffusion and Innovation Incentives

R&D externalities can cause a trade-off between the social benefits associated with increased knowledge flows and those arising due to the protection of innovation incentives.[87] Literature refers to this dual effect as a 'twin effect' of incomplete

81 Cohen and Levinthal (1989), pp. 575–576.

82 Cohen (2010), p. 186. See also Spence (1984), p. 103; Cohen and Levinthal (1989), p. 576.

83 Arrow (1962), p. 146; den Hertog J (2010) Review of economic theories of regulation. Utrecht School of Economics Discussion Paper Series No 10-18, p. 16; Bergstrom et al. (1986), p. 25.

84 Gruber (2016), p. 188; den Hertog J (2010) Review of economic theories of regulation. Utrecht School of Economics Discussion Paper Series No 10-18, p. 16.

85 Jaffe (1998), p. 14.

86 Below at Sect. 7.2.3.3, subheading 'The (Controversial) Role of Patents as a Means to Coordinate Research Efforts'.

87 Cohen (2010), p. 186.

appropriability,[88] the 'knowledge appropriability trade-off',[89] and 'a difficult trade-off between incentives for innovation and the need to encourage diffusion'.[90] Such trade-off is likely to arise, especially where knowledge is created in industrial R&D. At the firm level, the high degree of cumulativeness correlates with the high level of appropriability of returns on R&D,[91] whereby cumulativeness is understood as 'the continuity of innovative activities [that] strongly depends on the competencies of specific firms'.[92] At the sector level, the high degree of cumulativeness correlates with the low level of appropriability due to knowledge spillovers,[93] whereby cumulativeness refers to the diffusion of knowledge across firms.[94] In other words, there is an inherent tension between the appropriability of R&D benefits at the firm level and the cumulativeness of research and innovative activity at the industry level. Accordingly, a policy dilemma arises as to how to reconcile the objectives to provide sufficient incentives to private firms to undertake R&D and create knowledge, on the one hand, while ensuring dissemination and efficient use of knowledge once it is created, on the other hand.[95]

Patent Rights as a Means to Prevent the Disincentive Effect of Knowledge 'Spillovers'

The disincentive effect of knowledge 'spillovers' has been viewed as a standard economic justification for creating exclusive rights in knowledge. The proposition is that knowledge 'spillovers' transmitted through imitation can enhance price competition and increase gains in static efficiency, which can be, in a long-term perspective, offset by a loss in dynamic efficiency due to the diminished incentives to innovate. Thus, by internalising knowledge 'spillovers', patent rights create a symmetric *static-dynamic efficiency trade-off*, whereby the prospective gains in dynamic efficiency due to the protection of innovation incentives offset a loss in static inefficiency due to the restricted price competition.[96]

Patent law, however, does not intend to internalise R&D externalities fully—nor would the full internalisation of social benefits generated by knowledge externalities

[88] Antonelli (2017), p. 4.

[89] ibid p. 98 ff.

[90] Denicolo and Franzoni (2011), p. 112.

[91] ibid.

[92] Breschi and Malerba (2005), pp. 135–136.

[93] ibid.

[94] ibid.

[95] Foray (2004), pp. 116–119. See also Stiglitz and Wallsten (1999), p. 56.

[96] Among the fields of IP, the static-dynamic efficiency trade-off is characteristic of patents and copyright. See Parchomovsky and Siegelman (2002), p. 1458.

be possible or socially desirable.[97] The pursuits to derive an 'optimal measure' of patent protection have produced a rich body of economic research.[98] Conceptually, optimality can be viewed as a balance between the rules on appropriability and the rules on knowledge diffusion.[99] On the one hand, patents are granted to mitigate the disincentive effect of knowledge 'spillovers' by internalising external benefits.[100] On the other hand, provisions such as the requirement for enabling disclosure ('best mode')[101] and, to some extent, the experimental use exception can be viewed as acknowledging the positive aspect of knowledge externalities.[102]

At the core, patent protection restricts externalities that emanate through the imitation of products that embody innovative knowledge.[103] More uncertain are cases where patents allow to internalise R&D externalities where a patented invention can be used as research input. In this regard, the justification of patent protection as a means to resolve the static-dynamic efficiency trade-off falls short as it takes a 'static or noncumulative perspective [...] primarily concerned with providing the

---

[97] European Commission (28 Nov 2013) Impact assessment accompanying the document proposal for a Directive of the European Parliament and of the Council on the protection of undisclosed know-how and business information (trade secrets) against their unlawful acquisition, use and disclosure. SWD(2013) 471 final, p. 139 (noting that '[s]pillovers and diffusion of knowledge are considered important determinants of dynamic economic efficiency as innovations spread through industries and economies over time'). See also Lemley (2005), p. 1032; Foray (2004), p. 114 (noting that 'social returns may be so substantial that remunerating the inventor accordingly is unthinkable').

[98] For a literature review on the optimal design of patent protection, see e.g. Rockett (2010), pp. 333–361; Hall (2007), pp. 576–577.

[99] Drahos and Braithwaite (2002), p. 13 (referring to this task as a 'difficult trick for any legislature').

[100] Patent rights can create a market for technological knowledge and allow innovators to internalise R&D benefits through *ex ante* cooperative research arrangements or *ex post* licensing agreements. Menell (2000), p. 139.

[101] On the incentive-to-disclose theory of patent law, see e.g. Mazzoleni and Nelson (1998a), p. 1038 (arguing that 'patents encourage disclosure and, more generally, provide a vehicle for a quick and wide diffusion of the technical information underlying new inventions'). See also Landes and Posner (2003), p. 304 ('Invention is a matter of adding to the stock of useful knowledge and so of reducing uncertainty.'). The qualifier 'earlier' should be added to the premise of promoting disclosure. See The disclosure function of the patent system (or lack thereof), p. 2016 (noting that '[m]ost patented inventions can be uncovered through reverse engineering, and the patent system is therefore of limited value in promoting R&D spillovers and cumulative innovations'). However, even though 'the patent system as a whole does not reduce the overall level of wasteful research, the disclosure function is still socially desirable to the extent that it reduces duplicative research *after* a patent has been published'. ibid 2010 (emphasis added). Furthermore, it is worth noting that economic research finds 'little empirical evidence as to the extent of disclosure and its economic impact'. See Hall and Harhoff (2012), p. 549.

[102] Both early disclosure and experimental use exception can be conceptually viewed as promoting the 'standing-on-shoulders' effect associated with knowledge externalities. As for the experimental use exception, its scope tend to be rather narrow. For a comparative overview, see e.g. WIPO (2009) Exclusions from patentable subject matter and exceptions and limitations to the rights. SCP/13/3.

[103] Landes and Posner (2003), p. 294; Shavell (2004), p. 138.

best incentive mechanism to develop a primary invention that has no follow-ons'.[104] In other words, it does not factor in the potentially offsetting effects of exclusive control over the use of knowledge on dynamic efficiency in the context of cumulative R&D. The debate regarding the 'anticommons' effect of patents for research tools illustrates ambivalent implications of patents for innovation.[105]

Overall, economic analysis of patent law is inconclusive as to how exclusive control over the use of technological knowledge—as a means of internalising knowledge externalities and improving appropriability conditions—ultimately impacts innovation,[106] especially where innovation is sequential and cumulative.[107]

7.2.3.3 The Issue of Excessive Incentives Due to Knowledge Spillovers

The 'Exhaustion' Externality or the 'Stepping-on-Toes' Effect

Paradoxically, the non-excludability of technological knowledge can reduce the risk of duplicative research and, at the same time, induce duplicative efforts and 'racing' behaviour.[108] The combination of factors makes an environment conducive to duplicative research, such as the existence of a commonly accessible resource, the prospect of receiving a prize (e.g. obtaining patents in research outcomes) and the lack of information exchange and cooperation among the entities making independent decisions to invest in R&D.[109] A shared research resource can refer to the common pool of knowledge, which usually stems from publicly funded research. While such knowledge can point to a promising technological opportunity, a lack of

---

[104] American Bar Association (2015), p. 102. See also Scotchmer (1991), p. 30 (noting that '[m]ost economics literature on patenting and patent races has looked at innovations in isolation, without focusing on the externalities or spillovers that early innovators confer on later innovators', and that 'the cumulative nature of research poses problems for the optimal design of patent law that are not addressed by that perspective'); Mazzoleni and Nelson (1998b), p. 280 (observing that 'whenever an invention is understood as contributing to further invention potential as well as creating a new or improved product or process of immediately final use, a question can be raised as to whether strong patents enhance or hinder technical advances in the long run'); Sena (2004), p. 324.

[105] For an overview of this debate, see Chap. 8 at Sect. 8.2.2.2.

[106] Cahoy (2006), p. 589 (summarising that 'the degree to which current patent systems promote innovative behavior remains surprisingly unclear'); Posner (2005), p. 59 (noting that, '[u] nfortunately, economists do not know whether the existing system of intellectual property rights is, or for that matter whether any other system of intellectual property rights would be, a source of net social utility, given the costs of the system and the existence of alternative sources of incentives to create such property').

[107] Bessen and Maskin (2009), p. 611 (arguing that where 'innovation is "sequential" (so that each successive invention builds in an essential way on its predecessors) and "complementary" (so that each potential innovator takes a different research line), patent protection is not as useful for encouraging innovation as in a static setting').

[108] Foray (2004), pp. 166, 169.

[109] ibid.

coordination can prompt actual or potential competitors to pursue research projects in parallel. Competitive zest can generate redundancy in research[110] resulting in decreased R&D productivity on average.[111] Literature on the economics of innovation refers to such situations as a 'depletion externality',[112] 'rent dissipation',[113] the 'stepping-on-toes'[114] or 'fishing out' effect.[115]

In the context of patents, this phenomenon is known as 'patent races' where the prospect of obtaining the 'patent premium' can stimulate 'excessive' R&D. Social costs of innovation races can be defined as the diminishing quality of innovation if firms are motivated 'to structure their R&D programs for speed, rather than quality or cost minimization',[116] which leads to 'too many little fish [being] captured'.[117] Besides, duplicative research involves the opportunity cost in terms of unexplored research paths and unrealised research opportunities.

Duplicative Research v Multiplicity and Diversity of Experimentation

Literature on 'patent races' features divergent views: on the one hand, parallel R&D can generate duplicative costs and inefficiency; on the other hand, it can encourage investment by maintaining competitive pressure[118] and contribute to technological progress by stimulating alternative solutions.[119] At the outset, it might not be clear whether parallel research projects are wastefully duplicative or value-enhancing. Hence, one needs to understand the factors contributing to the likelihood of similar research projects producing redundant outcomes.

In the settings where knowledge is cumulative, the diversity and competitiveness of innovating agents are regarded to be 'socially preferable' compared to the centralised structure of inventive activity.[120] The same can be said about the fields characterised by high uncertainty.[121] Research on evolutionary economics

---

[110] ibid.

[111] Jones and Williams (2000), pp. 66, 69.

[112] Cockburn and Henderson (1994), p. 508.

[113] ibid

[114] Jones and Williams (1998), p. 1125.

[115] ibid.

[116] Foray (2004), p. 169.

[117] ibid.

[118] Menell (2000), p. 138 (with further references). For an overview of literature on this subject, see ibid 146–148.

[119] ibid 147. This argument is particularly relevant in the case of drug R&D. For a discussion, see Chap. 8 at Sects. 8.2.3.1 and 8.2.3.2.

[120] ibid.

[121] On uncertainty in drug R&D, see Chap. 8 at Sect. 8.2.3.3.

emphasises the importance of parallel experimentation,[122] favouring the view that 'diversity of individual capabilities, learning efforts, and innovative activities [lead to] growing, distributed knowledge in the economy'.[123] Overall, the benefits of parallel research are associated with the potential to generate multiple valuable solutions,[124] strengthen the adaptation to exogenous and endogenous shocks, broaden the 'search space' through multiple feedbacks on the hypotheses tested in parallel and enhance knowledge diffusion.[125] Therefore, parallel R&D is viewed as 'a much better social bet than a regime where only one or a few organizations control the development of any given technology'.[126]

The (Controversial) Role of Patents as a Means to Coordinate Research Efforts

Patent rights have been rationalised as an instrument of coordinating R&D efforts due to the disclosure requirement. In theory, the disclosure through patent applications can 'signal' the investment decisions and discourage entities from undertaking overlapping projects, thereby coordinating research efforts and minimising wasteful spending.[127] In practice, information about an ongoing research project can dissuade firms from working on the same problem only if they perceive themselves 'lagging in the race'.[128] However, companies can likely continue pursuing parallel research under uncertainty as long as they can expect commercial success.[129] Such outcome can benefit society in situations where the chance of developing a valuable invention is slim, such as in drug R&D.

The theory of patents coordinating R&D efforts at the sector level concerns the so-called 'prospect opening' inventions that can open a range of follow-on technological opportunities. The proposition is that patents for such inventions can efficiently allocate resources to their development and subsequent exploitation through the usage rights. The 'prospect theory' of patents is attributed to Edmund Kitch,[130] who viewed the patent system's goal in allocating resources efficiently by granting

[122] For an overview of the literature on evolutionary innovation economics with a focus on the interrelation between diversity, competition, and technological progress, see Linge (2008), pp. 81–114.

[123] Witt (2018), p. 4090.

[124] Menell (2000), p. 147; Kerber (2010), p. 184.

[125] Linge (2008), pp. 200–202.

[126] Merges and Nelson (1990), p. 908.

[127] For an overview of the literature on the coordination role of the patent disclosure, see Rockett (2010), pp. 350–354.

[128] Rockett (2010), p. 353.

[129] Andrade et al. (2016), p. 48 (noting, in the context of drug R&D, that '[p]rices and volumes are amongst the main factors that will bring benefits and returns for firms to recover the amount of R&D expended on drug development').

[130] Kitch (1977). See also Grady and Alexander (1992), p. 310 (arguing that the grant of patent rights early in the development process can reduce the likelihood of rent-dissipating patent races,

exclusive rights in developing a technological 'prospect'.[131] The assumption is that broad patent rights can attract investment for developing a resource,[132] facilitate the exchange of information among innovators[133] and avoid duplicative R&D.[134]

Regarding the latter premise, one should distinguish two functions of patent rights: the one enabling the patent holder to charge for using patented technology, and the other allowing to coordinate the development of a 'prospect' efficiently. Licensing the usage rights in developing a 'prospect' does not necessarily imply that such development would proceed most efficiently.[135] Theoretically, Kitch's argument can be true under the conditions that the prospect holders have the capacity and motivation either to 'extract the value' from the 'prospect' themselves or can efficiently allocate the usage rights in developing the 'prospect'.[136] However, in reality, the prospect's value can remain unrealised[137] due to imperfect information or a lack of willingness to cooperate.[138]

The centralised management of the exploration of a technological field broadly claimed by a patent can be especially problematic in two situations. First, where multiple components are required to develop and commercialise a technology,[139] bringing an innovative product or service can require a complex licensing scheme and substantial transaction costs.[140] Second, where the initial invention is 'a far distance from practical application, and its principal value is in providing *clues* as to how to proceed',[141] multiple independent entities will likely produce a 'wider and diverse set of explorations than when the development is under the control of one

---

especially 'a nascent invention that "signals" many different, possibly patentable, improvements should be given a broad scope so as to avoid the possibility of races to patent these improvements').

[131] Kitch (1977), p. 266. A 'prospect' is understood as a 'nascent' invention presenting a promising technological opportunity, which development requires further investment.

[132] ibid p. 276 (noting that 'the patent owner has an incentive to make investments to maximize the value of the patent without fear that the fruits of the investment will produce unpatentable information appropriable by competitors').

[133] ibid p. 266.

[134] ibid p. 278.

[135] Merges and Nelson (1990), p. 907 (warning that '[h]olders of broad patents would be operating as tollkeepers, not coordinators' and, even though 'the ability to charge a toll may add to the incentives facing an inventor, it does not ensure more efficient development').

[136] Menell (2000), pp. 146–147.

[137] ibid p. 871 (noting that 'proprietary control of technology tend[s] to cause "dead weight" costs').

[138] Merges and Nelson (1990), p. 872 (referring to 'many instances when a firm that thought it had control over a broad technology rested on its laurels until jogged to action by an outside threat').

[139] Mazzoleni and Nelson (1998b), p. 280 (referring to such technologies as 'cumulative system technologies').

[140] ibid p. 280 (referring to research on genes and gene expression as an example that 'is also running into systems problems, particularly insofar as patents are being granted piecemeal on various parts of the puzzle').

[141] ibid (with further references).

mind or organization'.[142] This argument goes along the lines of the evolutionary view of innovation as a process of parallel research and experimentation outlined earlier.[143] Besides, evidence suggests that patents might not be an effective instrument of 'dissuading' competitors from undertaking overlapping R&D projects.[144]

7.3 The Summary of Theoretical Propositions and Implications for Further Analysis

7.3.1 A Systematic View on R&D Externalities

This chapter intended to outline the views on R&D externalities that emerged from research on law and economics of innovation and served as a justification for exclusive control over knowledge resources. The Table 7.1 summarises the key propositions outlining a theoretical framework for the subsequent analysis of the 'access-incentives dilemma'.

7.3.2 General Caveats

Several reservations regarding the relationship between R&D externalities and innovation need to be taken into account. First, economic analysis is uncertain as to the net effect of R&D externalities on innovation. Upon reviewing empirical economic research on innovative activity over the past 50 years, Cohen concludes that the overall effect of R&D externalities on the innovative performance at the sector level remains ambiguous. While 'industry R&D intensity can rise with appropriability (fall with spillovers), industries' innovative output may decline (increase with spillovers)'.[145] According to some commentators, 'externalities do not necessarily undermine private incentives to invest in R&D',[146] and 'positive externalities associated with knowledge spillovers dominate the rent-stealing

[142] Merges and Nelson (1990), p. 873 (emphasis added). This argument is particularly relevant for the discoveries of generic nature that can be a catalyst for developing multiple technological applications embodied in diverse innovative products.

[143] Above (n 122–125) and the accompanying text.

[144] Drug R&D is a pertinent example. See below (Chap. 8, nn 87–88) and the accompanying text.

[145] Cohen (2010), pp. 185–186. Further, Cohen notes that 'a key question for understanding R&D incentives and innovation at the industry level is, in addition to considering the efficiency effect of spillovers [...], *what factors* condition the tradeoff between spillovers' negative appropriability incentive effect and their positive complementarity effects'. ibid p. 186 (emphasis added).

[146] Economic incentives to business R&D, p. 5. https://europa.eu/epc/sites/epc/files/docs/pages/annexd_en.pdf. Accessed 26 Mar 2021.

Table 7.1 Synthesis of theoretical assumptions about R&D externalities and their implications for the allocation of resources to R&D

The problem	Insufficient innovation incentives (under-investment in R&D) due to imperfect appropriability of R&D benefits	Excessive innovation incentives (over-investment in R&D); innovation 'races' resulting in wastefully duplicative research
Contributing factors	• The public-good nature of R&D results (non-excludability of knowledge) • Uncertainty regarding R&D outcomes and commercial success	• A 'common pool' knowledge resource • A viable technological opportunity identified by multiple actors • The prospect of a reward (e.g. supra-competitive profits including due to patent protection)
The types of externalities involved and implications for competition dynamics	• *Imitation* externalities: effects on price competition due to unrestrained competition by imitation • *Research* externalities: effects on competition in innovation if knowledge spillovers can benefit competitors' R&D	• The '*fishing out*' externality: once one firm realises a technological opportunity, the efforts of other innovators involved in a 'race' can be rendered wasteful
The assumed role of exclusive control over the use of knowledge	• A means of internalising R&D benefits • Protection of intermediate research results	• A means of coordinating research efforts
Opportunity costs of exclusive control over knowledge	The underutilisation of knowledge resources/missed research opportunities	Inhibited multiplicity and diversity of experimentation and research outcomes
Efficiency trade-offs	The *diffusion-appropriability* trade-off: on the one hand, knowledge spillovers allow for greater knowledge diffusion; on the other hand, they can diminish returns on R&D, hence, innovation incentive.	The *efficiency-multiplicity* trade-off: on the one hand, parallel research can result in higher quality and diversity of outcomes; on the other hand, parallel R&D might yield wastefully duplicative outcomes.

effect'.[147] The finding worth highlighting is that economists acknowledge that they 'have little empirical understanding [...] of the tradeoff between the negative

[147] Jaffe et al. (2005), p. 167 (with further references). See also Cohen and Levinthal (1989), pp. 592–593 (concluding that 'the positive absorption incentive associated with spillovers may be sufficiently strong in some cases to more than offset the negative appropriability incentive'); Jeffrey I. Bernstein, M. Ishaq Nadiri, 'Research and Development and Intra-industry Spillovers: An Empirical Application of Dynamic Duality' p. 249 (noting that 'the trade-off between the cost-reducing (or productivity) effect and the incentive effect of R&D investment may be exaggerated' (with further references)).

appropriability incentive effect and the positive complementarity effects of R&D spillovers'.[148]

Second, the question of whether parallel research can be regarded as wastefully duplicative or beneficially complementary can hardly be answered in general. As summarised by Kerber,

> multiplicity and diversity in competition have always a quantitative as well as a qualitative dimension, which both influence the effectiveness of competition as a process of parallel search and experimentation. Since the additional effects of more experimenting firms are presumably decreasing, it can be suggested that there might be an *optimal number* of parallel experimenting firms (parallel research projects) as well as an *optimal extent* of diversity. The main problem is that both the theoretical and empirical research about these questions is still in its infancy.[149]

If this insight—or rather an open-ended question—is the state-of-art economic understanding of R&D externalities, how should the analysis of the 'access-incentives dilemma' regarding clinical trial data proceed?

7.3.3 The 'Access-Incentives Dilemma' Revisited

What can be elicited from the preceding review is that, while R&D externalities are generally desirable given their role in fostering cumulative research,[150] their negative effect on innovation incentives ought to be mitigated.[151] Neither the improvement of appropriability conditions nor the internalisation of R&D externalities should be viewed as a goal in itself[152] but rather as a means to achieve 'a superior objective [of providing] proper organisational and institutional conditions to maximise innovative performances'.[153] Identifying such conditions calls for a contextual

[148] Cohen (2010), p. 194. See also Linge (2008), p. 63 (summarising that 'the more firms invent successfully and share their R&D results, the more learning opportunities are provided and thus from a total welfare perspective the negative effect of imitation on innovation incentives is outweighed by rapid diffusion' (with further references)).

[149] Kerber (2010), p. 185 (emphasis added) (with further references).

[150] Davidson and Spong (2010), p. 370 (recommending that industrial policy 'should abandon the Pigovian view that spillover benefits from R&D discourage innovation' and that it would benefit by following the approach of economists Alfred Marshall and Adam Smith and the view of knowledge externalities as 'the free exchange of ideas').

[151] Cohen (2010), p. 192 (recommending that 'when considering the impacts of such knowledge flows, one needs to be attentive to the associated tradeoffs for R&D and innovation between the appropriability incentive effects of such flows, on the one hand, and their complementarity and efficiency effects, on the other hand').

[152] Foray (2009), p. 34 (arguing that it is 'important to decouple the objective of limiting spillovers and the objective of securing rents from the innovation; the latter being the fundamental appropriability objective while the former is likely to serve this fundamental objective well in certain situations but not so well in others').

[153] ibid pp. 35–36 (emphasis added).

analysis of how a compromise between the protection of innovation incentives and knowledge diffusion can be achieved in a particular setting.

In the case of clinical trial data, it is prima facie clear that data aggregation, broad access and robust multiple data analyses can allow for the fuller realisation of the research potential of data. What appears less clear is whether the policy trade-offs might arise if data generated in industry-sponsored trials is treated as a non-excludable resource. The next chapter addresses this question by analysing how different types of knowledge externalities manifest and interact in the context of clinical trials.

References

American Bar Association (ed) (2015) Intellectual property and antitrust handbook, 2nd edn. American Bar Association, Chicago

Andrade LF, Sermet C, Pichetti S (2016) Entry time effects and follow-on drug competition. Eur J Health Econ 17(1):45–60. https://doi.org/10.1007/s10198-014-0654-9

Antonelli C (2017) Endogenous innovation. The economics of an emergent system property. Edward Elgar Publishing, Cheltenham

Arrow KJ (1962) Economic welfare and the allocation of resources for invention. In: Nelson RR (ed) The rate and direction of inventive activity: economic and social factors. Princeton University Press, Princeton, pp 609–626

Audretsch DB, Lehmann EE, Hinger J (2004) From knowledge to innovation. The role of knowledge spillover entrepreneurship. In: Antonelli C, Link AN (eds) Routledge handbook of the economics of knowledge. Routledge, London, pp 20–28

Bergstrom T, Blume L, Varian H (1986) On the private provision of public goods. J Public Econ 29 (1):25–49. https://doi.org/10.1016/0047-2727(86)90024-1

Bernstein JI, Nadiri MI (1989) Research and development and intra-industry spillovers: an empirical application of dynamic duality. Rev Econ Stud 56(2):249–267. https://doi.org/10.2307/2297460

Bessen J, Maskin E (2009) Sequential innovation, patents, and imitation. RAND J Econ 40 (4):611–635. https://doi.org/10.1111/j.1756-2171.2009.00081.x

Blumenthal D (2010) Characteristics of a public good and how they are applied to healthcare data. In: Grossmann C et al (eds) Clinical data as the basic staple of health learning: creating and protecting a public good. National Academy of Sciences, Washington DC, pp 137–142

Bohanon C (1985) Externalities: a note on avoiding confusion. J Econ Educ 16(4):305–307

Bohm P (2018) External economies. In: Macmillan Publishers (ed) The new Palgrave dictionary of economics, 3rd edn. Palgrave Macmillan, London, pp 4313–4317

Braunerhjelm P (2011) Entrepreneurship, innovation and economic growth: interdependencies, irregularities and regularities. In: Audretsch DB, Falck O, Heblich S (eds) Handbook of research on innovation and entrepreneurship. Edward Elgar, Cheltenham, pp 161–213

Breschi S, Malerba F (2005) Sectoral innovation systems: technological regimes, Schumpeterian dynamics, and spatial boundaries. In: Edquist C (ed) Systems of innovation. Technologies, institutions and organizations. Routledge, London, New York, pp 130–156

Cahoy DR (2006) An incrementalist approach to patent reform policy. N.Y.U. J Legis Public Policy 9:587–661

Callon M (1998) An essay on framing and overflowing: economic externalities revisited by sociology. Sociol Rev 46(1):244–269

Cameron G (2001) Scientific data, the electronic era, intellectual property. In: Workshop report on IPR (intellectual property rights) aspects of internet collaborations. European Commission, Luxemburg, pp 31–33
Cockburn I, Henderson R (1994) Racing to invest? The dynamics of competition in ethical drug discovery. J Econ Manag Strategy 3(3):481–519. https://doi.org/10.1111/j.1430-9134.1994.00481.x
Cohen WM (2010) Fifty years of empirical studies of innovative activity and performance. In: Hall BH, Rosenberg N (eds) Handbook of the economics of innovation, vol 1. Elsevier, Amsterdam, pp 129–213
Cohen WM, Levinthal DA (1989) Innovation and learning: two faces of R&D. Econ J 99 (397):569–596
Cornes R, Sandler T (1999) The theory of externalities, public goods, and club goods. CUP, Cambridge
Crown W (2010) Characteristics of the marketplace for medical care data. In: Grossmann C et al (eds) Clinical data as the basic staple of health learning: creating and protecting a public good. National Academy of Sciences, Washington DC, pp 143–150
Dasgupta P, David PA (1987) Information disclosure and the economics of science and technology. In: Feiwel GR (ed) Arrow and the ascent of modern economic theory. Palgrave Macmillian, London, pp 519–542
David P (2003) The economic logic of 'open science' and the balance between private property rights and the public domain in scientific data and information: a primer. In: Esanu JM, Uhlir PF (eds) The role of scientific and technical data and information in the public domain: proceedings of a symposium. The National Academies Press, Washington DC, pp 19–34
Davidson S, Spong H (2010) Positive externalities and R&D: two conflicting traditions in economic theory. Rev Polit Econ 22(3):355–372
Dedeurwaerdere T (2009) The role of law, institutions and governance in facilitating access to the scientific research commons: a philosopher's perspective. In: van Overwalle G (ed) Gene patents and collaborative licensing models. Patent pools, clearinghouses, open source models and liability regimes. CUP, Cambridge, pp 365–380
Denicolo V, Franzoni LA (2011) Weak intellectual property rights, research spillovers, and the incentive to innovate. Am Law Econ Rev 14(1):111–140
Drahos P, Braithwaite J (2002) Information feudalism. Who owns the knowledge economy? Earthscan Publications, London
Duffy DF (2005) Intellectual property isolationism and the average cost thesis. Tex Law Rev 83:1077–1095
Eagles I, Longdin L (2011) Refusals to license intellectual property. Testing the limits of law and economics. Hart, Oxford, Portland
Eisenberg RS (1989) Patents and the progress of science: exclusive rights and experimental use. Univ Chic Law Rev 56(3):1017–1086
Eisenberg RS (1997) Patenting research tools and the law. In: Intellectual property rights and research tools in molecular biology. The National Academies Press, Washington DC, pp 6–16
Foray D (2004) Economics of knowledge. MIT Press, Cambridge
Foray D (2009) Comments on innovation and appropriability: empirical evidence and research agenda. In: The economics of intellectual property. Suggestions for further research in developing countries and countries with economies in transition. WIPO, Geneva, pp 33–37
Frischmann BM (2005) An economic theory of infrastructure and commons management. Minn Law Rev 89:917–1030
Frischmann BM (2012) Infrastructure. The social value of shared resources. OUP, Oxford
Frischmann BM, Lemley MA (2007) Spillovers. Columbia Law Rev 107(1):257–301
Grady MF, Alexander JI (1992) Patent law and rent dissipation. Va Law Rev 78(1):305–350
Greenstein S (2010) Innovative conduct in computing and internet markets. In: Hall BH, Rosenberg N (eds) Handbook of the economics of innovation, vol 1. Elsevier, Amsterdam, pp 477–537

Griliches Z (1998) The search for R&D spillovers. In: R&D and productivity: the economic evidence. The University of Chicago Press, Chicago, pp 251–268
Gruber J (2016) Public finance and public policy, 5th edn. Worth Publishers, New York
Hall BH (2007) Patents and patent policy. Oxf Rev Econ Policy 23(4):568–587
Hall BH, Harhoff D (2012) Recent research on the economics of patents. Annu Rev Econ 4:541–565
Hall BH, Mairesse J, Mohnen P (2010) Measuring the returns to R&D. In: Hall BH, Rosenberg N (eds) Handbook of the economics of innovation, vol 2. Elsevier, Amsterdam, pp 1034–1082
Heal G (1999) New strategies for the provision of global public goods: learning from international environmental challenges. In: Kaul I, Grunberg I, Stern MA (eds) Global public goods: international cooperation in the 21st century. OUP, Oxford, pp 220–239
Henderson R, Cockburn I (1996) Scale, scope, and spillovers: the determinants of research productivity in drug discovery. RAND J Econ 27(1):32–59
Jaffe AB (1986) Technological opportunity and spillovers of R&D: evidence from firms' patents, profits, and market value. Am Econ Rev 76(5):984–1001
Jaffe AB (1998) The importance of 'spillovers' in the policy mission of the advanced technology program. J Technol Transfer 23(2):11–19
Jaffe AB, Newell RG, Stavins RN (2005) A tale of two market failures: technology and environmental policy. Ecol Econ 54(2/3):164–174
Jones CI, Williams JC (1998) Measuring the social return to R&D. Q J Econ 113(4):1119–1135
Jones CI, Williams JC (2000) Too much of a good thing? The economics of investment in R&D. J Econ Growth 5:65–85
Kerber W (2008) Should competition law promote efficiency? Some reflections of an economist on the normative foundations of competition law. In: Drexl J, Idot L, Moneger J (eds) Economic theory and competition law. Edward Elgar, Cheltenham, pp 93–120
Kerber W (2010) Competition, innovation and maintaining diversity through competition law. In: Drexl J, Kerber W, Podszun R (eds) Competition policy and the economic approach: foundations and limitations. Edward Elgar, Cheltenham, pp 173–201
Kerber W, Schwalbe U (2008) Economic foundations of competition law. In: Hirsch G, Montag F, Säcker FJ (eds) Competition law: European Community practice and procedure. Article-by-article commentary of the EC competition law. Sweet & Maxwell, London, pp 202–392
Kitch EW (1977) The nature and function of the patent system. J Law Econ 20(2):265–290
Laffont JJ (2018) Externalities. In: Macmillan Publishers (ed) The new Palgrave dictionary of economics, 3rd edn. Palgrave Macmillan, London, pp 4318–4321
Landes WM, Posner RA (2003) The economic structure of intellectual property law. Harvard University Press, Cambridge
Lauer MS (2010) Data primarily collected for new insights. In: Grossmann C et al (eds) Clinical data as the basic staple of health learning: creating and protecting a public good. National Academy of Sciences, Washington DC, pp 90–99
Lee P (2008) Contracting to preserve open science: lessons for a microbial research commons. Emory Law J 58:889–975
Lemley MA (2005) Property, intellectual property, and free riding. Tex Law Rev 83:1031–1075
Linge G (2008) Competition policy, innovation, and diversity. Tectum-Verlag, Marburg
Lucas RE (2002) Lectures on economic growth. Harvard University Press, Cambridge
Martin S (2002) Spillovers, appropriability and R&D. J Econ 75(1):1–32
Mazzoleni R, Nelson RR (1998a) Economic theories about the benefits and costs of patents. J Econ Issues 32(4):1031–1052
Mazzoleni R, Nelson RR (1998b) The benefits and costs of strong patent protection: a contribution to the current debate. Res Policy 27(3):273–284
Menell PS (2000) Intellectual property: general theories. In: Bouckaert B, de Geest G (eds) Encyclopedia of law & economics, vol 2. Edward Elgar, Cheltenham, pp 129–188
Merges RP, Nelson RR (1990) On the complex economics of patent scope. Columbia Law Rev 90 (4):839–916

Merton RK (1974) The sociology of science: theoretical and empirical investigations. University of Chicago Press, Chicago
Murray F, Stern S (2007) Do formal intellectual property rights hinder the free flow of scientific knowledge? An empirical test of the anti-commons hypothesis. J Econ Behav Organ 63:648–687. https://doi.org/10.1016/j.jebo.2006.05.017
Nelson RR (1969) The simple economics of basic scientific research. J Polit Econ 67(3):297–306
Nelson RR (2009) Building effective 'innovation systems' versus dealing with 'market failures' as ways of thinking about technology policy. In: Foray D (ed) The new economics of technology policy. Edward Elgar Publishing, Cheltenham, pp 7–16
Oakland WH (1987) Theory of public goods. In: Auerbach AJ, Feldstein M (eds) Handbook of public economics. Elsevier, Amsterdam, pp 485–535
OECD (2015) Data-driven innovation: big data for growth and well-being. OECD Publishing, Paris. https://doi.org/10.1787/9789264229358-en
Orsenigo L, Dosi G, Mazzucato M (2006) The dynamics of knowledge accumulation, regulation, and appropriability in the pharma-biotech sector: policy issues. In: Mazzucato M, Dosi G (eds) Knowledge accumulation and industry evolution: the case of pharma-biotech. CUP, Cambridge, pp 402–431
Parchomovsky G, Siegelman P (2002) Towards an integrated theory of intellectual property. Va Law Rev 88(7):1455–1528
Platteau JP (2008) Common property resource. In: Durlauf SN, Blume LE (eds) The new Palgrave dictionary of economics, vol 2, 3rd edn. Palgrave Macmillan, London, pp 20–22
Posner RA (2005) Intellectual property: the law and economics approach. J Econ Perspect 19 (2):57–73
Razzolini L (2004) Public goods. In: Rowley CK, Schneider F (eds) The encyclopedia of public choice. Kluwer Academic Publishers, Dordrecht, pp 457–459
Reichman JH (2009) Rethinking the role of clinical trial data in international intellectual property law: the case for a public goods approach. Marquette Intellect Prop Law Rev 13(1):1–68
Reichman JH, Uhlir PF, Dedeurwaerdere T (2016) Governing digitally integrated genetic resources, data, and literature. Global intellectual property strategies for a redesigned microbial research commons. Cambridge University Press, Cambridge
Rockett K (2010) Property rights and invention. In: Hall BH, Rosenberg N (eds) Handbook of the economics of innovation, vol 1. Elsevier, Amsterdam, pp 315–380
Rose CM (2003) Romans, roads, and romantic creators: traditions of public property in the information age. Law Contemp Probl 69:89–110
Sandmo A (1972) Optimality rules for the provision of collective factors of production. J Public Econ 1(1):149–157. https://doi.org/10.1016/0047-2727(72)90023-0
Schall LD (1971) Technological externalities and resource allocation. J Polit Econ 79(5):983–1001. https://doi.org/10.1086/259810
Scotchmer S (1991) Standing on the shoulders of giants: cumulative research and the patent law. J Econ Perspect 5(1):29–41
Sena V (2004) The return of the prince of Denmark: a survey on recent developments in the economics of innovation. Econ J 114:312–332
Shavell S (2004) Foundations of economic analysis of law. Harvard University Press, Cambridge
Smith K (2005) Economic infrastructures and innovation systems. In: Edquist C (ed) Systems of innovation. Technologies, institutions and organizations. Routledge, London, pp 86–106
Spence M (1984) Cost reduction, competition, and industry performance. Econometric Soc 52 (1):101–122
Stiglitz JE (1999) Knowledge as a global public good. In: Kaul I, Grunberg I, Stern MA (eds) Global public goods: international cooperation in the 21st century. OUP, Oxford, pp 308–325

Stiglitz JE, Wallsten SJ (1999) Public-private technology partnerships. Am Behav Sci 43(1):52–73
van Overwalle G (2014) Governing genomic data: plea for an 'open commons'. In: Frischmann BM, Madison MJ, Strandburg KJ (eds) Governing knowledge commons. OUP, Oxford, pp 137–154
Wadley JH III (2011) Economic theory. Author House, Bloomington
Witt U (2018) Evolutionary economics. In: Macmillan Publishers (ed) The new Palgrave dictionary of economics, 3rd edn. Palgrave Macmillan, London, pp 1873–1879

Chapter 8
IPD as a Research Resource: Exclusively Controlled or Readily Accessible?

Abstract This chapter applies the above-outlined conceptual framework to examine whether and to what extent de facto exclusive control over IPD can be justified as a means of internalising R&D externalities. The analysis seeks to define and qualitatively weigh up innovation-related benefits and costs of exclusive control over vis-à-vis unrestricted access to IPD as a knowledge resource for research and innovation. The policy conclusion is drawn as to whether regulatory intervention by access measures can be justified on the grounds of promoting drug innovation.

8.1 Examining a Potential Disincentive Effect of Clinical Trial Data Disclosure

8.1.1 The Relationship Between Clinical Trial Data and the Problem of Incentives in Drug Innovation

First, let us clarify the relationship between clinical trial data and the public-good problem in drug R&D. As discussed earlier, non-summary clinical trial data can be characterised as an *inherently* public good and an impure public good.[1] However, the public-good problem of underinvestment does not arise concerning generating clinical trial data *as such*. Once the decision regarding a drug R&D program is made, drug sponsors have limited discretion to decide whether to conduct trials or not and how much data to gather. The scope of evidence from trials is ultimately determined by the regulatory requirements for safety and efficacy that address the market failures of insufficient testing[2] and information asymmetries.[3]

[1] See Chap. 7 at Sect. 7.1.1.1.

[2] Kwerel (1980), p. 506; Lemmens (2004), p. 647.

[3] The purpose of enforcing the regulatory standards for drug efficacy and safety is to mitigate 'harmful "externalities" generated by an industrialized market economy' and the risk of misrepresentation of product information due to information asymmetry in the marketplace. Stewart (1981), p. 1260. See also Katz (2007), p. 11. The 'market-for-lemons' metaphor for the information

D. Kim, *Access to Non-Summary Clinical Trial Data for Research Purposes Under EU Law*, Munich Studies on Innovation and Competition 16,
https://doi.org/10.1007/978-3-030-86778-2_8

At the same time, as the most cost- and time-intensive part of drug R&D, clinical trials contribute to the overall public-good problem in drug innovation.[4] Hence, it is important to consider how data disclosure can impact the ability of trial sponsors to recover R&D costs.

8.1.2 Protection of the Competitive Advantage as an Innovation Incentive

As discussed earlier, the notions of innovation incentives and competitive advantage are inherently related.[5] Thus, to understand the potential disincentive effect of disclosure of non-summary clinical trial data, one needs to consider how it can affect the competitive position of the drug sponsor in line with the distinction between three forms of competition in the pharmaceutical sector: competition by imitation, product improvement and a new product.

8.1.2.1 The Structure of the Pharmaceutical Market

The drug market is segmented by the areas of therapeutic indications.[6] Besides, the market structure is defined by two types of companies: originator companies engaged in developing innovative drugs often protected by patents[7] and generic companies entering competition once patent rights and other types of regulatory protection expire.[8] After the generic launch, prices start declining; the price reduction accelerates as more generic competitors enter the market.[9]

asymmetry-induced market failure was introduced by George Akerlof. See Akerlof (1970), p. 488 (referring to 'lemons' as a phenomenon of the diminishing quality of goods due to the information asymmetry between buyers and sellers). See also Eisenberg (2007), p. 367 (referring to the 'original function' of the USFDA as 'protecting the public from snake oil').

[4]Chapter 5 at Sect. 5.2.3.

[5]Chapter 5 at Sect. 5.2.2.2.

[6]WHO collaborating centre for drug statistics methodology, ATC, structure and principles. https://www.whocc.no/atc/structure_and_principles. Accessed 26 Mar 2021.

[7]European Commission (8 Jul 2009) Pharmaceutical sector inquiry report. Final report, p. 22. https://ec.europa.eu/competition/sectors/pharmaceuticals/inquiry/staff_working_paper_part1.pdf. Accessed 26 Mar 2021.

[8]The business model of generic companies is based on producing drugs identical or equivalent to successful originator treatments and their commercialisation upon the loss of exclusivity of the originator product. ibid p. 35.

[9]On the strategies deployed by originator companies to delay the generic entry, see ibid p. 351 ff.

8.1.2.2 New Medicinal Products: First-in-Class v Follow-on Drugs

The concept of innovation includes 'entirely new goods and services [and] significant improvements [of] existing products'.[10] In the context of pharmaceutical innovation, 'entirely new' products are known as 'first-in-class' (or 'breakthrough') drugs. The improvements of existing drugs are referred to as 'follow-on' products, while substantial improvements are sometimes called 'best-in-class' drugs.

While there is no unanimously accepted definition of a 'follow-on' drug,[11] in broad terms, it can refer to any new entrant in the established therapeutic class approved for the same indication and underlying the same or a different chemical structure as the first-in-class drug.[12] Therapeutic benefits of follow-drugs can vary considerably, from featuring 'greater regimen convenience, higher efficacy, or reduced side effects'[13] to presenting only minor modifications of the pioneer drug.[14] While the research-based pharmaceutical industry has been criticised for directing R&D resources to 'me-too' drugs, studies show that a substantial share of follow-on drugs has been approved with priority rating based on significant advances in treatment.[15]

From a clinical perspective, even slight differences among the treatments within a therapeutic class can be of high value, especially if they cater to individual patient needs and characteristics.[16] Sometimes, a follow-on product can establish a new market segment within a therapeutic area due to the marked differences in the therapeutic properties.[17] Personalised medicine is a pertinent example.[18]

---

[10] OECD and Eurostat (2005), p. 17.

[11] For definitions and an overview of the literature on follow-on drugs, see e.g. Andrade et al. (2016), pp. 45–47.

[12] DiMasi and Paquette (2004), p. 2; Ahn (2014), p. 56.

[13] Petrova (2014), p. 67. Sometimes a follow-on drug 'may even surpass the pioneer drug through enhanced effectiveness, greater convenience, or weaker side effects'. ibid pp. 34–35.

[14] Some authors differentiate between 'me-too' and follow-on drugs: the former result from the parallel R&D by competing drug companies, while the latter are developed and launched by the same drug company that sponsored the pioneer drug. See Petrova (2014), p. 33 ff.

[15] DiMasi and Paquette (2004) and DiMasi and Faden (2011).

[16] See e.g. Petrova (2014), p. 24; Pharmaceutical sector inquiry report (n 7), p. 187.

[17] Petrova (2014), p. 23.

[18] In general terms, personalised medicine refers to targeted diagnostics and treatments based on an individual patient's medical history and genetic profile. More specifically, it is defined as 'a medical model using molecular profiling technologies for tailoring the right therapeutic strategy for the right person at the right time, and determine the predisposition to disease at the population level and to deliver timely and stratified prevention'. European Commission (2010) Stratification biomarkers in personalised medicine. Workshop report. http://ec.europa.eu/research/health/pdf/biomarkers-for-patient-stratification_en.pdf. Accessed 26 Mar 2021. See also Jain (2015), pp. 1–2.

8.1.2.3 Competitive Strategies of Drug Companies

The types of drugs discussed above correspond to the following competitive strategies of drug companies:

(i) competition by *imitation* (generic drugs);[19]
(ii) competition by *improvement* (the 'best-in-class strategy') that can include new formulations, modes of administration, combinations of active ingredients with known therapeutic activity, new derivatives of the pioneer product;[20] and
(iii) competition by *new drugs* (the 'first-in-class strategy').[21]

Competition by imitation is also known as price competition, while competition in innovation is referred to as non-price or dynamic competition. Competition by improvement occupies an intermediate position as introducing improved products can cause price effects in the existing product market.

8.1.2.4 Innovation as the Main Parameter of Competition in the Pharmaceutical Sector

Research-based pharmaceutical companies compete by investment and efforts directed at the development and commercialisation of new medicinal products.[22] This form of competition 'takes place outside and before the emergence of markets'[23] and is also known as Schumpeterian competition.[24] From a long-term perspective, competition in innovation, as a promoter of technological progress, is regarded more socially valuable than the static competition.[25] At the same time, economists point out that there is a limited understanding of how firms compete in innovation,[26] how the 'optimal intensity' of competition in innovation should be

[19] Andrade et al. (2016), p. 48.

[20] Pharmaceutical sector inquiry report (n 7), p. 49; Ahn (2014), p. 56.

[21] This strategy corresponds to competition *for* the market—R&D efforts directed at developing new products that can either replace the existing ones or create a new market.

[22] Reinganum (1981), p. 21 ff.

[23] Drexl (2012), p. 507.

[24] Schumpeter (1950), pp. 81–86. On competition in innovation and dynamic efficiency in high technology industries and markets, see e.g. Jones and Sufrin (2016), p. 13 ff; van den Bergh and Camesasca (2001), p. 36 ff.

[25] See e.g. Kerber and Schwalbe (2008), para 1-8-071 (with further references); Fatur (2012), p. 71. For an overview of the theories of dynamic competition, see Elling and Lin (2001), pp. 16–44.

[26] Cohen (2010), p. 156.

defined,[27] and how R&D intensity correlates with innovation performance[28] and social welfare.[29]

In the pharmaceutical sector, competition in innovation is generally viewed as the main form of competition.[30] The sector is characterised by the fierce pre-market rivalry among originator companies to be the first to identify a promising molecule or a new target and 'to win the race to discover new products'.[31]

8.1.2.5 The Hypothesis Regarding the Effects of Data Disclosure on Innovation Incentives

In line with the distinction between the types of competition in the pharmaceutical market, the hypothesis regarding the potential disincentive effects of data disclosure can be formulated as follows. Mandatory disclosure of non-summary clinical trial data can hinder innovation incentives of research-based companies

(i) in the context of *price competition*, to the extent to which third-party access and use of the disclosed data can facilitate the loss of exclusivity in the existing drug market;

---

[27] Kerber and Schwalbe (2008), paras 1-8-097, 1-8-207 (observing that there might be 'only a weak relationship between the degree of concentration, R&D expenditure and innovation in a market', which 'means that no unambiguous conclusions can be drawn regarding the effects of changes in R&D activities on the rate of innovation' (with further references)). See also Glader (2006), p. 85 (noting that, even though 'scholars and policy makers have been sensitive to the possibility that a more permissive attitude could be beneficial to spur international competition and R&D incentives, the problem has been that the links between concentration and R&D or concentration and innovation have always been murky'). See also below (n 242).

[28] As summarised by Carrier, '[a]fter a half-century of debate and innumerable studies, the overwhelming consensus is that there is no clear answer to the question [regarding the relationship between market structure and innovation]'. Carrier (2008), p. 396. See also Rapp (1995), p. 27 (stating that '[t]he leap of faith is to believe that there is a positive functional relationship between the rate of R&D expenditure (or the amount of R&D capacity) and the quantum of innovation produced by a firm').

[29] Henderson and Cockburn (1996), p. 36.

[30] Orsenigo et al. (2006), p. 407. See also OECD (2013) The role of efficiency claims in antitrust proceedings. DAF/COMP(2012)23, p. 15 (noting that, in the pharmaceutical industry, 'competition mainly occurs through races to innovate rather than through price setting, in a process known as Schumpeterian rivalry'); Gambardella (1995), p. 142 (concluding that 'research and innovation are the most important determinants of competitive performance and growth among the largest US drug companies'); Cockburn and Henderson (1994), p. 483 (referring to competition in the pharmaceutical industry as 'a prime example of the types of strategic racing behavior [whereby] firms invest heavily in research and development since successful research is a key contributor to commercial success').

[31] Charles River Associates (2004) Innovation in the pharmaceutical sector, p. 63. https://www.crai.com/insights-events/publications/innovation-pharmaceutical-sector/. Accessed 26 Mar 2021.

(ii) in the context of *dynamic competition*, to the extent to which third-party data analysis can offset the advantage of the data holder in developing new including improved drugs.

The next section explores these propositions.

8.1.3 Implications of Non-summary Clinical Trial Data Disclosure for Competition by Imitation

8.1.3.1 The Relevance of Access to IPD for Generic and Biosimilar Drug Development and Manufacturing

Generic competitors usually do not need to access and re-analyse IPD to reproduce the originator drug as the composition and formulation of small-molecule drugs can be 'reverse engineered' relatively easily without access to the source data.[32] However, biosimilar drugs are different in this regard. The development of biosimilar drugs is 'more complex and costly due to the fact that the active ingredient is based on live tissue'.[33] Competitors' access to the non-summary clinical trial data can reveal the detailed information related to the bioassays and the originator's manufacturing and analytical methods that might facilitate the development of a biosimilar drug. This argument was raised, for instance, in the *AbbVie v EMA* case. When objecting to the EMA's decision to grant third-party access to the clinical study reports related to the authorisation of a biological drug Humira, AbbVie argued that those reports

> describe the manner in which [AbbVie] planned and implemented the clinical trials necessary for obtaining the [marketing authorisation] for the medicinal product for the indication of Crohn's Disease. Those reports therefore provide a *very specific road map* for a company wishing to *develop* a TNF antagonist for the therapeutic use in question, by enabling it to *develop a similar 'biologics/biosimilar'* strategy in order to produce a *follow-on* medicinal product or to add *new therapeutic indications* to an existing medicinal product. The reports also provide information about some of the hurdles the applicants had to overcome, which could reduce the development process for a medicinal product by *two to three years*.[34]

One can only speculate to what extent such concerns can be justified. What is known is that the first biosimilar of AbbVie's Humira Amgevita was approved by the EMA on 26 January 2017.[35] The timing of Amgevita's market launch correlates

---

[32] Bansal AK, Koradia V (2005) The role of reverse engineering in the development of generic formulations. Pharmaceutical Technology 28(9). http://www.pharmtech.com/role-reverse-engineering-development-generic-formulations. Accessed 26 Mar 2021.

[33] Pharmaceutical sector inquiry report (n 7), p. 8.

[34] Case T-44/13 R *AbbVie v EMA* [2013] ECLI:EU:T:2013:221, para 60 (emphasis added).

[35] EMA (26 Jan 2017) CHMP assessment report. Amgevita. EMA/106922/2017, p. 9. Besides, several biosimilar products were authorised by the EMA in 2017–2018, for which Humira was designated as a reference medicine, including Imraldi, Hyrimoz, Hefiya, Cyltezo, Halimatoz, Hulio.

with the expiration of patent protection[36] rather than the development of the biosimilar version. Data submitted by Amgen was obtained in clinical studies on efficacy and safety[37] conducted in 2013–2015.[38] In other words, by the time the request for CRSs related to Humira was filed on 12 November 2012,[39] the development of the competing biosimilar product was already in the advanced stage and could not have been used for exploratory, hypothesis-generating analysis.[40]

8.1.3.2 Relevance of IPD Disclosure for the Protection Against the Competition by Imitation

As concluded in Chap. 5, third-party access to IPD is unlikely to interfere with the regulatory exclusivities that protect innovation incentives of research-based companies by delaying the entry of generic products. Where such instruments are applied and enforced properly, competitors would not be able to take undue advantage of the disclosed non-summary data by re-submitting it to other regulatory authorities.[41] Neither can access to IPD allow competitors to circumvent patent protection covering the products for which IPD was generated or compromise the patentability of follow-on inventions if the timing of data disclosure is adjusted accordingly.[42]

Accordingly, as long as access to IPD for research purposes does not affect the protection against imitation, access measures would not cause a disincentive effect in the sense of undermining the capacity of trial sponsors to earn time-limited supra-competitive profits in the market for the drug for which the accessed data was generated. However, the research potential of IPD can extend beyond the original

[36] Amgen (15 Oct 2018) Amgen launches AMGEVITA™ (biosimilar adalimumab) in markets across Europe. https://www.amgen.com/media/news-releases/2018/10/amgen-launches-amgevita-biosimilar-adalimumab-in-markets-across-europe/. Accessed 26 Mar 2021. See also Davio K (15 Oct 2018) On the eve of Humira's patent expiry, Europe prepares for biosimilar Adalimumab (stating that 'October 16 marks European patent expiry for AbbVie's blockbuster anti-tumor necrosis factor drug, adalimumab (Humira), and multiple competitors stand ready to launch their biosimilar products on, or shortly after, that date'). https://www.centerforbiosimilars.com/view/on-the-eve-of-humiras-patent-expiry-europe-prepares-for-biosimilar-adalimumab. Accessed 26 Mar 2021.

[37] EMA (26 Jan 2017) CHMP Assessment Report. Amgevita. EMA/106922/2017, p. 36 ff.

[38] ibid pp. 53, 70.

[39] In that case, access to CSRs was requested by a university student. Case T-44/13 R *AbbVie v EMA* [2013] ECLI:EU:T:2013:221, para 20. AbbVie argued that, 'even if access to the disputed reports were granted only to one student, the confidential information could be disclosed to anybody, including current or potential competitors'. ibid para 46.

[40] The EMA publication policy provides for the reservations to mitigate competitive concerns regarding competitors' use of the trial-related information and data in R&D. For instance, information regarding bioassays and analytical methods can be removed from a CSR before it can be disclosed. EMA publication policy 0070, p. 18.

[41] For an analysis, see Chap. 5 at Sect. 5.4.

[42] For an analysis, see Chap. 5 at Sect. 5.3.

hypothesis and research objectives addressed by the trial in which that IPD was collected.[43] Hence, implications of third-party access to IPD for actual and potential competition in *innovation* need to be considered further.

8.1.4 Implications of IPD Disclosure for Competition in Innovation

8.1.4.1 New Research Hypothesis as a Technological Opportunity

Companies compete in innovation by seeking and pursuing new *technological opportunities*, the prospects of developing new technologies and commercialising products embodying innovative features. In the context of drug innovation, a technological opportunity refers to a *viable hypothesis* regarding the structure-activity relationship that may become a starting point of a new R&D project. The latter corresponds to an 'R&D pole' defined as R&D efforts directed at developing a new treatment.[44]

Knowledge stemming from the 'upstream' basic research often becomes a catalyst of technological opportunities and commercial applications that are explored and developed in the 'downstream' industrial R&D.[45] In this regard, secondary analysis of IPD from past trials has a unique potential to contribute to both academic research (e.g. by improving understanding of disease factors and pathways) and industrial R&D.[46]

8.1.4.2 Concerns About the Impact of Data Disclosure on Competitive Advantage in Drug R&D

Drug companies have argued that clinical trial data disclosure can facilitate the development of new, potentially competing medicinal products. For instance, AbbVie alleged that disclosure of the clinical study reports submitted to the EMA would allow competitors to 'improve their competitive position with (actually or

[43] As discussed in Chap. 3.

[44] Pharmaceutical sector inquiry report (n 7), p. 9. The notion of R&D pole is often applied in the competition law analysis, e.g. in the assessment of mergers and horizontal co-operation agreements. See European Commission (14 Jan 2001) Communication from the Commission. Guidelines on the applicability of Article 101 of the Treaty on the Functioning of the European Union to horizontal co-operation agreements. OJ C 11, paras 119–122.

[45] Cockburn and Henderson (1994), p. 490 ff (discussing the example of the discovery of an angiotensin-converting enzyme inhibitor). See also Merges and Nelson (1990), p. 908.

[46] See e.g. Institute of Medicine of the National Academies (2015), p. 20 (noting that clinical trial data analysis can facilitate 'the identification and validation of new drug targets [and] new indications for use').

potentially) competing products'[47] and that such reports could provide competitors with the information, 'which could reduce the development process for a medicinal product by two to three years'.[48] Some commentators argued that confidential treatment of non-summary clinical trial data could allow originator companies 'to protect original laboratory method used for drug development or to protect exploratory endpoints or biomarkers that do not support the current label claims and benefit-risk evaluation but represent the basis for *future potential development* of the medicine by the sponsor'.[49]

If third-party exploratory data analysis can facilitate the development of new, potentially competing drugs, data protection against disclosure cannot be considered 'redundant' as it might seem.[50] Instead, factual confidentiality and exclusive control over non-summary trial data can enable drug sponsors to *control new research paths* that could be revealed through secondary data analysis. In so doing, control over non-summary test data provides an additional layer of protection allowing drug sponsors to retain a competitive advantage in competition in innovation. This way, it differs substantively from patents and regulatory exclusivities that protect innovator drugs against the competition by imitation. If secondary data analysis could facilitate competitors' R&D projects leading to new medicinal products[51] or otherwise strengthen competitors' research capacity,[52] the idea of data-sharing would run counter the business logic. In this view, the reservation for a potential 'conflict of interest' under the companies' data-sharing policies[53] can be understood as referring not only to the potential (mis)use of the disclosed data for generic drug approval but also to exploratory IPD analyses in competitors' R&D. Such strategic, competition-based considerations can explain why the data-sharing policies adopted by the drug companies presumptively deny access to actual and potential competitors.[54] In this regard, the intention of the EMA to 'establish[] a *level* playing field [for] all medicine developers'[55] by public disclosure of non-summary trial data appears controversial.

To understand more specifically how IPD disclosure can affect competitive dynamics in drug R&D, let us consider what new drugs can be developed based on exploratory data analysis. Several scenarios are conceivable.

[47] Case T-44/13 R *AbbVie v EMA* [2013] ECLI:EU:T:2013:221, para 60.

[48] ibid.

[49] Bonini et al. (2014), p. 2454 (emphasis added).

[50] See Eisenberg (2011), pp. 468–469 (observing that 'with regulatory exclusivity to protect against free riders, it is difficult to justify the continuing treatment of data submitted in pursuit of regulatory approval as trade secret or confidential information', which creates 'redundancy in protection').

[51] Case T-44/13 R *AbbVie v EMA* [2013] ECLI:EU:T:2013:221, para 60 (emphasis added).

[52] For a review of these arguments, see Chap. 5 at Sect. 5.1.

[53] See Chap. 6 at Sect. 6.3.2.

[54] ibid.

[55] EMA publication policy 0070, p. 4 (emphasis added).

8.1.4.3 Potential Scenarios

Follow-on Improvements of the Investigational Product

Suppose company A generated clinical trial data while developing drug X. Secondary analysis of IPD related to drug X could contribute to developing drug X^1 that would treat the same condition as drug X but would be clinically superior. For instance, secondary IPD analysis could identify the subgroup populations in which drug X^1 might have higher efficacy or a more optimal benefit-risk profile than drug X.[56] Drug X^1 could underlie the same or a different chemical structure. The latter is possible because the correlation between the chemical structure and the pharmacological characteristics of the tested compound combined with the detailed efficacy and safety test data can guide researchers to *alternative* chemical structures.[57]

New Medical Use of the Initial Investigational Product

Secondary analysis of data related to drug X could contribute to developing drug Y that would feature a different therapeutic indication[58] of the same active ingredient as drug X. IPD could reveal additional effects observed during the trial that were not defined *ex ante* as the primary endpoints. For instance, in *AbbVie v EMA*, it was argued that the reports at issue could 'provide a *very specific road map* for a company wishing to [...] add *new therapeutic indications* to an existing medicinal product'.[59] New uses of known substances are often discovered during clinical trials.[60] Nevertheless, the discovery of new indications of known active ingredients is serendipitous and, in some cases, can take over half a century.[61]

---

[56] On the subgroup IPD analysis, see Chap. 3 at Sect. 3.2.2.4.

[57] Response to the questionnaire by a chemist who works in a pharmaceutical company (on file with the author). The respondent also noted that such 'correlation can be done only if the full [individual patient-level] data are available' and that the clinical study reports and the respective publications do not contain such data.

[58] See e.g. Institute of Medicine of the National Academies (2015), p. 20.

[59] Case T-44/13 R *AbbVie v EMA* [2013] ECLI:EU:T:2013:221, para 60 (emphasis added). See also Eisenberg (2007), p. 383.

[60] See e.g. Correa (2015), p. 44; Pharmaceutical sector inquiry report (n 7), p. 187 (observing that 'clinical trials may reveal new medical uses'). The often-cited example of a new use discovered during clinical trials is the drug Viagra first tested to treat angina pectoris. Another example is minoxidil, initially developed to treat ulcers that demonstrated effectiveness as a vasodilator and was subsequently 're-purposed' for the severe hypertension condition. Watkins et al. (1979). Later on, the antihypertensive agent minoxidil showed a side effect of hair growth during efficacy trials and was repositioned for treating alopecia. Clissold and Heel (1987).

[61] For instance, thalidomide was approved as a treatment for morning sickness in pregnant women in the 1950s and repositioned to treat multiple myeloma in 2006. Thioguanine was initially

A Different Chemical Structure and a Different Therapeutic Indication

Exploratory analysis of IPD related to drug X could contribute to the subsequent research, eventually leading to drug Z that would treat a different condition and have a different chemical structure than drug X. This scenario is most plausible in the context of 'big data' analysis, where exploratory analysis of large volumes of the aggregated health data might reveal unforeseen correlations and insights that can further guide the discovery and validation of new drug targets and molecules.[62] While data related to drug X could then be one 'piece of a puzzle' among numerous knowledge inputs, it might be impossible to discern its contribution to the development of drug Z.

8.1.4.4 Analysis

Theoretical Assumptions

Let us assume that, as a result of IPD disclosure, competitors could analyse it in their research projects and that such analysis—as alleged by drug companies[63]—could facilitate the development of drug improvements. From a market competition perspective, even though the original investigational product (drug X) and the follow-on products (drugs X^1, X^2, X^3) could be viewed as different products,[64] they would compete in the same therapeutic market.[65] The timing of the launch of drug X^1 by company B would depend on several factors, including how long it would take to develop and obtain marketing authorisation[66] for drug X^1, the scope of patent protection covering drug X, as well as whether company A might have developed and patented improvements over drug X.[67] The impact of drug X^1 entry on

developed to treat leukemia in children; it took 65 years before it was approved for inflammatory bowel disease. Simsek (2018), pp. 17–18.

62 Gauch (2009), p. 285 (defining data-mining as 'searching a database and looking for every kind of a relationship between variables that have not been previously discovered').

63 See Chap. 5 at Sect. 5.1.

64 Pharmaceutical sector inquiry report (n 7), p. 150 (stating that '[d]ifferent dosages or different forms of administration of the same prescription medicine have been considered as different products').

65 See Case T-718/15 *PTC Therapeutics International v EMA* [2018] ECLI:EU:T:2018:66, para 92.

66 Dir 2001/83/EC, art 10(2)(b) (stipulating that 'different salts, esters, ethers, isomers, mixtures of isomers, complexes or derivatives of an active substance shall be considered to be the same active substance, unless they differ significantly in properties with regard to safety and/or efficacy'; if so, 'additional information providing proof of the safety and/or efficacy of the various salts, esters or derivatives of an authorised active substance must be supplied by the applicant'.) See also Dir 2001/83/EC, annex I, pt II(3).

67 Patents related to such improvements are also known as 'second-generation' or 'secondary' patents. See Pharmaceutical sector inquiry report (n 7), p. 381 (explaining that originator companies usually file for secondary patents (e.g. directed at the modifications, improvements, combinations

competition and the market share of company A in the respective therapeutic market would be determined by factors such as the degree of the substitutability of drugs X and X1 and the presence of other therapeutic alternatives.[68] Even if drug X^1 can represent a minor modification of drug X, the impact on A's market share can be substantial because, as discussed earlier, small improvements 'may elicit big therapeutic benefits'.[69]

In effect, secondary data analysis could reduce the time lag between the launch of the first-in-class and follow-on drugs.[70] The longer time and the greater investment it might take the competitors to develop improvements, the longer is the *lead time* during which the originator company can earn returns on R&D (if, during that period, the innovator drug would also be protected against generic competition by patents and regulatory exclusivities). Alternatively, the greater data is available from previously conducted trials, the more targeted and efficient can be subsequent R&D across the firms. In theory, if the time lag between the launch of the first-in-class drug and its improvements is viewed as a key driver of return on investment for R&D', the reduction of such time advantage can have a discouraging effect on incentives.[71] Let us consider next how IPD disclosure might affect the lead time in practice.

The Case of Drug Improvements

First, one should bear in mind that the ad hoc exploratory analysis results are 'merely suggestive'[72] and highly probabilistic and, at best, can hint at a hypothesis for future research.[73] Commentators refer to exploratory data analysis as 'the practice of studying data without preconceived notions as a means of obtaining ill-conceived ones',[74] '[t]he art of seeing a Rembrandt in a Jackson Pollock',[75] a 'fishing expedition'[76] and 'data dredging'.[77] From a methodological perspective, it is recommended

---

with other molecules, etc.) in order to maintain the freedom to operate when conducting further research and improve first-generation products 'without interference from competitors').

68 Andrade et al. (2016), pp. 51–53.

69 Reichman (2009), p. 39. See also above (n 16) and the accompanying text.

70 Institute of Medicine of the National Academies (2015), p. 63.

71 See Scotchmer (2004), p. 147 (pointing out that the faster the improvements come, 'the shorter is the market incumbency of each innovator', and that the prospect of selling an innovative product 'at a price established in competition with its successor [. . .] might easily discourage investment').

72 Brody (2016), p. 88. See also EMA (23 Jan 2014) EMA guideline on the investigation of subgroups in confirmatory clinical trials. EMA/CHMP/539146/2013, p. 9.

73 Moyé (2003), p. 119.

74 Senn (2007), p. 43.

75 ibid.

76 Browner et al. (2007), p. 61.

77 Kirwan (1997), pp. 822–825. See also Merson et al. (2016), p. 2414 (referring to 'repositories of data without metadata, data dictionaries, or documentation needed for meaningful or correct reanalysis' as 'data dumpster'); Sydes et al. (2015) (pointing out that 'data dredging is likely to

that 'unanticipated associations' and hypotheses suggested by *ad hoc* data analysis 'should be viewed with interest but scepticism'.[78]

Second, even if secondary data analysis might guide researchers to a new drug candidate, such hypothesis would require validation in confirmatory studies. Given that the drug development process from identifying a promising target to the market launch takes on average 12–15 years,[79] there will be a substantial time lag between exploratory IPD analysis and the commercialisation of its results. Even in cases where a follow-on drug is based on the same active ingredient as the originator drug, it would need to undergo new trials before it can be authorised for marketing, provided it differs significantly in properties.[80] By the time drug X^1 can reach the market, the market share of the sponsor of drug X is likely to be eroded by the substitutes.[81]

Third, competition by improvement is often intense already during the pre-market stage. For instance, the study by DiMasi and Faden[82] examined the sample of 94 drug classes[83] and found that

> [f]or all drug classes introduced since the late 1980s, at least one follow-on drug was *synthesized before the approval* of the first-in-class drug and the first pharmacological testing of some follow-on drugs occurred before the first-in-class drug was approved, in all but one class, since the late 1980s. Indeed, *initial clinical testing* of at least one follow-on drug in a class occurred *before* the approval of the first-in-class drug for at least 80% of the classes in each of the periods since the late 1980s. At least half of all classes had at least one follow-on drug for which the investigational new drug application (InD) was *filed before* the approval of the first-in-class drug for each period since the early 1980s, with this being the case for 89% of the classes since the early 1990s. At least half of the classes in any period since the early 1980s had at least one follow-on drug with *Phase II testing initiated before* the first drug in the class was approved. Since the late 1980s, 64% of the classes had at least

---

provide some false positive findings and lead to over-interpretation'). But see Meinert (2012), p. 446 (finding that '*ad hoc* data analysis aimed at finding statistically significant differences among different groups, especially when leading to a presentation or publication heralding differences [is] real and important').

[78] Browner et al. (2007), p. 61.

[79] Hughes (2011), p. 1239.

[80] Above (n 66).

[81] Studies show that pharmaceutical companies pursue the 'best-in-class' strategy aimed at developing drugs 'with a particularly attractive clinical or economic profile' improving the existing medicines, instead of the 'first-in-class' strategy. See Lanthier et al. (2013), pp. 1434–1436 (observing that the time-lag between the marketing of a first-in-class drug and the marketing of similar products has decreased significantly over time); Pharmaceutical sector inquiry report (n 7), p. 49 (stating that 'incremental innovations constitute a parameter for competition between originator medicines in the same therapeutic class'); Andrade et al. (2016), p. 48; DiMasi and Paquette (2004), p. 12.

[82] DiMasi and Faden (2011).

[83] Within those classes, the first-in-class compound was approved in the US between 1960 and 2003 and 287 follow-on drugs were authorised in the US between 1960 and 2007.

one follow-on drug in the class with *Phase III testing initiated before* first-in-class approval, with the share at 88% since the late 1990s.[84]

In other words, evidence suggests that, to a large extent, the development of drugs that are subsequently considered as 'follow-on' versions occurs *before* the pioneer drug is authorised for marketing.[85] In this regard, the term 'follow-on' can be confusing as it suggests that such drugs result from research that *follows* the introduction of the first-in-class drug.[86] Another indication of the high concentration of R&D efforts in the pharmaceutical sector is the evidence on the overlapping R&D programs conducted by the originator companies[87] and patent litigation arising from such overlaps.[88]

The high intensity of pre-market competition can suggest that follow-on drugs would likely be in the advanced stage of development by the time the IPD related to the first-in-class drug can become accessible (i.e. after the first-in-class drug is approved). Hence, it is unlikely that the lead time between the market launch of the first-in-class drug and the introduction of the follow-ons could be reduced due to exploratory data analysis of data related to an investigational product if the disclosure takes place after the investigational product is authorised.

Furthermore, one should also consider the incentives of drug companies to undertake secondary analysis of IPD. Drug sponsors are motivated to introduce second-generation drugs to take over the market shares of the first-generation products.[89] In many cases, the originator company would also be able to patent improvements.[90] Before embarking on a new R&D program, research-based companies normally carry out the 'freedom-to-operate' analysis to determine whether the prospective project may violate the existing patent rights of other companies.[91] Usually, companies would avoid head-to-head competition if it is perceived to be

[84] ibid p. 25 (emphasis added).

[85] ibid.

[86] See Petrova (2014), p. 33 (explaining that, since competing drug companies 'work in parallel on similar targets, often applying the same fundamental knowledge sourced from open science, the solutions they come up with may not be all that different', and that 'the vast majority of me-too drugs are not the product of brazen, deliberate imitation [as most] of them have been in clinical development prior to the approval of the pioneer drug' (emphasis added)).

[87] Pharmaceutical sector inquiry report (n 7), p. 514 (identifying over 1100 instances, where patents of one originator company could be infringed by the overlapping R&D programmes and/or patents of another originator company).

[88] According to the report by the European Commission, the majority (75.8%) of the examined patent infringement cases had been litigated between the originator companies pursuing R&D programs in the same ATC3 class. Pharmaceutical sector inquiry report (n 7), pp. 410.

[89] ibid pp. 184–185. See also Petrova (2014), p. 24.

[90] Creating 'patent clusters' through secondary patents is reportedly one of the so-called 'lifecycle management' strategies often employed by research-based drug companies to protect exclusivity in the market. Pharmaceutical sector inquiry report (n 7), pp. 60, 173 ff.

[91] ibid p. 381 ff.

'inherently unproductive'.[92] Thus, the motivation of drug developer B to undertake secondary analysis of IPD related to drug X to develop its improved version might be low, especially if IPD becomes available after the marketing authorisation of drug X.

In light of the foregoing, concerns that post-MA exploratory analysis of IPD could affect the market share of the initial investigational product significantly and the competitive position of its sponsor appear exaggerated. This conclusion is consistent with the findings of economic analysis of R&D externalities,[93] which distinguishes the following factors as mitigating, in general, the disincentive effect of knowledge externalities.

(i) The 'spilled over' knowledge is not used to *imitate* the product that embodies innovative knowledge but can facilitate the development of a new product commercialised in a market other than the product of the knowledge producer.[94]
(ii) The *absorption* of the externally created knowledge involves high costs.[95]
(iii) There is likely to be a substantial *time lag* between the knowledge 'spillover' and the commercialisation of a new product to which development the 'spilled over' knowledge may have contributed.[96]

First, exploratory data analysis, by definition, is conducted not to imitate the original investigational product but to generate new research hypotheses.[97] Second, IPD analysis is time- and cost-intensive. In this view, it is not quite accurate to refer to benefits that third parties can derive through secondary IPD analysis as 'knowledge spill-overs' or 'free-riding' as such benefits do not simply 'spill over' but will require substantial efforts on behalf of the data user. Third, the time lag between exploratory data analysis and the marketing authorisation of a product to which development exploratory analysis might contribute is likely to be substantial.

---

[92] Cockburn and Henderson (1994), p. 495. See also Pharmaceutical sector inquiry report (n 7), p. 57 (finding that the competitive environment (product differentiation) is one of the key determinants of R&D investment decision making).

[93] If third-party research can benefit through the analysis of disclosed IPD, IPD disclosure can be viewed as a catalyst of R&D externalities.

[94] Antonelli (2017), p. 97.

[95] See e.g. Hall et al. (2010), p. 1065; Antonelli (2017), p. 5.

[96] Jaffe (1998), p. 14.

[97] Small-molecule drugs can be 'reverse-engineered' without accessing and re-analysing IPD related to the originator drug. Bansal AK, Koradia V (2005) The role of reverse engineering in the development of generic formulations. Pharmaceutical Technology 28(9). http://www.pharmtech.com/role-reverse-engineering-development-generic-formulations. Accessed 26 Mar 2021.

Protection of Exploratory Endpoints as Intermediate Research Results

As noted earlier, IPD re-analysis can guide researchers to alternative chemical structures and indications. Thus, the competitive effects of secondary data analysis could extend beyond the market for the initial investigational product. Even when third-party exploratory analysis of data related to drug X might eventually contribute to the development of drugs Y and Z, it may well be that the sponsor of drug X is also researching the same direction. In other words, de facto exclusive control over the primary research data can protect the priority of the investigational drug sponsor in deriving new insights and drug discovery opportunities. Under such circumstances, a policy intervention by access measures could diminish the advantage of the data holder in competition in innovation and, thus, interfere with innovation incentives.

Data on exploratory endpoints is an apt example where control over IPD can protect the competitive advantage of the trial sponsor in R&D. Even though confirmatory trials are conducted primarily for obtaining drug marketing authorisation, they often include exploratory aspects.[98] Exploratory endpoints are variables not usually predefined in a trial protocol that can form the basis for generating a new research hypothesis (e.g. the treatment comparison or subgroup analysis[99]).[100] For instance, trials for an anti-inflammatory drug might investigate an exploratory lipid profile to inform future studies, while the results of the exploratory analysis might be included in the CSRs submitted to a drug authority for the marketing authorisation of the investigational product.[101] In this situation, there is little doubt that exploratory endpoints should be protected against premature disclosure.[102]

The normative justification for protecting undisclosed information representing 'intermediate research results'[103] can be invoked in this context. Quite obviously, disclosure of non-summary clinical trial data would cause a disincentive effect if it exposes to the competitors a promising hypothesis regarding a new treatment effect that the trial sponsor has already identified.[104]

---

[98] The International Council for Harmonisation of Technical Requirements for Pharmaceuticals for Human Use (ICH) (1998) ICH harmonised tripartite guideline. Statistical principles for clinical trials. E9, p. 5.

[99] See e.g. Huque and Röhmel (2010), p. 3.

[100] Bonini et al. (2014).

[101] EMA publication policy 0070, pp. 19–20.

[102] Academic researchers also consider exploratory biomarker studies as commercially-sensitive information that needs to be maintained confidential. See e.g. Multi-Regional Clinical Trials Center at Harvard University (2014) Overview of data disclosure initiatives: current and ongoing data transparency activities in the pharmaceutical industry. https://www.regulations.gov/comment/FDA-2013-N-0271-0031. Accessed 26 Mar 2021.

[103] Bone (1998), pp. 270–271.

[104] According to the EMA, for instance, disclosure of data on exploratory endpoints could allow 'competitors to gain insights into additional future study plans and/or indications for the product' and, in some situations, affect the patentability of future inventions. Hence, such information can qualify as CCI and be deleted. EMA publication policy 0070, pp. 19–20.

However, one would doubt whether the protection of data on exploratory endpoints can provide a sufficient justification for allowing drug sponsors to maintain, as a general rule, exclusive control over research paths that could be explored through secondary IPD analysis. The above-outlined scenarios can be characterised as competition in innovation only if both company A (the original data holder) and company B (the potential data user) are indeed conducting research directed at developing drugs X^1, X^2 ... Y, Z. This might not always be the case. The least constructive outcome would be when the initial data holder neither analyses data in own research (beyond the benefit-risk assessment of the investigational product) nor allows third parties to conduct secondary IPD analysis. The unrealised value of data in terms of additional knowledge gained through secondary data analyses would then constitute a welfare loss[105] and justify concerns regarding missed research opportunities.

8.1.4.5 The Interim Conclusion

The preceding analysis suggests that exclusive control over IPD cannot be justified as a means of protecting the competitive advantage of trial sponsors (the initial data holders) in competition by imitation. As far as competition in innovation is concerned, it appears unclear to what extent drug companies' concerns over the diminished competitive advantage due to the unrestricted third-party access and secondary analyses of IPD might be warranted, given that the results of exploratory IPD analysis are highly probabilistic and distant in time. The only situation where control over IPD might be justified on innovation grounds is if control over data can provide the initial trial sponsor with the priority in deriving new insights and research hypotheses through IPD analysis, as illustrated by the example of data collected on exploratory endpoints. Apart from that, control over third-party exploratory analysis of IPD can hardly be rationalised as a means of protecting innovation incentives of drug sponsors.

8.2 The Issue of the Underutilised Research Potential of Data

8.2.1 Concerns Regarding Lost Research Opportunities

Proponents of access to non-summary clinical trial data argued that data sharing

> might accelerate the *drug discovery* and development process, reducing and facilitating the identification and validation of *new* drug targets or surrogate endpoints. In short, there are

[105] See e.g. Ben-Asher (2000), p. 279 (noting that if the additional knowledge is not discovered in the course of the initial or follow-on projects, 'there is likely to be a welfare loss').

> today many *missed opportunities* to gain scientific knowledge from clinical trial data that could strengthen the evidence base for the treatment decisions of physicians and patients. In economic terms, these missed opportunities result in a *suboptimal return* on the altruism and contributions of clinical trial participants, the efforts of clinical trialists and research staff, and the financial resources invested by study funders and sponsors.[106]

Concerns regarding missed research opportunities due to the exclusive control of drug sponsors over IPD can evoke an analogy with the 'tragedy of anticommons'—the topic that has been debated intensely, especially concerning patents for research tools in the context of biopharmaceutical innovation.

8.2.2 A 'Tragedy of Anticommons' Due to Exclusive Control Over IPD?

8.2.2.1 The Notion of Anticommons

As a social phenomenon, an 'anticommons' refers to a cooperation failure, a collective action problem where individually rational, gain maximising behaviour can lead to a collectively suboptimal outcome.[107] In the law-and-economics literature, a 'tragedy of anticommons'[108] is defined as a situation where multiple independent actors control separate, complementary inputs that are 'collectively [...] necessary in order to utilize a resource or generate a product or make a decision deemed to have positive social value',[109] irrespective whether such inputs can be protected by ownership rights.[110] The central anticommons concern is inefficiency due to the unrealised value or underuse[111] of resources. In this regard, it is an antipode of the 'tragedy of commons' positing that rivalrous finite goods

---

106 Institute of Medicine of the National Academies (2015), p. 18 (emphasis added) (with further references).

107 Ostrom (2008), p. 5. See also King et al. (2016), p. 67 (defining an anticommons as a situation, where '[b]ehavior that is individually rational and maximizing [...] results in outcomes that are collectively perverse and systematically suboptimal').

108 Heller (1998).

109 King et al. (2016), p. 70.

110 ibid (further arguing that 'there is no apparent necessity to insert the concept of legal property as a critical element' of the anticommons definition because 'separate, necessary yet complementary inputs [are the sufficient] preconditions to the anticommons problem, consistent with the non-cooperative game model').

111 Heller and Eisenberg (1998), p. 698; Hess and Ostrom (2007), p. 11 (defining the 'tragedy of the anticommons in the knowledge arena [as] the potential underuse of scarce scientific resources caused by excessive intellectual property rights and over-patenting in biomedical research'); Wang (2008), p. 253 (noting that the 'core problem with the anticommons is underuse'); Frost and Morner (2010), p. 178 (noting that resources are 'underused, because too many "knowledge empire builders" have the right to exclude').

(e.g. natural resources) will be overused in the absence of the incentives for their preservation or enforceable rules rationing their use.[112]

8.2.2.2 'Anticommons' as the Problem of Unrealised Value Due to the Failure to Cooperate

While the anticommons hypothesis has been studied in different contexts, the debate has been especially intense in life sciences with regard to patents for research tools.[113] Concerns have been raised that such 'upstream' resources might be underutilised due to the failure to 'bundle' fragmented exclusive rights[114] or 'stacking' patent licences.[115] While the argument that patents for research tools can inhibit scientific research and cumulativeness of innovation might seem intuitively appealing, the question is whether such effects can be proved and assessed.

The problem of anticommons has been demonstrated mathematically as a non-cooperative, game-theory type of model.[116] Buchanan and Yoon[117] showed the symmetry between the over-usage of resources due to multiple use rights (the 'commons') and the under-usage of resources due to multiple exclusion rights (the 'anticommons'). In both cases, individual decision makers, who exercise their rights independently, can 'impose external diseconomies'[118] (generate inefficiencies).

The model by Parisi, Schulz and Depoorter shows that 'the private incentives of excluders do not capture the external effects of their decisions'[119] and predicts that '[a]nticommons losses increase monotonically in [...] the foregone synergies and complementarities between the property fragments'.[120] They argue that legal rules should be designed taking into account potential deadweight losses due to property fragmentation[121] and cautioned against 'the current trend of vesting individual property right protection in an increasingly broader scope of [knowledge] resources'.[122] Even more critically, Zhou alleges that 'intellectual property *per se* is the very source of the tragedy of the anticommons in knowledge and that the

---

[112] Hardin (1968).

[113] See e.g. Murray and Stern (2007), p. 654; Heller and Eisenberg (1998), p. 698 (arguing that the anticommons problem in the case of patents for biotechnological research tools is 'distinct from the routine underuse inherent in any well-functioning patent system').

[114] Heller and Eisenberg (1998), pp. 698–699.

[115] ibid.

[116] Major et al. (2016), Zhou (2015), Schulz et al. (2001) and Parisi et al. (2004).

[117] Buchanan and Yoon (2000).

[118] ibid p. 4.

[119] Parisi et al. (2004), p. 183.

[120] ibid.

[121] ibid p. 184.

[122] ibid.

tragedy is unavoidable so long as an intellectual property regime exists'.[123] His model forecasts that, even though society might gain from more private investment in R&D due to IP protection, 'it loses to a greater extent from a lower rate of knowledge spillover and diffusion'.[124]

More recently, King, Major and Cosmin challenged the conventional assumption that the 'anticommons tragedy' can be avoided, even if the actors could act rationally, had perfect information, property rights were clearly defined, and no transaction costs were involved.[125] According to their model, '[t]he failure to achieve efficient bundling among the fragments [. . .] emerges as a *systematic* rather than accidental outcome',[126] which implies that even under the *ideal* bargaining conditions, the outcome is likely to be *suboptimal* due to the 'intentional maximizing behaviour of *fully rational* actors with *perfect* information'.[127] The authors argue that the problem of anticommons is 'inherent to any game comprised from autonomous actors *controlling complementary inputs*, where separate but *bundled consent* is necessary to permit use and achieve an external gain'.[128]

Even though the propensity for the underutilisation of resources due to the failure to cooperate can be demonstrated as a mathematical model, an empirical inquiry should determine whether, in a particular setting, the conditions can be conducive for the efficient conversion of rights. In the case of patents for research tools, licensing has been viewed as an optimal solution as it can, on the one hand, protect incentives by allowing innovating firms to internalise benefits of R&D and, on the other hand, enable knowledge diffusion.[129] However, while several surveys have attempted to test the anticommons hypothesis concerning patents for research inputs,[130] it would

[123] Zhou (2015), p. 3.

[124] ibid p. 14.

[125] King et al. (2016), pp. 72–74. The authors also argue that

> Ronald Coase was not correct when he asserted, given rational and costless transactions, that the negotiated outcome will always lead to the Pareto efficient use of resources. Anticommons provides a direct counter-example. It is not merely that multiple owners complicate transactions, or even that they have extra incentive to exaggerate costs and provide misleading information. Rather, the logic of anticommons maximization itself entails equilibrium at a suboptimal location relative to what would have occurred had the property right been unified. ibid p. 72.

[126] ibid p. 77.

[127] ibid (emphasis added).

[128] ibid p. 72 (emphasis added).

[129] As argued by Walsh, Arora and Cohen in the context of patent rights for research tools, '[f]rom a social welfare perspective, nothing is wrong with restricted access to IP for the purpose of subsequent discovery as long as the patent holder is as able as potential downstream users to *fully* exploit the potential contribution of that tool or input to subsequent innovation and commercialization'. Walsh et al. (2003), pp. 290–291 (emphasis added). See also Scotchmer (2004), p. 161 ff; Scotchmer (1991), p. 32 ff.

[130] Walsh et al. (2003), pp. 285–340; Walsh et al. (2005). These empirical studies are not reviewed here in detailed as they addressed the anticommons hypothesis specifically in the case of patented research tools and, therefore, might not be applicable to research data (for a critical view of the study

be safe to say that their results are fragmented and inconclusive and that 'the impact of IPR on future progress in the broader scientific community remains open to debate'.[131]

8.2.2.3 The Analogy Between Exclusive Control Over IPD and Patents for Biotechnological Research Tools

The dilemma over access to IPD as research input is reminiscent of the 'tragedy of anticommons' hypothesised with regard to patents for research tools in the context of biopharmaceutical innovation. In particular, several common features can be identified.

i. Research inputs

Both biotechnological research tools[132] and IPD constitute research inputs that can be used in 'upstream' scientific research and 'downstream' drug R&D directed at developing commercial applications of scientific knowledge.[133]

ii. Exclusive control

Patent holders can ration third-party access to biotechnological research tools; trial sponsors are usually in the position to exercise quasi-exclusive control over third-party access to non-summary clinical trial data.[134]

---

methodology, see David (2003), p. 31 ff). However, one finding is worth highlighting: based on the survey among 414 biomedical researchers in universities, government and non-profit institutions, Walsh, Cho and Cohen found 'little empirical basis for claims that restricted access to IP is currently impeding biomedical research'; at the same time, they submit that 'there is evidence that access to material research inputs [such as research data not protected by IP rights] is restricted more often'. Walsh et al. (2005), p. 2003. According to the authors, the welfare effects of restricted access to such research materials are inconclusive as the restrictions can both impede scientific progress and benefit it (in particular, by reducing duplicative research and enhancing project diversity).

[131] Murray and Stern (2007), p. 649 (emphasis added) (with further references). See also Long (2000), p. 239 (stating that the question of whether the broad scope of patent protection can be an efficient instrument of stimulating downstream innovation in the field of biomedical research as 'an unsettled issue in need of more research'); Wang (2008), p. 253 ff.

[132] Biomedical research tools refer to research resources such as cell lines, monoclonal antibodies, reagents, DNA libraries, clones, etc. See the US Department of Health and Human Services (23 Dec 1999) Principles and guidelines for recipients of NIH research grants and contracts. Fed. Reg. 64 (246), p. 72092. https://grants.nih.gov/grants/intell-property_64FR72090.pdf. Accessed 26 Mar 2021.

[133] On scientific data as research input, see Chap. 7 at Sects. 7.1.1.2 and 7.1.1.3.

[134] Such control can be enabled through technical measures of protection and non-property forms of legal protection. See Chap. 4 at Sect. 4.2.

iii. The 'bundling' of multiple authorisations

The 'bundling' of the necessary authorisations is a prerequisite for realising the value of research inputs for medical research and clinical practice. It can be achieved through patent licensing agreements in the case of patents for research tools and data-sharing agreements in the case of IPD.[135]

iv. The failure to cooperate

The anticommons hypothesis posits that multiple holders of complementary rights in an asset, who operate and make decisions autonomously, are likely to fail to 'bundle' the necessary authorisations to extract the value from that asset.[136] Scientific research and innovation can be particularly prone to this problem as they require numerous heterogeneous knowledge inputs by their very nature. As a result, the 'bundling' of the necessary authorisations can be complex and entail substantial transaction costs, especially due to the uncertainty regarding the prospective results of using the research inputs and their value.

v. Stacking licences

In the case of patents for research tools, the so-called 'reach-through' licensing agreements can allow the initial patent owner to claim benefits generated in the downstream development;[137] as will be discussed later, clinical trial data sharing agreements demonstrate a striking similarity in this regard.[138]

vi. The underutilisation of research inputs as a socially suboptimal outcome

The problem of anticommons 'takes the form of *underusage* rather than overusage of the resource [. . .] measured in nonrealized economic value'.[139] In the case of non-rivalrous knowledge resources, such unrealised economic value is

[135] As discussed earlier, many data-sharing policies adopted by pharmaceutical companies contain a standard clause explicitly stating that data shall not be shared with actual or potential competitors. See Chap. 6 at Sect. 6.3.2. See also Mattioli (2017), p. 179.

[136] King et al. (2016). On the game-theoretic models of anticommons, see above (nn 116–128) and the accompanying text.

[137] Stacking licences present another anticommons mechanism, whereby 'too many upstream patent owners [can] stack licenses on top of the future discoveries of downstream users'. Heller and Eisenberg (1998), p. 699 (further arguing that such agreements can provide the patent holder with a 'continuing right to be present at the bargaining table as a research project moves downstream toward product development'). See also Wang (2008), p. 282 (noting that '[t]he sheer number of relevant patents and patentees in a research or commercialization project spawns the burden of reviewing patent claims and negotiating necessary licenses').

[138] Chapter 9 at Sect. 9.3.2.4, subheading 'Concerns Regarding 'Stacking Licenses''.

[139] Buchanan and Yoon (2000), p. 4 (emphasis added). According to the authors, anticommons is 'a useful metaphor for understanding how and why potential economic value may disappear into the "black hole" of resource underutilization, a wastage that may be quantitatively comparable to the overutilization wastage employed in the conventional commons logic'. ibid p. 2.

contemplated as missed research opportunities[140] and 'significantly inhibited' knowledge externalities.[141] Patents for research tools were argued to impede the processes of discovery and invention.[142] Similarly, restricted access to clinical trial data is viewed to 'slow the rate of scientific discovery and advancement'.[143] Such concerns arise, in particular, due to the role of research tools in enabling and promoting the cumulativeness of knowledge, research and innovation.

vii. Dual implications of exclusive control for innovation

The anticommons debate stems from the dual implications of exclusive control over 'upstream' inputs for innovation.[144] The ability to control access to both IPD and patents for research tools posits a policy dilemma between the need to protect innovation incentives, conventionally associated with exclusive control over R&D results as a means of earning returns on R&D, and the desirability of broad access and utilisation of knowledge.[145]

Given the above-outlined similarities, let us consider how the concept of anticommons could inform the normative analysis of access to IPD.

8.2.2.4 The Relevance of the Concept of Anticommons for Clinical Trial Data

Even without applying a game-theoretic model, there is ample evidence of a lack of cooperation among IPD holders. As discussed above, competitive concerns urge companies to pursue a non-sharing strategy,[146] even though such concerns could be rationalised only in limited cases.[147] Research on anticommons further provides a conceptual framework for evaluating the social cost of exclusive control over data as a source of knowledge.

The notion of anticommons epitomises a social dilemma where an individually rational strategy can lead to a collectively suboptimal outcome.[148] In the context of access to IPD, companies' 'individually rational strategy' of protecting competitive

---

[140] Käseberg (2011), p. 13 (observing that 'it has been recognised [. . .] that broad patent protection in cumulative innovation settings may impede follow-on innovation and lead to opportunity losses in terms of dynamic efficiency').

[141] Zhou (2015), p. 4.

[142] David (1993), p. 55 (emphasis added).

[143] Institute of Medicine of the National Academies (2015), p. 141 (with further references). See also Brandt-Rauf (2003), p. 66 (highlighting an analogous argument with regard to scientific data in general, namely, that the resistance to share such data can 'slow the progress of science because scientists cannot easily build on the efforts of others or discover errors in completed work').

[144] See above (nn 129–131) and the accompanying text.

[145] On the 'access-incentives paradox' concerning access to IPD, see Chap. 7 at Sect. 7.1.2.

[146] Chapter 6 at Sect. 6.3.2.

[147] Chapter 5 at Sects. 5.3 and 5.4.

[148] See above (n 107) and the accompanying text.

advantage is posited to lead to a collectively suboptimal outcome—'missed opportunities to gain scientific knowledge'[149] and, ultimately, 'suboptimal return on the altruism and contributions of clinical trial participants, the efforts of clinical trialists and research staff, and the financial resources invested by study funders and sponsors'.[150] The scope of 'missed opportunities' can hardly be appraised in the absence of a realistic counterfactual—an understanding of what discoveries could have been made and what innovative products could have been developed through exploratory IPD analysis, but for the trial sponsors' control over IPD.[151]

The main distinguishing factor between patents for research tools and exclusive control over IPD can be seen in the *excludability* of the subject matter at issue. Technical teaching embodied in an innovative product can often be reproduced once the product is commercialised. In contrast, IPD cannot be 'reverse-engineered' from a marketed drug while trial sponsors are usually in the position to exercise exclusive control over third-party access even in the absence of property rights in data as such. In other words, in the case of IPD, there is no 'spillover' effect to begin with that would necessitate the protection by exclusive rights by analogy with patents for research tools. As discussed earlier, the sponsors' control over all data from trials was rationalised by reasons other than solving the 'public-good' problem of innovation incentives.[152] Hence, the balancing exercise in the case of access to IPD is different. First, there is no need for (additional) protection of innovation incentives; second, concerns due to trial sponsors' control over IPD—including the potential impact on medical research and drug R&D—need to be factored into the equation, as considered below.

[149] Institute of Medicine of the National Academies (2015), p. 18 (emphasis added) (with further references).

[150] ibid.

[151] See e.g. Hall and Harhoff (2012), p. 554 (observing that the hypothesis that the overall effect of patents can be negative, especially in a setting, where innovation is cumulative, 'is difficult to test because of the absence of a true counterfactual'); Heller (2011), p. 73 (noting that it is 'hard to know how to quantify gridlock, in part because it involves testing a counterfactual: what cures would we have it people could work together more easily?'). Conversely, one could also question whether research tools would have been invented but for the incentives provided by patent law, which locks the debate in a circular argumentation. See Murray and Stern (2007), p. 684 (finding support for the 'anticommons' effect and also acknowledging that such evidence 'captures only one aspect of the impact of IP on dual knowledge', while patent rights might have 'enhanced the incentives for (unobserved) research [and] led to more effective (or rapid) commercialization [. . .], or allowed for cumulative innovation through patents and future patent citations').

[152] See Chap. 4 at Sect. 4.2.1.2, subheading 'The Obligation to Protect Data Against Unauthorised Access as the Source of de facto Exclusive Control'.

8.2.3 *Foregone Efficiencies in Drug Development as a Distinct Social Cost*

8.2.3.1 *Duplicative* v *Cumulative Research*

Duplicative research efforts result from the inability of researchers to build on the knowledge developed in earlier studies, in particular, due to the barriers to search, access or exchange of the relevant information. Duplication of efforts generates inefficiency in R&D resource allocation at the sector level. The opposite is the *cumulativeness* of knowledge and the associated efficiency gains. Given that knowledge accumulation is an inherent feature and a fundamental principle of scientific research and technological innovation,[153] the cost of restricted access to resources enabling cumulativeness is likely to be appreciable.

Furthermore, inefficiency due to redundant research can be defined as opportunity costs—expenditures that could have been either saved (without diminishing social benefits[154]) or directed at bridging knowledge gaps. In the context of medical research and drug innovation, 'knowledge gaps' refer to unmet needs and clinical uncertainties that often serve as the starting point of drug R&D programs.[155]

Such considerations can explain why facilitating access to knowledge resources is generally considered a policy priority. For instance, in its Recommendation on access to and preservation of scientific information,[156] the European Commission emphasises that access to scientific data 'enhances data quality, reduces the need for duplication of research, speeds up scientific progress and helps to combat scientific fraud'.[157] It also explicitly states that the conditions for conducting research can be improved 'by reducing duplication of efforts and minimising the time spent searching for information and accessing it'.[158] In medical research, the elimination of duplicative trials is viewed as a distinct goal of the EMA's publication policy.[159] In this regard, it is worth emphasising that promoting cumulativeness and efficiency in research through secondary exploratory analysis of IPD should be distinguished from eliminating duplication of trials in the case of generic drug approval.

---

[153] Johnson (2005), pp. 42–43; Edquist (2005), p. 19; David (1993), p. 28 (arguing that scientific and technological knowledge is 'cumulative and interactive [and] grows by increments, with each advance building on [. . .] previous findings in complicated and often unpredictable ways').

[154] OECD (2017), p. 11 (defining wasteful spending as costs that could have been avoided by cheaper substitutes with identical or better benefits).

[155] Pharmaceutical sector inquiry report (n 7), p. 395.

[156] European Commission (21 Jul 2012) Commission recommendation of 17 July 2012 on access to and preservation of scientific information, 2012/417/EU. OJ L 194/39.

[157] ibid rec 10.

[158] ibid rec 6, 10.

[159] EMA. Clinical data publication. http://www.ema.europa.eu/ema/?curl=pages/special_topics/general/general_content_000555.jsp. Accessed 26 Mar 2021.

8.2.3.2 The Role of Secondary Analysis of Clinical Trial Data in Fostering Cumulativeness of Drug R&D

Conventionally, drug innovation was characterised by low cumulativeness.[160] Innovative capacity and experience of drug companies in one therapeutic area were assumed to be of limited relevance for research in other therapeutic fields because 'it is difficult [for firms] to use the knowledge accumulated in a particular research project in subsequent innovative efforts'.[161] Knowledge and experience acquired in the earlier R&D were not considered advantageous for drug discovery even within the same therapeutic category.[162] However, such views appear too narrow. Drug innovation is science-driven[163] and cumulative in the sense that the discovery and development of new molecules and targets are guided by knowledge gained in earlier research, including clinical trials.[164] In this regard, the expertise accumulated by a firm in a particular therapeutic area can play a crucial role for its competitiveness in R&D.[165]

Furthermore, pharmacological knowledge gained in clinical trials can support cumulativeness of research beyond the boundaries of a therapeutic indication tested in the initial trial. The human organism is a complex system with several inherently interrelated sub-systems. A treatment that targets one condition is likely to affect other sub-systems. In this regard, it is helpful to distinguish between the primary intended (on-target) effects of an intervention and secondary unintended (off-target) effects. Off-target effects are also known as side effects that can be harmful, neutral or beneficial.[166] Research on one molecule, or a drug based on a particular chemical structure, is likely to generate data and knowledge about its pharmacological effects on other sub-systems. Such knowledge can, among others, advance the understanding of other medical conditions, lead to new hypotheses regarding structure-activity correlations and predict biological responses to medical interventions.[167] In other words, the cumulative nature of research cannot be restricted to one particular therapeutic area. Consequently, IPD analysis can enable the cumulativeness of knowledge and innovation across therapeutic fields.

The potential to foster cumulativeness of innovative activity can be regarded as a general definitional feature of research inputs (research tools) as 'intermediate'

---

160 Levin (1988), p. 427 (noting in the chemical and drug industries 'innovations [. . .] stand alone as isolated discoveries').

161 Cefis et al. (2006), p. 164.

162 Garavaglia et al. (2006), p. 244.

163 See Orsenigo et al. (2006), p. 416 (emphasising that the role of science is 'more direct and immediate in pharmaceuticals than in most other technologies').

164 See e.g. British Pharmacological Society. Pharmacology skills for drug discovery, p. 4. https://thebiologist.rsb.org.uk/images/Pharmacology_Skills_for_Drug_Discovery.pdf. Accessed 26 Mar 2021; Achilladelis and Antonakis (2001), p. 571.

165 Henderson and Cockburn (1996).

166 Taniguchi et al. (2008), p. 63. On the types of unintended effects, see Kim et al. (2016), p. 399.

167 Khan (2014), p. 397 ff.

goods. Walsh, Arora and Cohen observe that the 'progress in biomedical research is now more cumulative; it depends more heavily than heretofore on prior scientific discoveries and previously developed research tools'.[168] As discussed earlier, aggregated IPD as research input can be used in multiple analyses directed at the research questions beyond those addressed in the initial trial. More specifically, secondary IPD analysis could promote cumulativeness of medical research and drug R&D in the following ways.

First, comprehensive datasets from past trials can be sufficient for analysing new research questions and, thus, exempt from conducting new trials. As noted by Gøtzsche, '[m]uch research could be performed, at almost no cost, on existing data, making it unnecessary to collect new data',[169] while '[a]n incomplete knowledge base [...] leads to redundant research'.[170] The concept of 'extrapolation' discussed earlier[171] exemplifies the idea that secondary analysis of data from past trials could make the design of subsequent trials more efficient.[172] This proposition relates to the much-debated issue of wasteful trials.[173] According to Chalmers and Glasziou, up to 85% of clinical trials can be cumulatively considered wasteful.[174] By 'waste', the authors broadly refer to the defects in the ways research is designed, conducted, analysed, reported, regulated and managed. *Redundant* trials—trials addressing research questions that could be 'answered satisfactorily with existing evidence'—represent one category of 'waste'.[175] Several empirical studies provide evidence on wasteful trials due to the insufficient analysis of the available data.[176] Apart from obvious ethical concerns, such trials constitute misallocation of economic resources and entail the opportunity cost of unaddressed 'knowledge gaps'.[177]

Second, the re-analysis of data from earlier trials can support the cumulativeness of research by optimising the design of new trials.[178] Usually, research teams thoroughly examine literature and available evidence from earlier studies relevant to the research hypothesis of the planned trial. The more detailed evidence can be accessed and analysed, the more precisely research questions, endpoints, effect size,

---

168 Walsh et al. (2003), p. 281.

169 Gøtzsche (2012), p. 237.

170 ibid.

171 Chapter 3 at Sect. 3.4.2.

172 EMA (19 Mar 2013) Concept paper on extrapolation of efficacy and safety in medicine development. EMA/129698/2012.

173 Research: increasing value, reducing waste. http://www.thelancet.com/series/research. Accessed 26 Mar 2021.

174 Chalmers and Glasziou (2009), p. 88.

175 ibid p. 87. See also Flohr and Weidinger (2016), p. 1930 (pointing out that an 'unnecessarily large number of vehicle-controlled studies' have been conducted in the field of atopic eczema (with further references)).

176 For an overview and discussion, see Kim and Hasford (2020).

177 Flohr and Weidinger (2016), p. 1931.

178 See Stoney and Johnson (2018), p. 251 (emphasising that understanding the state of scientific knowledge in a particular therapeutic area is key in trial design).

control groups, analytic strategies and other elements of trial design can be defined.[179] Furthermore, IPD meta-analyses can identify inter alia the prognostic factors, risk groups and disease characteristics that can modify the treatment effect and be further explored in new trials.[180]

Third, knowledge about the failed pursuits is an important factor of the sector-level R&D productivity and the prompt dissemination of negative outcomes is indispensable for preventing unnecessary trials.[181] From an economic perspective, this corresponds to the *cost-reduction* effect of knowledge externalities, whereby 'one firm's abandonment of a particular research line signals to others that the line is unproductive and hence saves them the expense of learning this themselves'.[182] While a 'failed' trial means that a hypothesis regarding the treatment safety or efficacy could not be supported, the treatment effect might still be confirmed under the modified conditions. Exploratory data analysis of IPD from unsuccessful trials could reveal additional insights and direct the subsequent research[183] if researchers could access such data.[184] Notably, drug companies also acknowledge that secondary analysis of trial data allows to 'avoid some of the trial-and-error development'.[185]

Finally, it should be emphasised that the above-outlined efficiency gains can be realised only when IPD is analysed. In this regard, it is worth highlighting that IPD meta-analysis is 'resource demanding, time consuming, and methodologically challenging',[186] which might explain why scaling up secondary analysis of even

---

[179] See Massaro (2009), p. 46 (noting that one must have

> a *reasonable assumption* [. . .] to ensure an adequate sample size and sufficient power. *Hopefully* there are previous exploratory or pilot studies or previously published data that the researchers can use to make a reasonable assumption of the true effect size. If not, the researchers are left with no choice but to make a "*best guess*" at the effect size (emphasis added).

See also The International Council for Harmonisation of Technical Requirements for Pharmaceuticals for Human Use (ICH) (1998) ICH harmonised tripartite guideline. Statistical principles for clinical trials. E9, p. 5 (recommending to use 'a reliable and validated variable with which experience has been gained either in *earlier studies* or in published literature' (emphasis added)); Stoney and Johnson (2018), p. 215; Cleophas et al. (2006), p. 1.

[180] Tierney et al. (2015).

[181] Gøtzsche (2012), p. 237 (observing that '[w]hen failures with previous drugs or devices are kept secret, expensive development programs for similar drugs or devices can continue for years after they would have been stopped if the data had been known').

[182] Jaffe (1998), p. 11.

[183] See Chap. 3 at Sect. 3.3.4.

[184] See e.g. Gustafsson (2010), p. 941 (referring to situations where 'commercial sponsors [. . .] have lost interest supporting academic trialists in pursuing posttrial studies or even have attempted to discourage them from doing so, particularly if the original hypothesis was rejected').

[185] Case T-718/15 R *PTC Therapeutics International v EMA* [2016] ECLI:EU:T:2016:425, para 92 (emphasis added).

[186] Nevitt et al. (2017).

accessible data can be a challenge.[187] Thus, factors enabling secondary IPD analysis are decisive for the effectiveness of policy measures intended to promote the cumulativeness of medical research and drug development through access to data.

8.2.3.3 Reducing Uncertainty

Reduction of uncertainty in research is another important implication of secondary IPD analysis. Drug R&D is generally characterised by high uncertainty and serendipity and is often compared to a 'lottery'.[188] As the term itself suggests, clinical trials are subject to trial and error, while the risk of failure persists till the very late stage.[189] On average, only a small number of compounds become marketable products through the 'iterative cycles of testing, understanding, modifying, and retesting potential solutions'.[190] While the pharmaceutical industry has been criticised for directing research efforts at incremental innovation and 'me-too' drugs, there is also evidence to the contrary. For instance, the study by Pammolli, Magazzini and Raccaboni shows that the decline in drug R&D productivity in the past decades is associated with the increasing concentration of R&D investments in the areas of unmet therapeutic needs and unexplored biological mechanisms where uncertainty and the risk of failure are especially high.[191]

Over the past decades, drug discovery transformed from the large-scale 'random screening' (heuristic drug discovery) towards a more 'rational', science-guided drug design.[192] Such shift is attributed to the improved understanding of the biological basis of diseases allowing researchers and drug developers to design research strategies in a more targeted way.[193] The more knowledge on disease processes and determinants is available, the more precisely the new research hypotheses and

[187] Below at Sect. 8.3.5 in this chapter.

[188] Mitscher (2002), p. 31.

[189] See e.g. USGAO (2006), p. 87 (reporting the failure rates in human clinical trials due to the lack of safety or efficacy: 82% during the period between 1996 and 1999 and 91% during the period between 2000 and 2003).

[190] Nightingale and Mahdi (2006), p. 81. See also National Research Council of the National Academies (2010), p. 126 (defining the process of treatment discovery as 'an iterative process of studying a disease, hypothesizing and developing treatments, evaluating those treatments, and, for successful treatments, further refining the indication to account for lack of efficacy or toxicities (or both) in particular subgroups of patients'). Furthermore, the report points out that 'the scientific development of a particular treatment indication is [as a rule] connected with that of other treatments, and thus it may be difficult to identify the exact process that led to the adoption of some treatment'. ibid.

[191] Pammolli et al. (2011), p. 428. See also Pharmaceutical sector inquiry report (n 7), p. 395 (finding that the overlaps in drug R&D projects are common because most companies direct R&D efforts at the unmet needs).

[192] See e.g. Dosi and Mazzucato (2006), p. 3; Garavaglia et al. (2006), p. 238. See generally Drews (2000).

[193] OECD (2004), p. 44. See also Dosi and Mazzucato (2006), p. 3.

questions can be formulated, and the more optimally the subsequent trials can be designed. Conversely, the less is known about the drug mechanism of action and the disease characteristics, the harder it is to predict the prospective on- and off-target effects.[194]

The types of IPD analysis surveyed in Chap. 3 play a crucial role in fostering cumulativeness of research and, thus, reducing uncertainty. The importance of IPD concerning on- and off-target effects should be highlighted in this regard. Given that the binding of a drug to a target 'largely depends on the structural correspondence of the drug molecule and the binding cavity of the target molecule',[195] the analysis of correlations between the drug structure and *individual* biological responses is key.[196] The spectrum of the individual biological responses can be 'overwhelmingly diverse'[197] and depend on the cell types, cellular states, individual genetic characteristics, etc.[198] For overcoming such challenge, the authors recommend the centralised and systematic collection of measurements across different drugs, cell types and diseases that could enable extensive search for structure-activity patterns.[199]

Furthermore, confirmatory secondary data analysis should be viewed as equally important for enabling the cumulativeness of knowledge. After all, exploratory research needs to build on verified knowledge.[200]

8.3 The Issue of Wasteful Duplication of Research Efforts Due to Data Disclosure

8.3.1 The Hypothesis Regarding Wasteful Duplication of Research Efforts

The pharmaceutical industry has been criticised for 'misallocating resources'[201] by concentrating research efforts in developing follow-on products featuring minor improvements of the existing drugs.[202] One could hypothesise that disclosure of IPD would exacerbate such resource misallocation because multiple exploratory

---

[194]Khan (2014), p. 497.

[195]ibid.

[196]ibid.

[197]ibid.

[198]ibid.

[199]ibid.

[200]Complementarity between confirmatory and exploratory analyses constitutes a general principle of scientific research. See e.g. David (2003), p. 19.

[201]Mueller and Frenzel (2015), p. 73.

[202]In this regard, patents for drug improvement have derived criticism for reducing the incentive to invest in the development of pioneering drug. See Hollis A (13 Dec 2004) Me-too drugs: is there a

analyses of the same data would likely produce duplicative outcomes—similar modifications of the investigational products tested in the initial trial.

8.3.2 Factors Contributing to Duplicative R&D

The combination of the following factors is generally considered to make an environment conducive to the 'racing' behaviour:

(i) a shared (commonly accessible) knowledge resource that can ease a firm's entry into R&D;
(ii) a viable technological opportunity that can be identified and pursued by multiple actors;
(iii) the prospect of commercial success (supra-competitive profits, especially due to patent protection);
(iv) a *lack of coordination* among the entities who make independent decisions to invest in R&D.[203]

Drug R&D presents as a pertinent example where the key determinants of R&D investment decision making are an opportunity of discovering a new therapeutic indication anticipated by the unmet medical need, the demand size (estimated by the medical need) and the prospect of obtaining supra-competitive profits (usually due to patent protection).[204] Knowledge produced in fundamental biomedical research often becomes a 'commonly accessible resource' that can open technological opportunities in drug discovery and enable multiple companies to enter into competition in R&D directed at the development of the downstream applications of basic scientific knowledge.[205] In light of these factors, one could assume that IPD disclosure would contribute to the 'racing' behaviour. If secondary exploratory analysis of non-summary trial data can increase the chances of successful drug discovery, companies would be motivated to conduct secondary analyses of IPD once it becomes accessible, thereby increasing the likelihood of the suboptimal allocation of R&D resources (overinvestment).[206]

---

problem? p. 3. https://www.who.int/intellectualproperty/topics/ip/Me-tooDrugs_Hollis1.pdf. Accessed 26 Mar 2021.

[203] Foray (2004), p. 169; Cockburn and Henderson (1994), p. 484 (noting that 'free entry into R&D competition [can] result in overinvestment relative to both the private or social optima'). See also Chap. 7 at Sect. 7.2.3.3, subheading 'The 'Exhaustion' Externality or the 'Stepping-on-Toes' Effect'.

[204] Cockburn and Henderson (1994), p. 482; Pharmaceutical sector inquiry report (n 7), p. 57. As noted earlier, patents are generally viewed as a strong incentive in the drug industry. See Chap. 5 Sect. 5.2.4.

[205] Cockburn and Henderson (1994), p. 486. See also Foray (2004), p. 171.

[206] Cockburn and Henderson (1994), p. 484 (highlighting the finding of the literature on firms' strategic interaction that 'free entry into R&D competition [can] result in overinvestment relative to both the private or social optima').

However, it should be emphasised that parallel R&D, as such, should not be equated with the wasteful duplication of research efforts. As highlighted above, the relationship between R&D intensity and innovation is not straightforward: In some cases, the reduction of experimentation can lead to the improved research quality;[207] in other situations, a high concentration of R&D in a particular area can cause excessive investment (the 'stepping-on-toes' effect).[208] Thus, the question is under what conditions parallel IPD analyses are likely to produce duplicative results.

8.3.3 *The Rivalry of R&D Benefits*

The rivalry of R&D benefits is one factor correlating with the propensity of parallel research to generate redundant outcomes. In general, rivalry in consumption (the perfect divisibility of benefits) means that one's consumption of a good 'fully eliminates any benefits that others might have derived from that unit',[209] while non-rivalry (the indivisibility of benefits) implies that 'a unit of the good can be consumed by one individual without detracting [. . .] from the consumption opportunities still available to others from that same unit'.[210] Scientific data can be characterised as a non-rivalrous resource as it can be utilised (analysed) by multiple researchers in concurrent projects without depriving each other of deriving benefits (knowledge)[211] or imposing additional 'amortisation' costs on the trial sponsor.[212] However, there is a further nuance where *intermediate* goods, such as research tools, are used as inputs for developing knowledge and innovative products.[213] While such tools can be non-rivalrous in *use* in the sense that they can be utilised in parallel research activities, the *benefits* derived through such use can, to some extent, be rivalrous.[214] The decisive factor is whether one company's use of a research tool can diminish the opportunities for other entities to earn returns on innovative activity.[215]

---

[207] Linge (2008), p. 214.

[208] See Chap. 7 at Sect. 7.2.3.3, subheading 'The 'Exhaustion' Externality or the 'Stepping-on-Toes' Effect'.

[209] Cornes and Sandler (1999), p. 8.

[210] ibid.

[211] See e.g. Nelson (2009), p. 10 (observing that 'understandings won from any R&D effort have public good properties [. . .] in the sense that use by one party does not reduce the stock of understanding that might be used by another'); Reichman (2009), p. 51 (arguing that 'the information gleaned from the clinical testing of drugs and therapies is a public good in the sense that each individual citizen benefits from such information without reducing its value to others').

[212] David (2003), p. 20 (observing that the 're-use of the information will neither deplete it nor impose further costs').

[213] On intermediate non-rivalrous goods, see Chap. 7 at Sect. 7.1.1.2.

[214] Walsh et al. (2003), p. 332 (referring to the 'rival-in-use' research tools as tools 'primarily used to develop innovations that will compete with one another in the marketplace').

[215] ibid (referring to a receptor responsible for a particular disease as an example of 'rivalrous' research tools). Once one firm identifies a compound that blocks such receptor, 'it undermines the

In other words, the benefits derived from a research resource can be rivalrous where a limited number of technological opportunities can be identified and developed based on that resource. Once those opportunities are identified and realised by one research team, the efforts of others would be rendered futile if they are not able to earn returns on R&D. As discussed earlier, in the 'winner-take-all' setting, one competitor's success in R&D imposes an 'exhaustion externality' on other competitors researching the same direction.[216]

Firms conduct R&D in the pursuit of new technological opportunities. Innovation races are likely to entail duplicative R&D if 'the opportunities to discover are *finite* and firms vie for them'.[217] Hence, '[b]y making a discovery, a firm ends the race and deprives the other of the chance to claim priority'.[218] Applying this criterion to IPD, the benefits of exploratory IPD analysis would be rivalrous if one company's discovery of knowledge in data deprived other drug developers of realising economic benefits from an independent discovery of such knowledge. For example, suppose that secondary analysis of IPD related to investigational product X can reveal its new therapeutic indication. If those datasets are freely available, a new indication could, theoretically, be discovered by multiple research teams. In this sense, the use of IPD is non-rivalrous. However, the *results* of IPD analysis would be rivalrous in a situation where a new indication is discovered and commercialised by one company and where that company can protect its competitive advantage—for instance, by obtaining patents—thereby precluding competitors from realising commercial benefits of the independent discovery of the same indication, thus, rendering their research efforts wastefully duplicative. That is in theory. Let us consider next whether the benefits of IPD analysis are likely to be rivalrous in the sense that one company's findings would 'deplete' data as a source of medical (economically relevant[219]) knowledge.

8.3.4 Clinical Trial Data as a Rivalrous and Non-rivalrous Good

At the outset, it should be noted that drug R&D, in general, is not regarded as duplicative in the sense that research projects directed at a particular problem are

ability of another to profit from [the] compound that blocks the same receptor'. In contrast, non-rivalrous research tools tend to be of the 'general purpose' nature, in the sense of being utilised across various therapeutic areas, e.g. genomics databases, DNA chips, recombinant DNA technology, etc. ibid p. 323.

[216] Henderson and Cockburn (1996), p. 36. See Chap. 7 at Sect. 7.2.3.3, subheading 'The 'Exhaustion' Externality or the 'Stepping-on-Toes' Effect'.

[217] Denicolo and Franzoni (2012), p. 120 (emphasis added).

[218] ibid (also referring to such situation as 'a common pool problem in innovation races').

[219] The economic relevance of knowledge is arguably an integral characteristic of the rivalry of goods.

likely to yield identical results (perfectly substitutable drugs).[220] On the contrary, independently pursued research paths is viewed as an 'insurance value of parallel experimentation',[221] which is especially important in drug R&D, given the high rate of failure.[222] As a side note, for the same reason, it has been disputed under competition law analysis whether the reduction of parallel R&D tracks carried out by actually or potentially competing drug companies can be justified on the efficiency grounds, in particular, in the merger assessment,[223] even though the elimination of the overlapping R&D activities is generally viewed as a source of efficiency gains in cooperative R&D.[224]

In the case of exploratory IPD analysis, it appears unfeasible to define *ex ante* to what extent the analysis benefits, such as new opportunities in drug discovery, might be rivalrous. Such opportunities can vary from case to case.[225] Given that exploratory data analysis often requires IPD aggregation, it would be logical to assume that the larger the data pool, the lower the probability that multiple data analyses would

---

[220] See Blass (2015), p. 455 (defining duplicative efforts as the research programs 'focused on the same macromolecular target'). See also Ben-Asher (2000), p. 232 (noting, in the context of the assessment of drug mergers, that the 'straightforward and wasteful duplication in medical research leading to welfare loss is unlikely'); Cockburn and Henderson (1994), p. 484 (observing that 'competing [drug R&D] projects may well be complementary [and] similar research can lead to related but significantly different outcomes'); Carrier (2009), p. 305 ff.

[221] Kerber (2010), p. 184 (with further references).

[222] On the resource allocation to research in general, see Dasgupta and Maskin (1987), p. 582 (arguing that parallel research does not necessarily imply waste and, since 'the outcome of any research project is uncertain, it is generally in society's interest to hold a portfolio of active projects on any scientific or technological problem'). Concerning drug R&D, see Ben-Asher (2000), p. 317; Pammolli et al. (2011), p. 437 (arguing that 'parallel R&D along similar trajectories [...] should not necessarily be considered as wasteful duplication or imitation').

[223] As emphasised by Carrier,

> where the merging firms only have products in preclinical development, the staggering odds that either one would reach the market, let alone both, counsels the agencies not to challenge the merger. Nor does a merger between a firm with a product in advanced trials [...] and one in preclinical development raise concern. Even if [...] those two firms are 'closest' to the market, the improbability that the latter will ever reach the market reduces concern.

Carrier (2009), pp. 305–306.

[224] See Foray (2004), p. 57 (noting that efficiency gains arising from collaborative research include 'sharing research costs and avoiding duplicative projects; the benefit to be harnessed from creating larger pools of knowledge, which in turn generate greater variances from which more promising avenues of research can be selected; and the economic gains to be generated from division of labor in research activities'. See Kerber and Schwalbe (2008), para 1-8-368 (observing that 'it is not clear-cut that the reduction of parallel R&D tracks is always efficient'). They argue, in the context of merger assessment, that 'in some situations, where the firms prior to a merger might be involved in a patent race, a reduction in R&D efforts is likely to be justified on the efficiency grounds, while in other situations, it might be more beneficial if the merging firms keep own research activities, especially, if this increases the probability of making a certain invention'). See also Link (2007), p. 135; OECD (2013) The role of efficiency claims in antitrust proceedings. DAF/COMP(2012) 23, p. 15.

[225] Gustafsson et al. (2010).

yield duplicative outcomes. Besides, as shown in Chap. 3, IPD analysis can support research throughout the drug R&D cycle. In this sense, the aggregated anonymised IPD can be viewed as a 'general-purpose' knowledge resource. As noted earlier, the benefits of using 'general-purpose' research tools tend to be non-rivalrous.[226]

Overall, concerns regarding duplicative results of multiple secondary data analyses appear to be unwarranted, given potentially unlimited ways of integrating IPD with other types of health data,[227] perspectives from which secondary data analyses can be approached and research questions beyond those addressed in the 'parent' trial.[228] Thus, multiple secondary (exploratory) analyses of IPD should be viewed as contributing to multiplicity and diversity of research and experimentation[229] rather than causing wasteful duplication.

Finally, it is worth highlighting that the general assumption that control over 'upstream' knowledge resources can efficiently coordinate R&D efforts has been challenged. The criticism of the 'prospect theory' of patents[230] can well apply in the context of clinical trial data. As will be discussed later, negotiating access to clinical trial data for R&D purposes on an individual basis can hardly present a viable solution for realising its research potential.[231]

8.3.5 Evidence on Secondary Analysis of Clinical Trial Data in the Industry R&D

Empirical studies show that the potential of even the accessible historical data is not realised optimally in planning and designing new trials. For instance, the study by Tierney et al. identified several ways of how IPD meta-analysis can be used to inform the design, conduct, analysis, and interpretation of the subsequent trials that may not be possible based on the systematic reviews of the *summary-level* data.[232] Notably, they found that only about half of the reviewed IPD meta-analyses had indeed influenced the design and conduct of new trials, including by informing the choice of the comparators and trial participants, sample size calculations, and interpretation of the trial outcomes. Jones et al. examined how meta-analysis of data from the completed trials had been used to inform the design of the subsequent

---

[226] Above (n 215) and the accompanying text.

[227] As discussed in Chap. 3, exploratory data analysis often requires aggregating datasets from multiple trials.

[228] Gustafsson et al. (2010), p. 938. For an overview of the types of IPD analysis, see Chap. 3.

[229] On the role of multiplicity and diversity of experimentation in evolutionary economics, see Chap. 7 at Sect. 7.2.3.3, subheading 'Duplicative Research v Multiplicity and Diversity of Experimentation'.

[230] On this issue, see Chap. 7 at Sect. 7.2.3.3, subheading 'The (Controversial) Role of Patents as a Means to Coordinate Research Efforts'.

[231] For a detailed analysis, see Chap. 9 at Sect. 9.3.1.

[232] Tierney et al. (2015).

trials and found that only about half of the trial documentation in the analysed sample had used the referenced systematic reviews to select and define trial outcomes, calculating the sample size and other aspects of the trial design.[233] Goudie et al. report that only a small fraction of the examined sample of trials referenced the relevant prior studies and related the trial results to the previous research.[234] They argue that '[p]revious evidence is not used (or not reported to be used) as extensively as it could be in justifying, designing, and reporting RCTs'.[235] Based on the cumulative meta-analysis, the study by Storz-Pfennig[236] shows that, to a large extent, clinical trials are unnecessary because the available evidence from the past trials could be sufficient to 'reach a reliable conclusion'.[237]

These findings demonstrate that the mere accessibility of clinical trial data is insufficient to achieve the policy objective of making research more informed, efficient and targeted. Only when data is analysed can its value as a source of knowledge be realised.

8.4 On Balance

8.4.1 The Summary of Implications of IPD Disclosure for the Allocation of Resources to R&D

This chapter sought to determine whether a regulatory intervention by access measures could be rationalised on the grounds of promoting drug innovation. The problem of access to IPD was framed as a case on R&D externalities. The characterisation of IPD as an 'intermediate' good—the result of and input in R&D as a knowledge-production activity—poses divergent implications of data disclosure (non-excludability) for innovation incentives at the firm and sector level. Theoretically, the non-excludability of data as a knowledge resource could lead to two specific problems in R&D resource allocation: *insufficient* incentives for data generation as a knowledge production activity (conducting clinical trials) and *excessive* incentives for data consumption (data analysis) if IPD analysis can yield wastefully duplicative outcomes. In line with the theoretical framework outlined in Chap. 7, the Table 8.1 summarises the findings of the preceding analysis.

---

[233] Jones et al. (2013). Given the potential of systematic reviews to optimise the design of subsequent trials, the authors proposed that special guidelines for applicants and funders should be developed 'to optimise delivery of new studies informed by the most up-to-date evidence base and to minimise waste in research').

[234] Goudie (2010).

[235] ibid p. 984.

[236] Storz-Pfennig (2017).

[237] ibid p. 61.

Table 8.1 Summary of implications of IPD disclosure for resource allocation in R&D

	The relationship between IPD disclosure and the problem of *insufficient* incentives in R&D	The relationship between IPD disclosure and *excessive* incentives for data analysis
The initial hypotheses concerning the impact of IPD disclosure	Disclosure of (anonymised) IPD can negatively affect the appropriability conditions and, hence, innovation incentives of drug sponsors.	Disclosure of (anonymised) IPD can lead to wastefully duplicative secondary data analyses.
The revised hypotheses upon the analysis of the role of exclusive control over IPD	(i) Competition by *imitation*: exclusive control over IPD does not protect originator drugs against *generic* competition (provided that test data exclusivity protection is duly applied and enforced). (ii) Competition in *innovation*: exclusive control over IPD can protect intermediate research results (the priority of the trial sponsor in conducting exploratory IPD analysis).	Exclusive control over IPD is unlikely to perform the function of coordinating research efforts of research-based drug companies.
Findings on the effects of IPD disclosure on knowledge externalities	(i) Data disclosure is unlikely to cause *imitation* externalities (beyond the level achieved through reverse-engineering). (ii) Data disclosure is likely to generate *research* externalities (if competitors' R&D can benefit through secondary IPD analysis).	Data disclosure is unlikely to entail the '*fishing* out' externality due to the unlimited ways of aggregating data and approaching exploratory data analysis.
Opportunity costs of exclusive control over IPD	Under the trial sponsors' exclusive control, the research potential of IPD is likely to be underutilised.	Exclusive control over IPD can (i) inhibit *multiplicity* and *diversity* of experimentation and research outcomes; (ii) entail *foregone efficiencies* in terms of better informed and optimised subsequent drug discovery and design of new trials.
Potential policy trade-offs due to IPD non-excludability (if a policy measure removes trial sponsors' control over IPD)	The *diffusion-appropriability* trade-off is unlikely to arise due to IPD non-excludability in the presence of patents, SPCs, test data exclusivity and other regulatory exclusivities that protect the appropriability conditions and, thus, innovation incentives of drug sponsors.[a] Hence, exclusive	The *efficiency-multiplicity* trade-off is unlikely to arise due to IPD non-excludability as, in general, there is hardly a risk that multiple exploratory analyses of IPD aggregated from different trials would be wastefully duplicative. Thus, exclusive control over third-party IPD analysis cannot be

(continued)

Table 8.1 (continued)

	The relationship between IPD disclosure and the problem of *insufficient* incentives in R&D	The relationship between IPD disclosure and *excessive* incentives for data analysis
	control over IPD cannot be rationalised as a means of resolving the diffusion-appropriability trade-off. Rather, access measures can be justified as a means of promoting *cumulativeness* of research and knowledge.	rationalised as a means of resolving the efficiency-multiplicity trade-off. Rather, access measures can be justified as a means of promoting *diversity* and *multiplicity* of IPD data analysis and research outcomes.

[a]As a safeguard, policy measures enabling access to IPD could be designed so that they do not interfere with the protection afforded by these regulatory instruments

8.4.2 Conclusion on Policy Implications

From an innovation policy perspective, the most significant finding is that exclusive control over IPD can curb *research externalities*, while such control can be justified only in limited situations. Theoretically, exclusive control over third-party use of data could be justified based on efficiency considerations in two situations: first, where non-excludability of R&D results (such as data and knowledge) could hinder the appropriability conditions and, hence, innovation incentives; second, where it could stimulate wastefully duplicative research. The analysis in this chapter showed that neither justification is convincing in the case of IPD.

Furthermore, enabling access to source data from trials would be consistent with the innovation-based justification of test data exclusivity. An analogy with patents can be drawn in this regard: both test data exclusivity and patent rights are conventionally rationalised as innovation incentives. However, while disclosure of technical teaching is required as the *quid pro quo* for a patent, test data exclusivity does not envisage disclosure or access to the primary research data, even upon the expiration of the protection term. In this regard, data exclusivity does not seem to follow the logic of IP protection[238] of balancing the rules on appropriability with the rules on knowledge diffusion.[239]

From a broader perspective, the issue of access to IPD for research purposes relates to a fundamental question under what conditions a regulatory intervention to facilitate competition in innovation can be justified. In this regard, research on law and economics of innovation highlights uncertainty as to what constitutes the 'optimal amount' and the 'right direction' of R&D[240] and how a policy intervention

[238]See Eisenberg (2011), p. 487 (suggesting that regulatory exclusivity 'could follow the example of the patent system, providing innovators with the exclusive right to use submitted data for regulatory purposes for a period of time in exchange for disclosure').

[239]Drahos and Braithwaite (2002), p. 13.

[240]Ben-Asher (2000), p. 292. See generally Jones and Williams (2000).

should be designed to improve the allocation of R&D resources at the sector level. In the context of drug innovation, it would prudent to say that what constitutes 'the right amount of research [is] unknown'.[241] While it is obvious that 'excessive duplication of research effort [generates] social inefficiency',[242] it is unclear how 'excessive' or 'redundant' R&D can be identified *ex ante* given the uncertainty of research in life sciences.

What is clear is that two extremes should be avoided: wastefully duplicative research (where the overlapping research projects result in the perfectly substitutable treatments) and the underutilisation of the existing research resources, such as scientific data. Theoretically, R&D resources would be allocated to their optimal use if all possible research paths are explored through data analysis without generating duplicative research results. Furthermore, the policy would need to discriminate between the R&D programs based on the therapeutic value of the *future* research outcomes: encourage research directed at the significant improvements and unmet needs while discouraging R&D leading to minor modifications of the existing medicines.[243] In reality, such coordination of research efforts can hardly be implemented because the outcomes of drug R&D cannot be ascertained *ex ante*. At the time when exploratory analysis of aggregated IPD takes place, the structural characteristics and the clinical value of prospective drugs, to which development IPD analysis might contribute, cannot be predicted.

In practice, some degree of duplicative efforts and missed opportunities is perhaps unavoidable. The question is rather under which regime the social cost of duplicative research and missed opportunities can be minimised. Insights from economic literature suggest that, while R&D externalities have multiple implications for economic efficiencies,[244] the overall effect is likely to be

[241] Ben-Asher (2000), p. 232 (emphasis added). See also Carrier (2009), p. 298 (noting that more R&D 'does not necessarily result in more innovation' (with further references)); Scherer (1993), p. 111 (noting that 'the conditions for determining the socially optimal R&D program are too complex to reach a confident judgment as to whether the market has overshot or undershot'); Henderson and Cockburn (1996), p. 55 (noting that 'determining the size and shape of the optimal research portfolio requires solving a complex, nonlinear constrained optimization problem whose parameters are not fully known'); Cockburn and Henderson (1994), p. 487 (observing that economic models of competition by R&D investment 'are fundamentally indeterminate: competitive industries may invest too much, too little or just about the right amount in research'); Rapp (1995), p. 33 (submitting that '[i]n fact, there is no functional relationship between the level of R&D expenditure and the level of innovation the market level').

[242] Dasgupta and David (1987), p. 532.

[243] Hollis A (13 Dec 2004) Me-too drugs: is there a problem? p. 4 (noting that '[t]he more differentiated the me-too product is from the pioneer, the greater the likelihood that the social value of having more product diversity will compensate for the harm to the incentive for pioneering R&D'). https://www.who.int/intellectualproperty/topics/ip/Me-tooDrugs_Hollis1.pdf. Accessed 26 Mar 2021.

[244] Spence (1984), pp. 101–122.

efficiency-enhancing.[245] In the case of IPD, this implies that its value as a source of biomedical knowledge would be enhanced with the maximised use through confirmatory and exploratory analyses. In other words, from a normative perspective, treating IPD as a non-excludable knowledge resource should be justified as a means to promote knowledge diffusion and research cumulativeness.

References

Achilladelis B, Antonakis N (2001) The dynamics of technological innovation: the case of the pharmaceutical industry. Res Policy 30(4):535–588

Ahn H (2014) Second generation patents in pharmaceutical innovation. Nomos, Baden-Baden

Akerlof GA (1970) The markets for 'lemons': qualitative uncertainty and the market mechanism. Q J Econ 84(3):488–500

Andrade LF, Sermet C, Pichetti S (2016) Entry time effects and follow-on drug competition. Eur J Health Econ 17:45–60. https://doi.org/10.1007/s10198-014-0654-9

Antonelli C (2017) Endogenous innovation. The economics of an emergent system property. Edward Elgar Publishing, Cheltenham

Ben-Asher D (2000) In need of treatment? Merger control, pharmaceutical innovation, and consumer welfare. J Leg Med 21(3):271–349. https://doi.org/10.1080/01947640050174813

Blass B (2015) Basic principles of drug discovery and development. Elsevier, Amsterdam

Bone RG (1998) A new look at trade secret law: doctrine in search of justification. Calif Law Rev 86 (2):241–313

Bonini S et al (2014) Transparency and the European Medicines Agency--sharing of clinical trial data. N Engl J Med 371(26):2452–2455. https://doi.org/10.1056/NEJMp1409464

Brandt-Rauf S (2003) Biomedical research. In: Esanu JM, Uhlir PF (eds) The role of scientific and technical data and information in the public domain: proceedings of a symposium. The National Academies Press, Washington DC, pp 65–72

Brody T (2016) Clinical trials: study design, endpoints and biomarkers, drug safety, and FDA and ICH Guidelines, 2nd edn. Elsevier, Amsterdam

Browner WS, Newman TB, Hulley SB (2007) Getting ready to estimate sample size: hypotheses and underlying principles. In: Hulley SB et al (eds) Designing clinical research, 3rd edn. Wolters Kluwer Health, Philadelphia, pp 51–64

Buchanan JM, Yoon YJ (2000) Symmetric tragedies: commons and anticommons. J Law Econ 43:1–13

Carrier MA (2008) Two puzzles resolved: of the Schumpeter–Arrow stalemate and pharmaceutical innovation markets. Iowa Law Rev 93:393

Carrier MA (2009) Innovation for the 21st century. OUP, Oxford

Cefis E, Ciccarelli M, Orsenigo L (2006) Heterogeneity and firm growth in the pharmaceutical industry. In: Mazzucato M, Dosi G (eds) Knowledge accumulation and industry evolution: the case of pharma-biotech. CUP, Cambridge, pp 163–207

Chalmers I, Glasziou P (2009) Avoidable waste in the production and reporting of research evidence. Lancet 374(9683):86–89. https://doi.org/10.1016/S0140-6736(09)60329-9

Cleophas TJ, Zwinderman AH, Cleophas TF (2006) Statistics applied to clinical trials, 3rd edn. Springer, Dordrecht

[245] See Chap. 7 at Sect. 7.2.3.1. For a discussion in the context of drug R&D, see Cockburn and Henderson (1994), pp. 482 ff.

Clissold SP, Heel RC (1987) Topical minoxidil. A preliminary review of its pharmacodynamic properties and therapeutic efficacy in alopecia areata and alopecia androgenetica. Drugs 33 (2):107–122. https://doi.org/10.2165/00003495-198733020-00002

Cockburn I, Henderson R (1994) Racing to invest? The dynamics of competition in ethical drug discovery. J Econ Manag Strategy 3(3):481

Cohen WM (2010) Fifty years of empirical studies of innovative activity and performance. In: Hall BH, Rosenberg N (eds) Handbook of the economics of innovation, vol 1. Elsevier, Amsterdam, pp 129–213

Cornes R, Sandler T (1999) The theory of externalities, public goods, and club goods. CUP, Cambridge

Correa CM (2015) Guidelines for pharmaceutical patent examination: examining pharmaceutical patents from a public health perspective. UNDP, Geneva

Dasgupta P, David PA (1987) Information disclosure and the economics of science and technology. In: Feiwel GR (ed) Arrow and the ascent of modern economic theory. Palgrave Macmillian, London, pp 519–542

Dasgupta P, Maskin E (1987) The simple economics of research portfolios. Econ J 97 (387):581–595

David P (2003) The economic logic of 'open science' and the balance between private property rights and the public domain in scientific data and information: a primer. In: Esanu JM, Uhlir PF (eds) The role of scientific and technical data and information in the public domain: proceedings of a symposium. The National Academies Press, Washington DC, pp 19–34

David PA (1993) Intellectual property institutions and the panda's thumb: patents, copyrights, and trade secrets in economic theory and history. In: Wallerstein MB, Mogee ME, Schoen RA (eds) Global dimensions of intellectual property rights in science and technology. National Academy Press, Washington DC, pp 19–61

Denicolo V, Franzoni LA (2012) Weak intellectual property rights, research spillovers, and the incentive to innovate. Am Law Econ Rev 14(1):111–140

DiMasi JA, Faden LB (2011) Competitiveness in follow-on drug R&D: a race or imitation? Nat Rev Drug Discov 10(1):23–27. https://doi.org/10.1038/nrd3296

DiMasi JA, Paquette C (2004) The economics of follow-on drug research and development: trends in entry rates and the timing of development. Pharmaeconomics 22(2):1–14

Dosi G, Mazzucato M (2006) Introduction. In: Dosi G, Mazzucato M (eds) Knowledge accumulation and industry evolution. The case of pharma-biotech. CUP, Cambridge, New York, pp 1–18

Drahos P, Braithwaite J (2002) Information feudalism. Who owns the knowledge economy? Earthscan Publications, London

Drews J (2000) Drug discovery: a historical perspective. Science 17(287):1960–1964

Drexl J (2012) Anti-competitive stumbling stones on the way to a cleaner world: protecting competition in innovation without a market. J Compet Law Econ 8(3):507–542. https://doi.org/10.1093/joclec/nhs019

Edquist C (2005) Systems of innovation approaches – their emergence and characteristics. In: Edquist C (ed) Systems of innovation. Technologies, institutions and organizations. Routledge, London, pp 1–35

Eisenberg RS (2007) The role of the FDA in innovation policy. Mich Telecomm Tech Law Rev 13:345–388

Eisenberg RS (2011) Secrecy in the age of regulatory exclusivity. In: Dreyfuss RC, Strandburg KJ (eds) The law and theory of trade secrecy. Edward Elgar Publishing, Cheltenham, pp 467–491

Elling J, Lin D (2001) A taxonomy of dynamic competition theories. In: Elling J (ed) Dynamic competition and public policy. CUP, Cambridge, pp 16–44

Fatur A (2012) EU competition law and the information and communication technology network industries: economic versus legal concepts in pursuit of (consumer) welfare. Hart, Oxford

Flohr C, Weidinger S (2016) Research waste in atopic eczema trials-just the tip of the iceberg. J Invest Dermatol 136(10):1930–1933. https://doi.org/10.1016/j.jid.2016.06.614

Foray D (2004) Economics of knowledge. MIT Press, Cambridge
Frost J, Morner M (2010) Overcoming knowledge dilemmas: governing the creation, sharing and use of knowledge resources. Int J Strateg Change Manag 2(2/3):172–199
Gambardella A (1995) Science and innovation: the US pharmaceutical industry during the 1980's. CUP, Cambridge
Garavaglia C, Malerba F, Orsenigo L (2006) Entry, market structure, and innovation in a 'history-friendly' model of the evolution of the pharmaceutical industry. In: Mazzucato M, Dosi G (eds) Knowledge accumulation and industry evolution: the case of pharma-biotech. CUP, Cambridge, pp 234–265
Gauch RR (2009) It's great! Oops, no it isn't: why clinical research can't guarantee the right medical answers. Springer, Dordrecht
Glader M (2006) Innovation markets and competition analysis. EU competition law and US antitrust law. Edward Elgar, Cheltenham
Gøtzsche PC (2012) Strengthening and opening up health research by sharing our raw data. Circ Cardiovasc Qual Outcomes 5(2):236–237. https://doi.org/10.1161/CIRCOUTCOMES.112.965277
Goudie AC (2010) Empirical assessment suggests that existing evidence could be used more fully in designing randomized controlled trials. J Clin Epidemiol 63(9):983–991. https://doi.org/10.1016/j.jclinepi.2010.01.022
Gustafsson F et al (2010) Maximizing scientific knowledge from randomized clinical trials. Am Heart J 159(6):937–943. https://doi.org/10.1016/j.ahj.2010.03.002
Hall BH, Harhoff D (2012) Recent research on the economics of patents. Annu Rev Econ 4:541–565
Hall BH, Mairesse J, Mohnen P (2010) Measuring the returns to R&D. In: Hall BH, Rosenberg N (eds) Handbook of the economics of innovation, vol 2. Elsevier, Amsterdam, pp 1034–1082
Hardin G (1968) The tragedy of the commons. Science 162:1243–1248
Heller MA (1998) The tragedy of the anticommons: property in the transition from Marx to markets. Harv Law Rev 111(3):621–688
Heller MA (2011) The anticommons lexicon. In: Ayotte K, Smith HE (eds) Research handbook on the economics of property law. Edward Elgar Publishing, Cheltenham, pp 57–57
Heller MA, Eisenberg RS (1998) Can patents deter innovation? The anticommons in biomedical research. Science 280(5364):698–701. https://doi.org/10.1126/science.280.5364.698
Henderson R, Cockburn I (1996) Scale, scope, and spillovers: the determinants of research productivity in drug discovery. RAND J Econ 27(1):32–59
Hess C, Ostrom E (2007) Introduction: an overview of the knowledge commons. In: Hess C, Ostrom E (eds) Understanding knowledge as a commons: from theory to practice. MIT Press, Cambridge, pp 3–26
Hughes JP et al (2011) Principles of early drug discovery. Br J Pharmacol 162(6):1239–1249. https://doi.org/10.1111/j.1476-5381.2010.01127.x
Huque M, Röhmel J (2010) Multiplicity problems in clinical trials: a regulatory perspective. In: Dmitrienko A, Tamhane AC, Bretz F (eds) Multiple testing problems in pharmaceutical statistics. Taylor & Francis Group, Boca Raton, pp 1–34
Institute of Medicine of the National Academies (2015) Sharing clinical trial data: maximizing benefits, minimizing risk. The National Academies Press, Washington DC
Jaffe AB (1998) The importance of 'spillovers' in the policy mission of the advanced technology program. J Technol Transf 23(2):11–19
Jain KK (2015) Textbook personalized medicine, 2nd edn. Humana Press, Springer, New York, Heidelberg
Johnson B (2005) Systems of innovation: overview and basic concepts. In: Edquist C (ed) Systems of innovation. Technologies, institutions and organizations. Routledge, London, pp 36–49
Jones A, Sufrin B (2016) EU competition law: text, cases, and materials, 6th edn. OUP, Oxford

Jones AP et al (2013) The use of systematic reviews in the planning, design and conduct of randomised trials: a retrospective cohort of NIHR HTA funded trials. BMC Med Res Methodol 13:50. https://doi.org/10.1186/1471-2288-13-50

Jones CI, Williams JC (2000) Too much of a good thing? The economics of investment in R&D. J Econ Growth 5:65

Käseberg T (2011) Intellectual property, antitrust and cumulative innovation in the EU and the US. Hart Publishing, Oxford

Katz A (2007) Pharmaceutical lemons: innovation and regulation in the drug industry. Mich Telecomm Tech Law Rev 14:1–43

Kerber W (2010) Competition, innovation and maintaining diversity through competition law. In: Drexl J, Kerber W, Podszun R (eds) Competition policy and the economic approach: foundations and limitations. Edward Elgar, Cheltenham, pp 173–201

Kerber W, Schwalbe U (2008) Economic foundations of competition law. In: Hirsch G, Montag F, Säcker FJ (eds) Competition law: European Community practice and procedure. Article-by-article commentary of the EC competition law. Sweet & Maxwell, London, pp 202–392

Khan SA (2014) Identification of structural features in chemicals associated with cancer drug response: a systematic data-driven analysis. Bioinformatics 30(17):i497–i504. https://doi.org/10.1093/bioinformatics/btu456

Kim D et al (2016) Predicting unintended effects of drugs based on off-target tissue effects. Biochem Biophys Res Commun 469(3):399–404. https://doi.org/10.1016/j.bbrc.2015.11.095

Kim D, Hasford J (2020) Redundant trials can be prevented, if the EU clinical trial regulation is applied duly. BMC Med Ethics 21:107. https://doi.org/10.1186/s12910-020-00536-9

King RF, Major I, Marian CG (2016) Confusions in the anticommons. J Polit Law 9(7):64–79

Kirwan JR (1997) Making original data from clinical studies available for alternative analysis. J Rheumatol 24(5):822–825

Kwerel ER (1980) Economic welfare and the production of information by a monopolist: the case of drug testing. Bell J Econ 11(2):505–518

Lanthier M et al (2013) An improved approach to measuring drug innovation finds steady rates of first-in-class pharmaceuticals, 1987-2011. Health Aff (Millwood) 32(8):1433–1439. https://doi.org/10.1377/hlthaff.2012.0541

Lemmens T (2004) Leopards in the temple: restoring scientific integrity to the commercialized research scene. J Law Med Ethics 32(4):641–657. https://doi.org/10.1111/j.1748-720X.2004.tb01969.x

Levin RC (1988) Appropriability, R&D spending, and technological performance. Am Econ Rev 78(2):424–428

Linge G (2008) Competition policy, innovation, and diversity. Tectum-Verlag, Marburg

Link AN (2007) Public policy and entrepreneurship. In: Audretsch DB, Grilo I, Thurik AR (eds) Handbook of research on entrepreneurship policy. Edward Elgar, Cheltenham, pp 131–139

Long C (2000) Patents and cumulative innovation. Wash Univ J Law Policy 2:229–246

Major I, King RF, Marian CG (2016) Anticommons, the Coase Theorem, and the problem of bundling inefficiency. Int J Commons 10:244–264

Massaro J (2009) Experimental design. In: Robertson D, Williams GH (eds) Clinical and translational science: principles of human research: principles of human research. Elsevier, Amsterdam, pp 41–58

Mattioli M (2017) The data-pooling problem. Berkley Technol Law J 32(2):179–236

Meinert CL (2012) Clinical trials: design, conduct and analysis, 2nd edn. OUP, Oxford, New York

Merges RP, Nelson RR (1990) On the complex economics of patent scope. Columbia Law Rev 90 (4):839–916

Merson L, Gaye O, Guerin PJ (2016) Avoiding data dumpsters - toward equitable and useful data sharing. N Engl J Med 374(25):2414–2415. https://doi.org/10.1056/NEJMp1605148

Mitscher LA (2002) Drug design and discovery: an overview. In: Krogsgaard-Larsen P, Liljefors T, Madsen U (eds) Textbook of drug design and discovery, 3rd edn. CRC Press, Boca Raton, pp 1–36

Moyé LA (2003) Multiple analyses in clinical trials. Fundamentals for investigators. Springer, New York

Mueller MT, Frenzel A (2015) Competitive pricing within pharmaceutical classes: evidence on 'follow-on' drugs in Germany 1993-2008. Eur J Health Econ 16(1):73–82. https://doi.org/10.1007/s10198-013-0555-3

Murray F, Stern S (2007) Do formal intellectual property rights hinder the free flow of scientific knowledge? An empirical test of the anti-commons hypothesis. J Econ Behav Organ 63:648–687. https://doi.org/10.1016/j.jebo.2006.05.017

National Research Council of the National Academies (2010) The prevention and treatment of missing data in clinical trials. National Academies Press, Washington DC

Nelson RR (2009) Building effective ëinnovation systemsí versus dealing with ëmarket failuresí as ways of thinking about technology policy. In: Foray D (ed) The new economics of technology policy. Edward Elgar Publishing, Cheltenham, pp 7–16

Nevitt SJ et al (2017) Exploring changes over time and characteristics associated with data retrieval across individual participant data meta-analyses: systematic review. BMJ 357:j1390. https://doi.org/10.1136/bmj.j1390

Nightingale P, Mahdi S (2006) The evolution of pharmaceutical innovation. In: Mazzucato M, Dosi G (eds) Knowledge accumulation and industry evolution: the case of pharma-biotech. CUP, Cambridge, pp 73–111

OECD (2004) Innovation in the knowledge economy. Implications for education and learning. OECD Publishing, Paris

OECD (2017) Tackling wasteful spending on health. OECD Publishing

OECD, Eurostat (2005) Oslo Manual. Guidelines for collecting and interpreting innovation data, 3rd edn. OECD Publishing, Paris

Orsenigo L, Dosi G, Mazzucato M (2006) The dynamics of knowledge accumulation, regulation, and appropriability in the pharma-biotech sector: policy issues. In: Mazzucato M, Dosi G (eds) Knowledge accumulation and industry evolution: the case of pharma-biotech. CUP, Cambridge, pp 402–431

Ostrom E (2008) Governing the commons: the evolution of institutions for collective action. CUP, Cambridge

Pammolli F, Magazzini L, Riccaboni M (2011) The productivity crisis in pharmaceutical R&D. Nat Rev Drug Discov 10(6):428–438. https://doi.org/10.1038/nrd3405

Parisi F, Schultz N, Depoorter B (2004) Simultaneous and sequential anticommons. Eur J Econ 17:175–190

Petrova E (2014) Innovation in the pharmaceutical industry: the process of drug discovery and development. In: Ding M, Eliashberg J, Stremersch S (eds) Innovation and marketing in the pharmaceutical industry. Springer, New York, pp 19–81

Rapp RT (1995) The misapplication of the innovation market approach to merger analysis. Antitrust Law J 64(1):19–47

Reichman JH (2009) Rethinking the role of clinical trial data in international intellectual property law: the case for a public goods approach. Marquette Intellect Prop Law Rev 13(1):1–68

Reinganum JF (1981) Dynamic games of innovation. J Econ Theory 25(1):21–41

Scherer FM (1993) Prices, profits and technological progress in the pharmaceutical industry. J Econ Perspect 7(3):97–115

Schulz N, Parisi F, Depoorter B (2001) Fragmentation in property: towards a general model. J Inst Theor Econ 158:594–613

Schumpeter JA (1950) Capitalism, socialism and democracy. Harper, New York

Scotchmer S (1991) Standing on the shoulders of giants: cumulative research and the patent law. J Econ Perspect 5(1):29–41

Scotchmer S (2004) Innovation and incentives. MIT Press, Cambridge

Senn S (2007) Statistical issues in drug development, 2nd edn. John Wiley & Sons, Hoboken

Simsek M (2018) Finding hidden treasures in old drugs: the challenges and importance of licensing generics. Drug Discov Today 23(1):17–21. https://doi.org/10.1016/j.drudis.2017.08.008

Spence M (1984) Cost reduction, competition, and industry performance. Econometric Soc 52 (1):101–122

Stewart RB (1981) Regulation, innovation, and administrative law: a conceptual framework. Calif Law Rev 69(5):1256–1377

Stoney CM, Johnson LL (2018) Design of clinical trials and studies. In: Gallin JI, Ognibene FP, Johnson LL (eds) Principles and practice of clinical research, 4th edn. Academic Press, London, pp 250–268

Storz-Pfennig P (2017) Potentially unnecessary and wasteful clinical trial research detected in cumulative meta-epidemiological and trial sequential analysis. J Clin Epidemiol 82:61–70. https://doi.org/10.1016/j.jclinepi.2016.11.003

Sydes MR et al (2015) Sharing data from clinical trials: the rationale for a controlled access approach. Trials 16:104. https://doi.org/10.1186/s13063-015-0604-6

Taniguchi CM et al (2008) Drug toxicity. In: Golan DE et al (eds) Principles of pharmacology: the pathophysiologic basis of drug therapy, 2nd edn. Wolters Kluwer, Lippincott Williams and Wilkins, Baltimore, pp 63–74

Tierney JF et al (2015) How individual participant data meta-analyses have influenced trial design, conduct, and analysis. J Clin Epideiol 68(11):1325–1335. https://doi.org/10.1016/j.jclinepi.2015.05.024

USGAO (2006) New drug development: science, business, regulatory, and intellectual property issues cited as hampering drug development efforts. GAO, Washington DC

van den Bergh R, Camesasca PD (2001) European competition law and economics: a comparative perspective. Intersentia, Antwerpen

Walsh JP, Arora A, Cohen WM (2003) Effects of research tool patents and licensing on biomedical innovation. In: Cohen WM, Merrill SA (eds) Patents in the knowledge-based economy. National Academies Press, Washington DC, pp 285–340

Walsh JP, Cho C, Cohen WM (2005) Science and law. View from the bench: patents and material transfers. Science 309(5743):2002–2003. https://doi.org/10.1126/science.1115813

Wang RL (2008) Biomedical upstream patenting and scientific research: the case for compulsory licenses bearing research-through royalties. Yale J Law Technol 10(7):251–330

Watkins J et al (1979) Reduction of beta-blocking drugs in hypertensive patients treated with minoxidil. BMJ 1(6175):1400. https://doi.org/10.1136/bmj.1.6175.1400

Zhou Y (2015) The tragedy of the anticommons in knowledge. Rev Radic Polit Econ 48(1):1–18. https://doi.org/10.1177/0486613415586992

Chapter 9
Evaluating Legislative Options

Abstract As concluded in the previous chapter, the conventional innovation-based justifications of exclusive control over R&D results can hardly rationalise treating IPD as an excludable good. Rather, access to IPD and robust secondary analysis should be prioritised. This chapter contemplates legislative means of implementing access to IPD taking into account the pharmaceutical sector specificities. Its starts by revisiting the policy objectives and outlining the main aspects of the access regime. Next, three policy alternatives are examined: (i) no intervention whereby access and usage rights in IPD can be allocated on a contractual basis; (ii) creating a statutory right of access to IPD for research purposes; (iii) providing for an obligation on the trial sponsors to transfer IPD to a centralised repository whereby third-party access would be subject to terms and conditions implementing the necessary safeguards but not subject the authorisation by trial sponsors. The pros and cons of each option are evaluated relative to the policy goals.

9.1 General Aspects of the Access Regime

9.1.1 Policy Objectives

The problem analysis in Chap. 6 dissected interrelated yet distinct issues associated with the trial sponsors' de facto exclusive control over IPD. It showed that the allocation of access and usage rights in IPD as a research resource should be viewed not only as a matter of public health policy but also research and innovation policy. More specifically, the policy objectives of implementing access to IPD should be defined as:

(i) enabling secondary confirmatory IPD analysis;
(ii) leveraging the research potential of the existing IPD through secondary exploratory analysis.

D. Kim, *Access to Non-Summary Clinical Trial Data for Research Purposes Under EU Law*, Munich Studies on Innovation and Competition 16,
https://doi.org/10.1007/978-3-030-86778-2_9

9.1.2 Data-Sharing as a Matter of Regulation at EU Level

Measures directed at the above-stated objectives should be implemented at EU level. First, since confirmatory IPD analysis can contribute to research quality and reproducibility, it relates to the goal of ensuring the reliability and robustness of clinical trial data, which, as acknowledged by the European legislator, 'cannot be sufficiently achieved by the Member States but can rather, by reason of its scale, be better achieved at Union level'.[1] Second, insofar as exploratory analysis can maximise the research potential of IPD in drug R&D, it pertains to the EU agenda for facilitating data-driven innovation. For instance, the objective of leveraging the potential of digital data to enhance social benefits, including in public health, is realised by the European Commission through several policy initiatives, including 'Towards a thriving data-driven economy',[2] 'A Digital Single Market Strategy for Europe',[3] 'Building a European Data Economy'[4] and 'Towards a common European data space'.[5] Policy measures directed at enabling the free movement of data and facilitating data-driven innovation can be implemented at EU level within the framework of research and technological development policy (Articles 179–190 of the TFEU), industrial policy (Article 173 of the TFEU), the approximation of laws for improving the establishment and the functioning of the internal market (Article 114 of the TFEU) and the free movement of goods (Articles 28, 30 and 34–35 of the TFEU).

9.1.3 The Overarching Principles

The principles of necessity and proportionality of a policy measure vis-à-vis the pursued objective should be observed as the general principles of policymaking.[6] Besides, measures implementing access to IPD should duly account for and balance the stakeholders' interests that might be affected by third-party IPD analysis.

---

[1]Reg 536/2014/EU, rec 85.

[2]European Commission, COM/2014/0442 final.

[3]European Commission, COM(2015) 192 final.

[4]European Commission, COM(2017) 9 final.

[5]European Commission, COM(2018) 232 final.

[6]As discussed in Chap. 6 at 6.1.4.

9.1.4 *Main Parameters of the Access Regime*

Medical researchers elaborated various models, recommendations and 'best practices' of responsible and sustainable clinical trial data sharing.[7] In light of this work, the design of an access regime should consider the following parameters.

(a) The scope of the accessible clinical trial data: Which document types should be accessible? Should the rules on access to clinical trial data differentiate between commercial and non-commercial sponsorship?
(b) Eligible data users: Should access to IPD be conditioned on the data user's eligibility? If so, what should be the relevant criteria?
(c) Timing: When should IPD become accessible to third parties? Should the timing of access be conditioned on specific factors?
(d) Payment: Should access to IPD for research purposes be provided on a paid basis? If so, how should the fee be calculated?

These modalities can be tuned once the underlying legal instrument is chosen. Several policy alternatives are considered next.

9.2 Policy Options

9.2.1 *Arguments for the State Provision of Clinical Trials*

Even though the existing policy framework relies, to a significant extent, on commercial sponsors of clinical trials, it is worth considering arguments why the state provisioning of drug testing might be a more suitable regime. First, there is general uncertainty as to what policy approach to resolving the problem of public goods[8] is more optimal from a social welfare perspective—public funding, market-based regulatory instruments, or a combination thereof. Particularly in drug innovation, instruments such as public subsidies for medical research and clinical trials and impact-based rewards for innovative medicines should be considered as alternatives to exclusivity-based incentives.[9]

Second, information asymmetries should be taken into account. As argued by Kaul, 'state agencies may be the better-suited provider where the delivery of a good is difficult to observe [. . .] and hence, difficult to monitor, verify and contract out'.[10]

[7] See e.g. Institute of Medicine of the National Academies (2015), pp. 27–42; Miller et al. (2019); Hollis et al. (2016); Sudlow et al. (2016); Manamley et al. (2016); Fletcher et al. (2013).

[8] As discussed in Chap. 7 at 7.1.1.1, clinical trials can be viewed as an inherently public good prone to the market failure of insufficient incentives.

[9] Grootendorst et al. (2011), p. 681.

[10] Kaul I (2013) Public goods: a positive analysis. Discussion draft, UNDP Office of Development Studies, p. 17 (emphasis added).

In clinical trials, the 'difficult-to-observe' parameter is research quality. While many studies have attempted to prove the impact of trial sponsors' financial interests on the research quality, the extent of such impact remains unclear.[11] Given that the trial evidence needs to be verified by a drug authority in any case, it is uncertain whether entrusting drug testing to the private sector is, on balance, an efficient way of dealing with information asymmetries in the drug market. In this regard, one could recollect earlier proposals[12] for establishing an independent testing agency to mitigate the risk of biased research.

Third, as discussed earlier, aggregated trial data can be characterised as an infrastructural good.[13] Economists argue that such goods are more optimally provisioned and managed by the state than the private sector due to the economies of scale, indivisibilities and externalities.[14]

9.2.2 *More Feasible Policy Approaches*

If the state provision of clinical trials is not a feasible option—at least, for the foreseeable future, what could be an effective and proportionate measure to achieve the above-stated policy objectives? The spectrum of choices can be outlined as follows:

(i) no intervention ('do-nothing'[15]) whereby trial sponsors would maintain de facto exclusive control over IPD;
(ii) a regulatory intervention providing for
   a. a right to access IPD held by trial sponsors and use it for research purposes;
   b. an obligation on trial sponsors to transfer data to a data repository whereby IPD would not be released into the public domain and third-party access and use would be subject to contractual terms and conditions;
   c. an obligation on trial sponsors to disclose anonymised IPD publicly.

The remainder of the analysis examines these options.

---

[11] As discussed in Chap. 6 at 6.4.2.4.

[12] Shapiro (1978); Reichman (2009); Rodwin (2012). See also Lexchin (2012), p. 258 (finding 'no evidence that any measures that have been taken so far have stopped the biasing of clinical research' and concluding that '[w]hat will be needed to curb and ultimately stop the bias [...] is a paradigm change in the way that we treat the relationship between pharmaceutical companies and the conduct and reporting of clinical trials').

[13] Chapter 7 at 7.1.1.2.

[14] See e.g. Rose (1986), p. 719 (noting that 'a governmental body might be the most useful manager where many persons desire access to or control over a given property, but they are too numerous and their individual stakes too small to express their preferences in market transactions'); Smith (2005), p. 93.

[15] Mandelkern Group on Better Regulation (13 Nor 2001) Final report, p. 15 (recommending that the 'do-nothing' option should be considered among policy alternatives).

9.3 'Doing Nothing'

Potential costs and benefits of policy intervention should be assessed relative to the non-intervention baseline. If the foregone efficiencies due to the restricted access to IPD were defined as a distinct social cost,[16] one would first need to consider whether such cost can be efficiently internalised through the voluntary negotiations with data holders. Such approach would correspond to the view that externalities are a 'pervasive feature[] of human society'[17] and their mere presence does not justify regulatory intervention. Lawmakers should step in only where externalities cannot be overcome on a contractual basis.[18] Thus, the first question to consider is whether drug sponsors can allocate access and usage rights in IPD efficiently.

9.3.1 Factors of Efficient Allocation of Rights Through Negotiations

The Coase theorem[19] is often applied to analyse legal rules on the allocation of access and usage rights in resources,[20] including intangible goods.[21] Its central proposition is that, as a general rule, 'the rearrangement of legal rights [...] would be made through the market whenever this would lead to an increase in the value of production [if market transactions are] costless'.[22] The theorem was criticised for its 'heroic assumptions, including the existence of costless bargaining with no transaction costs'.[23] However, Coase himself considered the absence of transaction costs a 'very unrealistic'[24] circumstance. His point is that, given that the transaction costs bear directly on the parties' benefits, 'the rearrangement of rights will only be undertaken when the *increase* in the value of production consequent upon the rearrangement is *greater* than the costs which would be involved in bringing it about'.[25] However, where the contract formation and execution are 'extremely

[16] As shown in Chap. 8.

[17] Cornes and Sandler (1999), p. 10.

[18] Shavell (2004), p. 108.

[19] Coase (1960).

[20] de Meza (2002), p. 280 (noting that Coase's article 'does enough to show that the impact of the law on resource allocation is by no means as straightforward as it seems and provides many clues as to how to approach the issue'). According to de Meza, Coase's 'The Problem of Social Cost' 'is cited even more frequently in law journals than in economics journals'. ibid.

[21] Lemley (1997), p. 1048 ff; Merges (1994), p. 2664 ff; Rai (1999), p. 839 ff; Merges and Nelson (1990), pp. 876–877.

[22] Coase (1960), p. 15.

[23] Wallis and Dollery (1999), p. 18.

[24] Coase (1960), p. 15.

[25] Ibid pp. 15–16 (emphasis added).

costly [or] sufficiently costly[, they can] prevent many transactions that would be carried out in a world in which the pricing system worked without cost'.[26] In that case, legal rights might need to be rearranged.[27]

Accordingly, to assess the effectiveness and efficiency of the contractual allocation of access and usage rights in IPD, two key factors need to be examined:

(a) the parties' motivation to negotiate;
(b) the relevant determinants of the transaction costs.

9.3.2 Factor Analysis

9.3.2.1 Parties' Motivation for Negotiations

While both academic researchers and research-based drug companies showed considerable interest in access and analysis of non-summary clinical trial data, data holders' motivation to grant access might be low due to several reasons.

Personal Data Protection

Trial sponsors' cautious approach to IPD sharing could be, in part, attributed to the obligations to protect the personal data of trial participants. However, such obligation alone might not be sufficient to justify a refusal to grant access to IPD for research purposes. As discussed earlier, data protection law envisages derogations from individual rights, including for scientific research purposes, provided that appropriate safeguards are in place.[28]

The Lack of the Established Practice

IPD is generated as a 'by-product' in drug development, while its main 'business function' subsists in supporting drug marketing authorisation. After that, IPD is not utilised as a 'productive input', unlike technical know-how necessary for drug manufacturing, or as a commodity routinely transacted in a market.[29] When

---

[26] Ibid.

[27] See de Meza (2002), p. 270 (noting that Coase's 'conclusion is not that market processes always make regulation unnecessary, but that as transaction costs are normally present, it is necessary to investigate *case by case* to find the best solution' (emphasis added)). See also Shavell (2004), p. 108 (observing that the choice of the default legal rules matters as they can produce the socially advantageous outcome 'directly, reducing the need for parties to bargain and to incur associated transaction costs').

[28] GDPR, art 89. See also Chap. 6 at 6.5.2.1.

[29] In some cases, the originator company can license the marketing authorisation dossier to a generic manufacturer ('authorised generics').

companies share anonymised IPD, access is usually granted for non-commercial research purposes,[30] without payment obligations,[31] and is often presented as a philanthropic initiative to promote scientific progress and healthcare.[32] Such practice of voluntary data-sharing can hardly be viewed as a 'data market' where the originator company would expect to recoup the costs of conducting trials. Instead, returns on R&D are earned primarily in the drug market.

Loss Aversion

Drug companies raised competitive concerns regarding third-party access to and analysis of non-summary clinical trial data on multiple occasions.[33] Such concerns might be attributed to loss aversion, a cognitive phenomenon associated with a human inclination to perceive losses greater than gains.[34] However, cognitive biases are speculative and difficult to prove. In a highly competitive environment, such as the pharmaceutical sector, one can hardly ascertain to what extent drug sponsors' loss aversion due to competitive concerns might be justified.[35] On the one hand, it cannot be excluded that exploratory data analysis might facilitate competitors' R&D.[36] On the other hand, the results of the exploratory analysis are usually uncertain and might be too distant from the commercialisation to rationalise competitive concerns of data holders.[37]

---

[30] As discussed in Chap. 6 at 6.3.2.

[31] See e.g. Data use agreement, para 1.2. http://yoda.yale.edu/data-use-agreement. Accessed 26 Mar 2021. CSDR standard contract template for clinical trial data sharing (10 Apr 2017). https://www.clinicalstudydatarequest.com/Documents/CSDR%20DATA%20SHARING%20AGREEMENT%20Version%201%204.10.2017.pdf. Accessed 26 Mar 2021.

[32] For instance, the CSDR initiative aspires to become 'a leader in the data sharing community inspired to drive scientific innovation and improve medical care by facilitating access to patient-level data from clinical studies'. Our Mission. https://clinicalstudydatarequest.com/About/Mission.aspx. Accessed 26 Mar 2021. The recitals of the standard data-sharing agreement used by Johnson & Johnson, Janssen, Queen Mary University of London and SI-BONE state that access to data is provided 'for the purpose of promoting Research which will be used to create or materially enhance generalizable scientific and/or medical knowledge to inform science and public health'. Data use agreement. http://yoda.yale.edu/data-use-agreement. Accessed 26 Mar 2021.

[33] As discussed in detail in Chap. 5 at 5.1 and Chap. 6 at 6.3.2.

[34] Covey (2014), p. 647.

[35] See e.g. de Coninck (2011), p. 269 (observing that, 'even though loss aversion is often presented as one of the most successful explanatory constructs within behavioural economics, it does not allow for predicting under what circumstances these effects are likely to occur or to vary').

[36] Chapter 3 surveys the potential benefits of secondary IPD analysis for medical research and drug R&D.

[37] As discussed in Chap. 8 at 8.1.4.4.

9.3.2.2 Uncertainty About the Prospective Benefits

Knowledge regarding the transaction value is an essential condition for the efficient allocation of the usage rights through contracts.[38] In contrast, information asymmetries can constitute a barrier to contracting.[39] Negotiating the terms of access to IPD inevitably raises the issue of evaluating the prospective knowledge derived through data analysis and benefit-sharing. In the case of research inputs, the term 'information asymmetries' is not quite accurate if neither the data holder nor the data user might know *ex ante* what insights would be gained through exploratory data analysis and whether research findings would eventually generate commercial value. Such uncertainty is generally characteristic of research tools as they usually provide 'clues as to how to proceed'[40] at 'a far distance from practical application'.[41] Licensing such tools can be problematic, given that 'it may be virtually impossible for any party to estimate with any confidence the expected value of taking out a license to follow those clues'.[42] The case of IPD is illustrative in this regard: analysis of historical IPD can inform the decision making in research that may or may not lead to a promising hypothesis regarding the structure-activity relationship.[43]

Uncertainty is inherent in scientific research, especially exploratory studies. As vividly expressed by Cameron, research and discovery are akin to an 'unplanned journey through the information space'.[44] Drug R&D is serendipitous and involves numerous knowledge inputs. Even where exploratory IPD analysis might contribute to successful drug development, it would be unfeasible to pinpoint and evaluate the contribution of the particular datasets among numerous heterogeneous knowledge inputs.[45] Moreover, knowledge 'extracted' from data is not only the function of the datasets but also of researchers' knowledge and skills.[46]

[38] Hoffman et al. (2002), p. 120.

[39] Shavell (2004), p. 108; Polinsky and Shavell (2008), p. 22.

[40] Ibid.

[41] Mazzoleni and Nelson (1998), p. 280 (emphasis added).

[42] Ibid. See also Reichman et al. (2016), p. 151 (observing that, where the prospect of deriving financial gain through using a research resource is unknown, concluding contracts over their use can entail 'endless amounts of speculation resulting in even higher transaction costs for all the parties').

[43] As discussed in Chap. 3.

[44] Cameron (2001), p. 32.

[45] Pammolli et al. (2011), p. 429 (noting that 'measuring research inputs and outputs for pharmaceuticals is difficult, as the innovation process builds on multiple and heterogeneous sources of knowledge, involves significant knowledge spillovers and lasts several years').

[46] Given that the information gained through data analysis and its value are context-dependent, assessing the value of data prior to its analysis might be 'almost impossible'. OECD (2015), p. 186.

9.3.2.3 Transaction Costs

Transaction costs are defined as all resources expended to conclude a contract,[47] including the costs of searching for potential partners and overcoming obstacles, such as information asymmetries.[48] Sufficiently low transaction costs—costs that make a transaction worthwhile[49]—are critical for achieving a mutually beneficial negotiation outcome.

One could draw an analogy between obtaining datasets from multiple trial sponsors for meta-analysis and converging exclusive rights 'into a usable bundle'.[50] High transaction costs due to the fragmentation of complementary inputs held by independent parties can cause resource underutilisation.[51] Even though such problem could be, theoretically, resolved on a contractual basis, economists argue that contracts between 'users' and 'excluders' are 'impracticable [...] as the number of excluders tends to be small as opposed to [...] large numbers of common-resource users'.[52]

When seeking access to IPD for exploratory research, data requestors bear transaction costs mainly related to the search of IPD from past trials relevant to a research project and the respective data holders (trial sponsors), negotiations with the data holders and the preparation of a research proposal. Transaction costs borne by data holders would usually involve the costs of preparing data for sharing (in particular, data anonymisation and formatting[53]) and evaluating research proposals (in some cases, by setting up a review board). Furthermore, pooling data from different studies[54] can multiply the transaction costs by the number of data holders. All in all, the transaction costs of obtaining access to IPD can be appreciable.

[47] Niehans (2018), p. 13782.

[48] Shavell (2004), p. 84.

[49] Above (nn 22–27).

[50] Heller (1998), p. 640.

[51] Parisi et al. (2004), p. 184; Heller and Eisenberg (1998), p. 700 ff; Walsh et al. (2003), p. 314 ff (reporting that transaction costs related to licensing for patents on upstream research tools can be substantial).

[52] Buchanan and Yoon (2000), p. 4.

[53] On the costs of data de-identification, see Institute of Medicine of the National Academies (2015), p. 68.

[54] Exploratory data analysis often requires data aggregation. See Stewart and Tierney (2002), p. 91 (noting that '[a] key but time-consuming aspect of IPD meta-analysis is contacting trialists and persuading them to participate and provide data').

9.3.2.4 Concerns Regarding 'Stacking Licenses'

The so-called 'reach through license agreements' (RTLAs) might cause the underutilisation of research inputs.[55] An RTLA can enable the patent holder to partake in the benefits generated through a patented research tool, for instance, by stipulating a license of rights in the prospective discoveries or royalties on sales of a product that might be developed with the help of a research tool.[56] Implications of RTLAs for the follow-on research and innovation are ambivalent. On the one hand, they can alleviate the uncertainty regarding the value of a research tool as the duty to pay takes effect only if a product developed by applying a licensed input is eventually commercialised. On the other hand, the prospect of dealing with '[the] overlapping and inconsistent claims on potential downstream products'[57] might discourage researchers from seeking a license in the first place.

A notable similarity between RTLAs for patents and clinical trial data-sharing agreements can be observed. For instance, the standard agreement used by the ClinicalStudyDataRequest Consortium contains the provisions on 'New Intellectual Property' defined as

> all data, discoveries, developments, inventions (whether patentable or not), improvements, methods of use or delivery, processes, know-how, or trade secrets which are made by a Researcher as a result of the conduct of Analyses or as a result of the use of any information provided to Institution or a Researcher by a Study Sponsor under this Agreement.[58]

Notably, the definition of 'new IP' embraces *any* results of the data analysis, irrespective of their eligibility for IP protection. The clause on 'new IP' further reads:

> All New Intellectual Property shall be the sole property of Institution [ie data user]; however, Institution will notify each Study Sponsor, promptly and in writing, of any New Intellectual Property. Institution hereby *grants* to *each* Study Sponsor a *perpetual, non-exclusive, fully-paid up, royalty-free, irrevocable, worldwide, unrestricted license* under any New Intellectual Property for Study Sponsor Uses, with the *right to sublicense through multiple tiers*. Institution further grants an *exclusive option*, to be exercised within one hundred eighty (180) days from notice of the New Intellectual Property to negotiate in good faith *an exclusive, fee-bearing, worldwide license* with *the right to sublicense through multiple tiers* to any New Intellectual Property which Institution may have or obtain. If additional assistance from the Institution is requested beyond the rights provided by the non-exclusive license, Institution will provide reasonable assistance to each Study Sponsor, upon commercially reasonable terms that are at least as favourable to the Study Sponsor as the terms

[55] See e.g. Heller and Eisenberg (1998), p. 699 (stating that 'the use of RTLAs [effectively] gives each upstream patent owner a continuing right to be present at the bargaining table as a research project moves downstream toward product development'); Mueller (2001), p. 57 (observing that '[t] he royalty stacking problem in biotechnology [...] has escalated in severity').

[56] Heller and Eisenberg (1998), p. 699.

[57] Ibid.

[58] CSDR standard contract template for clinical trial data sharing (10 Apr 2017), para 1.7. https://www.clinicalstudydatarequest.com/Documents/CSDR%20DATA%20SHARING%20AGREEMENT%20Version%201%204.10.2017.pdf. Accessed 26 Mar 2021.

> agreed with any other licensee for such assistance, to facilitate Study Sponsor in *fully utilizing* any New Intellectual Property.[59]

Such far-reaching obligations might demotivate researchers from seeking access to IPD and undertaking data analysis. One might wonder whether the tendency to impose extensive obligations can be attributed to the cognitive phenomenon of overestimating the contribution of one's assets to others' success.[60] For instance, the propensity to overvalue own assets—also known as the attribution bias—was viewed as a potential hindrance to the efficient licensing of patents for research tools.[61] As noted by Heller, the Coase theorem does not account for cognitive biases that might prevent parties from reaching an efficient outcome even where no transaction costs are involved.[62]

9.3.2.5 A Reverse 'Information Paradox'

As mentioned above, the grant of access to IPD held by commercial trial sponsors is usually conditioned on the merits of a research proposal submitted by a data requestor.[63] Drug companies might be reluctant to seek access to data held by other drug companies not to reveal the research project, for which exploratory IPD analysis would be undertaken. Such reluctance could be viewed as a reverse version of the 'Arrow's Information Paradox', the proposition that, once information is revealed to a potential customer, it cannot be sold because the customer already knows it.[64] In the case of data, the potential data user might be unwilling to disclose the purpose of exploratory IPD analysis, assuming that the data holder, knowing such purpose, would prefer to analyse the data herself instead of licensing the datasets.

9.3.2.6 On Balance

The problem analysis in Chap. 6 already contemplated that, given multiple concerns associated with trial sponsors' control over IPD, the 'do-nothing' option should be foregone. The detailed factor analysis in this section identified specific reasons why the social cost associated with the restricted access to IPD are unlikely to be efficiently internalised through negotiations with the trial sponsors on an individual

[59] Ibid para 4.2 (emphasis added).

[60] Nadel (2003), p. 216.

[61] Heller and Eisenberg (1998), p. 701.

[62] Heller (1998), p. 625. Further, he observes that '[e]ven in a world without transaction costs, people would not necessarily bargain to put the anticommons resource to a unique use'. ibid pp. 673–674.

[63] See Chap. 6 at 6.3.2.

[64] Arrow (1962), pp. 614–616.

basis. Therefore, possible forms of regulatory intervention should be examined further.

9.4 IPD Disclosure as an Instrument of Access

The opposite of exclusive control over IPD is *erga omnes* disclosure. At the outset, it appears doubtful whether this option presents a proportionate or even necessary measure for achieving the policy goals of improving research reproducibility and maximising the research potential of IPD.

9.4.1 *Can* Erga Omnes *Disclosure of IPD Improve Research Reproducibility?*

By definition, reproducibility of research findings pre-supposes the accessibility of source data. However, it does not necessarily imply that IPD shall be publicly disclosed. IPD analysis requires expertise in medical statistics, epidemiology and other related scientific disciplines.[65] A layperson simply cannot evaluate the risk-benefit balance of medical intervention.[66] This may explain why the EU Clinical Trials Regulation mandates the publication of summary trial results to be 'accompanied by a summary written in a manner that is understandable to laypersons'.[67] Not surprisingly, the requests from the general public for access to data held by the EMA represent only a minor fraction.[68] If only medical professionals can carry out scientific analysis to assess the benefit-risk balance, *erga omnes* IPD disclosure cannot be viewed as a proportionate measure to promote the public interest in reproducible research.[69]

[65] Advice to the European Medicines Agency from the clinical trial advisory group on good analysis practice (CTAG4)—final advice (20 Mar 2013) https://www.ema.europa.eu/documents/other/ctag4-advice-european-medicines-agency-clinical-trial-advisory-group-good-analysis-practice-final_en.pdf. Accessed 26 Mar 2021. See also EFSPI (25 Apr 2013) European Federation of Statisticians in the Pharmaceutical Industry (EFSPI) position on European Medicines Agency (EMA) access to clinical trial data initiative, p. 6 (pointing out that IPD analysis requires 'advanced statistical expertise'). https://www.efspi.org/documents/publications/efspipositiononema250413.pdf. Accessed 26 Mar 2021. See also Skovlund (2009), p. 260 (noting that '[s]tatistics is essential to the design of clinical trials and the interpretation of results').

[66] See CIOMS (2005), p. 49 (wondering: 'How should access by the public to such complicated data be arranged? How can they, or even healthcare professionals, evaluate the quality of the study and interpret the statistics provided?').

[67] Reg 536/2014/EU, art 37(4).

[68] Annex A to this study.

[69] See Case C-513/16 *EMA v PTC Therapeutics International* [2018] ECLI:EU:C:2017:148, para 139 (holding that the President of the General Court did not err in law [. . .] in finding that the public

Furthermore, IPD disclosure alone cannot guarantee that data would be re-analysed by experts. IPD analysis is time-consuming and methodologically challenging.[70] Technical aspects, such as data formats, need to be addressed to achieve data usability.[71] Notably, evidence shows that researchers are more keenly interested in exploratory rather than confirmatory secondary analysis of IPD.[72] In other words, it would be irresponsible for the regulator to rely on an *ad hoc* analysis by external researchers to ensure that the trial design, data and conclusions are verified on a systematic basis. While broader access to IPD might increase the likelihood that more confirmatory analyses can be conducted, it can 'by no means replace any of the methodological assessment done so far in the European regulatory system'.[73] Neither should the validity of secondary analysis be taken for granted.[74] Thus, rather than providing for the *erga omnes* IPD disclosure, it should be examined first to what extent the existing regulatory practice might be deficient in safeguarding the methodological quality of clinical trials and data robustness.

9.4.2 Can IPD Public Disclosure Maximise the Research Potential of Data?

Like confirmatory data analysis, exploratory analysis requires specialised expertise. Consequently, for the same reasons as outlined above, public disclosure of IPD cannot be viewed as a proportionate measure for realising the research potential of IPD to its maximum.

If public disclosure of IPD is not an appropriate measure, two policy options can come into question: a statutory right of access and an obligation to aggregate data into a repository. Each is addressed in turn.

---

interest in transparency was *sufficiently satisfied* [. . .] by the publication of the *summary* of the characteristics of Translarna, the patient information leaflet and the [European Public Assessment Report]' (emphasis added)).

[70] Nevitt et al. (2017).

[71] Institute of Medicine of the National Academies (2015), p. 6; Yang et al. (2009), p. 151.

[72] Strom et al. (2016); Navar et al. (2016), p. 1284.

[73] Koenig (2015), p. 17.

[74] See Institute of Medicine of the National Academies (2015), p. 33.

9.5 Creating a Statutory Right to Access and Use IPD for Research Purposes

The access regime could be implemented through the right of access to IPD with a corresponding duty on the data holder to provide the datasets upon a third-party request. Such right could be vested in any person interested in undertaking IPD analysis. Exceptions and limitations would need to be carefully considered for protecting the affected rights and interests, such as individual rights in personal data of the trial subjects and the legitimate economic interests of the trial sponsors. The closest legislative analogy for implementing such sector-specific regime would be the right of access to chemical test data.

9.5.1 The Analogy with the Right of Access to Test Data Under the REACH Regulation

9.5.1.1 The REACH Model of Mandatory Data-Sharing

In the EU, the Regulation concerning the Registration, Evaluation, Authorisation and Restriction of Chemicals (the REACH Regulation)[75] established a system for sharing test data related to the previously registered chemical substances.[76] Before applying for registration, an applicant should ask the European Chemicals Agency whether the chemical substance was registered within the last 12 years.[77] If so, the prospective registrant has a duty to request from the previous registrant the results of the tests involving vertebrates[78] and the right to receive the full study report within two weeks following the payment to the entities who carried out the respective studies.[79] Hence, the protection of economic interests and industry

[75] Regulation (EC) No 1907/2006 of the European Parliament and of the Council of 18 December 2006 concerning the Registration, Evaluation, Authorisation and Restriction of Chemicals (REACH), establishing a European Chemicals Agency, amending Directive 1999/45/EC and repealing Council Regulation (EEC) No 793/93 and Commission Regulation (EC) No 1488/94 as well as Council Directive 76/769/EEC and Commission Directives 91/155/EEC, 93/67/EEC, 93/105/EC and 2000/21/EC (30 Dec 2006) OJ L 396 [hereinafter REACH Regulation].

[76] Data sharing obligations under the REACH Regulation do not apply to substances used in medicinal products for human or veterinary use and regulated under Directive 2001/82/EC and Regulation 726/2004/EC. REACH Reg, art 2(5)(a).

[77] REACH Reg, art 26(1).

[78] REACH Reg, rec 49, art 27(1)(a).

[79] REACH Reg, rec 50, 51, arts 27 and 30(2). Furthermore, payment procedures are detailed under Commission Implementing Regulation 2016/9/EU of 5 January 2016 on joint submission of data and data-sharing in accordance with Regulation (EC) No 1907/2006 of the European Parliament and of the Council concerning the Registration, Evaluation, Authorisation and Restriction of Chemicals (REACH) and the Commission Regulation (EC) No 340/2008 of 16 April 2008 on the fees and charges payable to the European Chemicals Agency pursuant to the REACH Regulation.

competitiveness[80] is ensured through the right to receive compensation determined on 'a fair, transparent and non-discriminatory'[81] basis.

9.5.1.2 Applying the REACH Model to Clinical Trial Data

One may wonder why the same model has not been, or might not be, adopted for clinical trial data, given the same policy rationale of eliminating duplicative testing[82] and the similarities between the regulatory procedures for marketing authorisation. Ethical considerations dominate when it comes to eliminating unjustified[83] repetition of studies involving humans and animals.[84] The principle of minimising the exposure of animals to testing is promoted at the international level[85] and in the EU,[86] particularly through the procedures for mutual recognition of test results. The idea of avoiding duplicative testing in humans was implemented through the abbreviated procedure of drug marketing authorisation.[87] Accordingly, if the REACH model of mandatory data-sharing were applied to clinical trial data, test data exclusivity would be replaced with the 'liability rule' (the 'cost-sharing approach'[88]). Instead of suspending a generic entry, the economic interests of the trial sponsors would be protected through a system of fair compensation.[89]

---

[80] REACH Reg, rec 51.

[81] REACH Reg, art 27 (3).

[82] Dir 2001/83/EC, rec 10; Dir 87/21/EEC, rec 4. See also Case C-368/96 *The Queen v The Licensing Authority* [1998] ECLI:EU:C:1998:583, paras 69–71.

[83] 'Unjustified' is an important qualifier as the concept of research reproducibility presupposes repetition of an experiment. In many cases, the distinction between wastefully duplicative research and research directed at clarifying genuine uncertainties might not be clear. Kim and Hasford (2020).

[84] REACH Reg, rec 49, art 25(1); Dir 2010/63/EU, rec 31 (stating that animal welfare 'should be given the highest priority in the context of animal keeping, breeding and use').

[85] See e.g. OECD (12 May 1981) Final decision of the Council amending the decision concerning the mutual acceptance of data in the assessment of chemicals. C(81)30; OECD (10 Feb 1989) Final decision-recommendation on compliance with principles of good laboratory practice. C(89)87.

[86] Dir 2010/63/EU, rec 13, 16, 42; arts 4, 13. Moreover, the compliance with the principles of good laboratory practice established by the OECD when testing chemical products is regulated under Directive 2004/10/EC of the European Parliament and of the Council of 11 February 2004 on the harmonisation of laws, regulations and administrative provisions relating to the application of the principles of good laboratory practice and the verification of their applications for tests on chemical substances. The principle of replacement and reduction of animal testing is reiterated under Regulation (EU) 2017/746 of the European Parliament and of the Council of 5 April 2017 on *in vitro* diagnostic medical devices (rec 74) and Regulation (EC) No 1107/2009 of the European Parliament and of the Council of 21 October 2009 concerning the placing of plant protection products on the market (rec 40, arts 61–62).

[87] On the abridged procedure for drug marketing authorisation, see Chap. 4 at 4.2.4.1.

[88] See above (n 69) and the accompanying text.

[89] By providing for the right to obtain fair compensation from test data users, the legislator intends to 'strengthen the competitiveness of Community industry' and 'respect the legitimate property rights of those generating testing data'. REACH Regulation, rec 51-52.

Notably, the regulation does not prohibit the subsequent applicant from conducting new clinical trials and generating safety and efficacy data.[90] Leaving this pathway open is at odds with the very rationale underlying the abbreviated registration procedure. If duplicative testing is deemed normatively unacceptable, it should not be allowed under any conditions, and the legislator should not rely on the practical implausibility[91] that generic companies would conduct trials. In this regard, the approach under the REACH Regulation is more coherent by explicitly prohibiting duplicative testing.[92]

On closer inspection, the analogy with the REACH Regulation appears of limited relevance. By providing for the referential use of the originator's test data,[93] the regulator already attended to the ethical considerations[94] underlying the REACH data-sharing model.[95] The key difference is that the main purpose of enabling access to IPD should be, as considered by this study, facilitating secondary data *analysis*, while the REACH Regulation provides for data-sharing for *regulatory* purposes.[96] In this regard, data on chemicals is not comparable with biomedical data that can be re-used in exploratory research and generate novel hypotheses regarding treatment effects. Hence, concerns regarding the diminished advantage in competition in innovation do not arise when data on chemicals is treated as a 'non-excludable good', which use is nevertheless subject to a fee.

9.5.2 The Pros and Cons of the Right of Access to IPD

The main advantage of the right of access to data is that it provides a legal basis on which a third party interested in undertaking IPD analysis can request access. Thus, it could balance the negotiation powers by countervailing the factual control over data by trial sponsors.

The complexity can be seen in delineating the contours of the right. In this regard, it appears unclear how the notions 'legitimate purpose' of access or 'legitimate research' should be defined and how the distinction between commercial and non-commercial research can be enforced in practice.[97] Secondary data analysis

[90] See Chap. 5 at 5.4.1.2.

[91] Above (n 148).

[92] REACH Reg, rec 49; art 26(3); Reg 1107/2009/EC, rec 40.

[93] On the referential use of test data for the generic drug approval, see Chap. 4 at 4.2.4.1.

[94] The dominance of the public interest in avoiding repetitive testing makes the right of access under the REACH Regulation unique and, thus, not generally applicable to industrial data. For a discussion of the relevance of the REACH data-sharing model for data-driven innovation, see Drexl (2017), paras 176–180.

[95] See Chap. 4 at 4.2.4.1.

[96] REACH Reg, arts 25 and 27.

[97] In this regard, the statistics on the data requests under the EMA publication policy is quite curious. See EMA (16 Jul 2018) Clinical data publication (Policy 0070) report Oct 2016-Oct 2017.

can take place in academic, industrial and mixed research. Research and development in the science-driven sectors often build on basic research results;[98] knowledge derived through exploratory IPD analysis in upstream medical research can facilitate downstream drug development.[99] In this view, it might be unfeasible to discriminate between commercial and non-commercial entities when enforcing the eligibility conditions of access.

Given that the outcome of data analysis is *ex ante* unknown,[100] trial sponsors and data requestors might not easily come to terms concerning what should be considered as 'legitimate research' or a 'fair' *quid pro quo*.

The main disadvantage of the right of access can be viewed in that it will unlikely reduce the transaction costs since access would remain subject to negotiations with the trial sponsors on an individual basis. In this regard, the above-discussed reasons[101] can explain why enabling secondary IPD analysis through the right of access can be rather inefficient. The right of access might be a more suitable option when access is sought to individual datasets for a clearly defined purpose. For instance, the REACH model of data-sharing relies on the negotiations between the parties, who 'shall make every effort to reach an agreement on the sharing of the information requested by the potential registrant(s)',[102] while such information is needed for a specific regulatory purpose. The parties' interests and the negotiation frame of reference are quite different when access to data is sought for exploratory analysis of aggregated data.

On balance, it appears unlikely that the creation of the right of access to IPD for research purposes can significantly change the status quo.

9.6 A Centralised Clinical Trial Data Repository

9.6.1 The Data Repository Model

The idea of aggregating IPD into a centralised repository managed as a commonly accessible research resource is not novel.[103] The theoretical analysis in Chap. 8 also supports such model as an optimal regime for IPD governance. Treating safety and

EMA/630246/2017, p. 1 (reporting that during the first year of implementing the EMA's publication policy, the released data comprised over 80,000 document downloads for *non-commercial* research purposes).

[98] Foray (2004), p. 51 (with further references).

[99] As discussed in Chap. 3.

[100] See above at Sect. 9.3.2.2.

[101] Above at Sect. 9.3.2.

[102] REACH Reg, art 27(2).

[103] See e.g. Institute of Medicine of the National Academies (2015), pp. 164–165; Kelly (2010), p. 212 ff; Multi-Regional Clinical Trials Center at Harvard University (2014) Overview of data disclosure initiatives: current and ongoing data transparency activities in the pharmaceutical

efficacy data as a non-excludable good implies that, as a default rule, access to IPD for research purposes should not be subject to the trial sponsors' (data holders') authorisation. However, it does not mean that IPD shall be released into the public domain or that the trial sponsor should not be compensated for the costs of sharing data.

9.6.2 The Pros and Cons of a Centralised Repository for IPD

The main advantage of a repository model is that it provides for data aggregation, which is an essential condition for many types of IPD analysis.[104] In particular, it allows achieving the totality of evidence[105] and adequate statistical power.[106] The need to aggregate health-related data for medical research and drug development purposes was recognised by the European Commission in the context of its initiatives supporting the digital transformation of the healthcare system.[107] In particular, it acknowledges that pooling health-related data can contribute to research on rare diseases,[108] 'open up possibilities for improved early warning and detection of infectious disease health threats, bolster the tracking and control of infectious diseases outbreaks, and enable rapid and personalised treatment of infected patients'.[109] In this regard, the Commission emphasised the need for coordinated action implemented at EU level.[110]

Setting up an organisational and financial framework for the model implementation would require further effort and investment. As noted earlier, ensuring the usability and analytical utility of IPD[111] can entail substantial costs, especially related to data de-identification.[112] Besides, the unification and standardisation of data formats enabling interoperability are the prerequisites for a meaningful data

industry (proposing a 'learned intermediary' model of data-sharing). https://www.regulations.gov/comment/FDA-2013-N-0271-0031. Accessed 26 Mar 2021.

[104] See Chap. 3 at 3.2.2.4.

[105] On the importance of totality of evidence in meta-analysis, see e.g. Dias et al. (2018), p. 3; Council of Europe (2012), para 6.C.20.2.

[106] Gustafsson et al. (2010), p. 939 ff; Institute of Medicine of the National Academies (2015), p. 212; Lauer (2010), p. 91.

[107] European Commission, COM(2018) 233 final 8; European Commission, SWD(2018) 126 final 9.

[108] Ibid p. 33.

[109] European Commission, SWD(2018) 126 final p. 26.

[110] Ibid pp. 9, 39 ff.

[111] See e.g. Geifman et al. (2015) (noting that the quality and usability of data need to be 'adequately addressed' and that 'the collaborative establishment of data standards and processes for data sharing and acquisition would greatly accelerate the progress of research based on this rich data source'); Yang et al. (2009), p. 151.

[112] See e.g. Institute of Medicine of the National Academies (2015), p. 68.

analysis.[113] At the same time, elaborating on the technical and organisational aspects does not need to start from scratch as the earlier proposals already laid the groundwork.[114]

9.6.3 The Legislative Implementation

9.6.3.1 Mandatory Data Transfer as a Default Rule

To implement the data repository model, IPD should be transferred to a centralised database upon the trial completion, or another pre-specified event, as a default rule. The duty should extend to data from all trials authorised under the EU Clinical Trials Regulation, including data from exploratory studies where the sponsor does not intend to proceed with further studies and data from unsuccessful trials.[115] Presently, only summary results from such trials are required to be submitted to the EU database.[116] Furthermore, the obligation to transfer IPD should not differentiate between commercial and non-commercial sponsorship as the problem of access to primary trial data is not unique to the industry-sponsored trials.[117]

In the EU, the proposal can be implemented based on the EU database established under the EU Clinical Trials Regulation[118] and administered by the EMA. While the database currently provides access to the summary-level data, access to the de-identified IPD could be granted through the authorisation system, whereby users would need to register and accept the terms and conditions of access and data use.

To implement the proposal under the EU framework, Article 37(4) of the EU Clinical Trials Regulation would need to be amended as follows.

[113] Institute of Medicine of the National Academies (2015), p. 15 (observing that the existing data-sharing platforms 'are not consistently discoverable, searchable, and interoperable'). See also Advice to the European Medicines Agency from the Clinical Trial Advisory Group on Clinical Trial Data Formats (CTAG2)—Final advice to EMA (30 Apr 2013). http://www.ema.europa.eu/docs/en_GB/document_library/Other/2013/04/WC500142850.pdf. Accessed 26 Mar 2021.

[114] Above (n 7) and the accompanying text.

[115] On the importance of secondary analysis of such data, see Chap. 3 at 3.3.4.

[116] Reg 536/2014/EU, art 37(4), first indent.

[117] See e.g. Nevitt et al. (2017) (pointing out the problem of 'lost data' in academic trials).

[118] Reg 536/2014/EU, art 81.

Current	Proposal
Article 37	*Article 37*
End of clinical trial, temporary halt and early termination of a clinical trial and submission of the results	**End of clinical trial, temporary halt and early termination of a clinical trial and submission of the results**
[…]	[…]
4. Irrespective of the outcome of a clinical trial, within one year from the end of a clinical trial in all Member States concerned, the sponsor shall submit to the EU database a summary of the results of the clinical trial.	4. Irrespective of the outcome of a clinical trial, within one year from the end of a clinical trial in all Member States concerned, the sponsor shall submit to the EU database a summary of the results of the clinical trial and individual patient-level datasets.
[…]	[…]
For cases where the sponsor decides to share raw data on a voluntary basis, the Commission shall produce guidelines for the formatting and sharing of those data.	~~For cases where the sponsor decides to share raw data on a voluntary basis, the Commission shall produce guidelines for the formatting and sharing of those data.~~ The Commission shall produce guidelines for the formatting and sharing of individual patient-level data submitted to the EU database through the authorisation system in accordance with Regulation (EU) 2016/679 of the European Parliament and of the Council of 27 April 2016 on the protection of natural persons with regard to the processing of personal data and on the free movement of such data, and repealing Directive 95/46/EC (General Data Protection Regulation).

In addition, Article 56 of the EU Clinical Trials Regulation should be supplemented with an additional paragraph.

Current	Proposal
Article 56	*Article 56*
Recording, processing, handling and storage of information	Recording, processing, handling and storage of information
1. All clinical trial information shall be recorded, processed, handled, and stored by the sponsor or investigator, as applicable, in such a way that it can be accurately reported, interpreted and verified while the confidentiality of records and the personal data of the subjects remain protected in accordance with the applicable law on personal data protection.	1. All clinical trial information shall be recorded, processed, handled, and stored by the sponsor or investigator, as applicable, in such a way that it can be accurately reported, interpreted and verified while the confidentiality of records and the personal data of the subjects remain protected in accordance with the applicable law on personal data protection.
2. Appropriate technical and organisational measures shall be implemented to protect information and personal data processed against unauthorised or unlawful access, disclosure, dissemination, alteration, or destruction or accidental loss, in particular where the processing involves the transmission over a network.	2. Appropriate technical and organisational measures shall be implemented to protect information and personal data processed against unauthorised or unlawful access, disclosure, dissemination, alteration, or destruction or accidental loss, in particular where the processing involves the transmission over a network.
	3. Obligations set out in paragraphs (1) and (2) shall apply until data is transferred to the EU database in accordance with Article 37(4) of this Regulation.

Furthermore, Article 94 of the EU Clinical Trials Regulation would need to be adjusted as follows.

Current	Proposal
Article 94 **Penalties** 1. Member States shall lay down rules on penalties applicable to infringements of this Regulation and shall take all measures necessary to ensure that they are implemented. The penalties provided for shall be effective, proportionate and dissuasive. 2. The rules referred to in paragraph 1 shall address, inter alia, the following: (a) non-compliance with the provisions laid down in this Regulation on submission of information intended to be made publicly available to the EU database; (b) non-compliance with the provisions laid down in this Regulation on subject safety.	*Article 94* **Penalties** 1. Member States shall lay down rules on penalties applicable to infringements of this Regulation and shall take all measures necessary to ensure that they are implemented. The penalties provided for shall be effective, proportionate and dissuasive. 2. The rules referred to in paragraph 1 shall address, inter alia, the following: (a) non-compliance with the provisions laid down in this Regulation on submission of information and data intended to be made publicly available and individually accessible through the authorisation system to the EU database; (b) non-compliance with the provisions laid down in this Regulation on subject safety.

Furthermore, safeguards for protecting interests that can be affected by the obligation to transfer IPD to a database should be considered.

9.6.3.2 Safeguards and Reservations

Personal Data Protection

As long as IPD remains within the ambit of personal data protection law, its analysis should comply with the processing obligations under the GDPR. The derogations envisaged under the GDPR, particularly for scientific research purposes,[119] can

[119] See Chap. 6 at 6.5.2.1.

enable both primary and secondary analysis of IPD,[120] irrespective of whether it takes place in academic research or industrial R&D.[121]

In principle, the proposal for a data repository concerns *de-identified* IPD; thus, it does not require an additional legal basis to enable third-party data analysis after its transfer to the EU database. The terms of access implemented through the authorisation system could stipulate that the datasets made available through the database shall not be subsequently re-identified.

Economic Interests of Trial Sponsors

As long as the regulatory system relies on the private sector for provisioning drug innovation, the protection of trial sponsors' economic incentives should, to some extent, be viewed as a matter of protecting the public interest in drug innovation. As follows from the analysis in Chaps. 7 and 8, legitimate economic interests of trial sponsors in relation to trial data should be defined in terms of protecting their advantage in two types of competition, namely, competition by imitation and competition in innovation, including by improved products. Accordingly, to mitigate the disincentive effect of access measures, the rules on access should be designed to avoid interference with, first, the protection against generic competition in a drug market and, second, the ongoing or planned research projects of a trial sponsor.

More specifically, the timing of making data accessible to third parties through the EU database should be adjusted to give the trial sponsors the priority to carry out additional data analyses and file for patent applications related to the treatment effects examined in a trial. As argued above, data on exploratory endpoints,[122] including exploratory biomarkers,[123] should be protected as the intermediate research results. In such situations, trial sponsors could be required to commit to transfer data subsequently and specify the event that would prompt the transfer.[124]

---

[120] See European Data Protection Board (23 Jan 2019) Opinion 3/2019 concerning the Questions and Answers on the interplay between the Clinical Trials Regulation (CTR) and the General Data Protection Regulation (GDPR) (art. 70.1.b)), pp. 8–9 (clarifying the legal basis for the processing of personal data of clinical trial participants for the primary and secondary uses).

[121] GDPR, rec 159.

[122] As discussed in Chap. 8 at 8.1.4.4.3. Under the EMA publication policy 0070, data collected on exploratory objectives of a study can qualify as CCI and be deleted from CSRs. See EMA publication policy 0070, p. 19.

[123] Institute of Medicine of the National Academies (2015), pp. 164–165; Kelly (2010), p. 212 ff; Multi-Regional Clinical Trials Center at Harvard University (2014) Overview of data disclosure initiatives: current and ongoing data transparency activities in the pharmaceutical industry. https://www.regulations.gov/comment/FDA-2013-N-0271-0031. Accessed 26 Mar 2021.

[124] Such proposal is in line with the view of the European Ombudsman that '[the] only [...] legitimate justification for redacting information from a clinical study report could relate to the ongoing development of new treatments or of new medicines'. European Ombudsman (8 June 2016) Decision on own initiative inquiry OI/3/2014/FOR concerning the partial refusal of the

Furthermore, the question arises whether third parties should compensate the trial sponsors for the use of IPD in subsequent research and, if so, on what basis such compensation should be calculated. This issue is reminiscent of the policy dilemma regarding the proper incentive structure in cumulative innovation,[125] namely, 'how to make sure that earlier innovators are compensated for their contributions, while ensuring that later innovators also have an incentive to invest'.[126] Three types of the relationship between the earlier and the later innovation efforts are distinguished:

(i) when the later innovation could not occur without the predecessor;[127]
(ii) when the earlier innovation *reduces the cost* of achieving the later innovation, whereby such cost reduction constitutes part of the social value generated by the earlier innovation;[128]
(iii) when the earlier innovation *accelerates* the development of the second, and thus the social value of the earlier innovation encompasses 'the value of getting the second innovation sooner'.[129]

The use of IPD for exploratory research purposes exemplifies the second and third types of innovation cumulativeness, suggesting that the data user should compensate the data provider. As discussed above, determining the compensation through individual negotiations can be problematic.[130] Allocating the usage rights on a flat-rate basis might be a more reasonable and feasible approach, whereby the compensation should defray the costs of preparing IPD for sharing. The relevant expenditures would be, for instance, the cost of data anonymisation and formatting where the trial sponsor bears such costs. At the same time, data users should not reimburse trials sponsors for the costs of conducting trials, given that multiple regulatory instruments can protect returns on drug R&D, including patents, SPCs, test data protection and market exclusivities.

Public Interest in Transparency

As argued earlier, *erga omnes* disclosure of IPD would not directly serve the public interest in transparency since a layperson cannot conduct the scientific benefit-risk assessment.[131] It is important to distinguish between transparency in regulatory decision-making (e.g. regarding drug marketing authorisation) and transparency in

European Medicines Agency to give public access to studies related to the approval of a medicinal product, para 72.

[125] For an overview of the economic literature on this topic, see Rockett (2010), p. 339 ff.

[126] Scotchmer (2004), p. 127.

[127] Ibid.

[128] Ibid.

[129] Ibid.

[130] Above at Sect. 9.3.2.1 in this chapter.

[131] Above at Sect. 9.4.1) in this chapter.

research directed at validating conclusions of the primary analysis. In both cases, concerns regarding accuracy and reproducibility of findings would be better addressed if secondary confirmatory data analysis occurs systematically rather than *ad hoc*. As a starting point, the effectiveness of the existing system of ensuring data reliability and robustness should be examined.

Transparency of secondary data analysis is a distinct issue. Pharmaceutical companies expressed concerns regarding (intentionally) invalid and misleading secondary IPD analysis.[132] The more clinical trial data is disclosed, the easier it might be to challenge the validity of the primary analysis and compromise the reputation of the marketed drugs.[133] Notably, the EMA acknowledges that it 'cannot guarantee that all secondary data analyses [...] enabled by the policy will be conducted and reported to the highest possible scientific standard'.[134] A reciprocal transparency obligation on data users could address concerns regarding scientific integrity. In particular, under the terms of access, data requesters could be required to notify the corresponding trial sponsors and drug regulators of 'any unexpected findings that inform the safe and effective use of a medicine'[135] before their publication and to take 'reasonable steps to explore [...] possible explanations for discrepancies'.[136] The results of the secondary IPD analysis could also be subject to disclosure in the EU database.

9.6.4 Conclusion on Chapter 9

This chapter intended to evaluate the legislative options vis-à-vis the dual policy objective of enabling confirmatory and exploratory IPD analysis, on the one hand, while protecting innovation incentives of drug sponsors, on the other hand. Ranking the options based on the preceding discussion, the centralised data repository implemented through the obligation on trial sponsors to submit IPD to the EU database should be the instrument of choice. Conditions and reservations are proposed to protect the interests of the affected parties. As for creating the right of access to IPD, it is doubtful whether it can constitute a 'second-best' solution since the exercise of such right through direct negotiations with IPD holders might not

[132] Institute of Medicine of the National Academies (2015), pp. 131, 143–144; EFSPI (n 65), pp. 3–4; Rathi et al. (2012).

[133] See e.g. Fletcher et al. (2013), p. 335 (explaining that 'there could be many reasons for the results not completely matching the results generated by the owners of the data'; 'the data sets will generally have complex data structures which a requester may not fully understand which could lead to an incorrect re-analysis; specific variables may be unavailable due to anonymising the data sets; and the requester will not have access to the computer software/code used to generate the analyses').

[134] EMA publication policy 0070, p. 4 (emphasis added).

[135] EFSPI (n 65), p. 7.

[136] Ibid.

change the status quo substantially. Neither the absence of intervention nor the *erga omens* IPD disclosure presents a viable policy option.

References

Arrow KJ (1962) Economic welfare and the allocation of resources for invention. In: Nelson RR (ed) The rate and direction of inventive activity: economic and social factors. Princeton University Press, Princeton

Buchanan JM, Yoon YJ (2000) Symmetric tragedies: commons and anticommons. J Law Econ 43:1–13

Cameron G (2001) Scientific data, the electronic era, intellectual property. In: Workshop report on IPR (intellectual property rights) aspects of internet collaborations. European Commission, Luxemburg, pp 31–33

CIOMS (2005) Management of safety information from clinical trials. Report of CIOMS working group VI. CIOMS, Geneva

Coase R (1960) The problem of social cost. J Law Econ 3:1–44

Cornes R, Sandler T (1999) The theory of externalities, public goods, and club goods. CUP, Cambridge

Council of Europe (2012) Guide for research ethics committee members. Council of Europe

Covey R (2014) Behavioral economics and plea bargaining. In: Zamir E, Teichman D (eds) The Oxford handbook of behavioral economics and the law. OUP, Oxford, pp 643–664

de Coninck J (2011) Behavioural economics and legal research. In: van Hoecke M (ed) Methodologies of legal research. Which kind of method for what kind of discipline? Hart, Oxford, pp 257–276

de Meza D (2002) Coase theorem. In: Newman P (ed) The new Palgrave dictionary of economics and the law, vol 1. Macmillan Press, London, pp 270–282

Dias S et al (2018) Network meta-analysis for decision-making. John Wiley & Sons, Hoboken

Drexl J (2017) Designing competitive markets for industrial data – between propertisation and access. JIPITEC 8, paras 1–190

Fletcher C et al (2013) European federation of statisticians in the pharmaceutical industry's position on access to clinical trial data. Pharm Statis 12(6):333–336. https://doi.org/10.1002/pst.1603

Foray D (2004) Economics of knowledge. MIT Press, Cambridge

Grootendorst P et al (2011) New approaches to rewarding pharmaceutical innovation. Can Med Assoc J 183(6):681–685. https://doi.org/10.1503/cmaj.100375

Gustafsson F et al. (2010) Maximizing scientific knowledge from randomized clinical trials. Am Heart J 159(6):937–943. https://doi.org/10.1016/j.ahj.2010.03.002

Heller MA (1998) The tragedy of the anticommons: property in the transition from Marx to markets. Harv Law Rev 111(3):622–688

Heller MA, Eisenberg RS (1998) Can patents deter innovation? The anticommons in biomedical research. Science 280(5364):698–701. https://doi.org/10.1126/science.280.5364.698

Hoffman E, McCabe K, Smith VL (2002) Experimental law and economics. In: Newman P (ed) The new Palgrave dictionary of economics and the law, vol 2. Macmillan Press, London, pp 116–123

Hollis S et al (2016) Best practice for analysis of shared clinical trial data. BMC Med Res Methodol 16(Suppl 1):76. https://doi.org/10.1186/s12874-016-0170-y

Institute of Medicine of the National Academies (2015) Sharing clinical trial data: maximizing benefits, minimizing risk. The National Academies Press, Washington DC

Kelly B (2010) Technical and operational challenges. In: Grossmann C et al (eds) Clinical data as the basic staple of health learning: creating and protecting a public good. National Academy of Sciences, Washington DC, pp 212–216

Kim D, Hasford J (2020) Redundant trials can be prevented, if the EU clinical trial regulation is applied duly. BMC Med Ethics 21:107. https://doi.org/10.1186/s12910-020-00536-9

Koenig F (2015) Sharing clinical trial data on patient level: opportunities and challenges. Biom J 57 (1):8–26. https://doi.org/10.1002/bimj.201300283

Lauer MS (2010) Data primarily collected for new insights. In: Grossmann C et al (eds) Clinical data as the basic staple of health learning: creating and protecting a public good. National Academy of Sciences, Washington DC, pp 90–99

Lemley MA (1997) The economics of improvement in intellectual property law. Tex Law Rev 75:989–1084

Lexchin J (2012) Those who have the gold make the evidence: how the pharmaceutical industry biases the outcomes of clinical trials of medications. Sci Eng Ethics 18(2):247–261. https://doi.org/10.1007/s11948-011-9265-3

Manamley N et al (2016) Data sharing and the evolving role of statisticians. BMC Med Res Methodol 16(Suppl 1):75. https://doi.org/10.1186/s12874-016-0172-9

Mazzoleni R, Nelson RR (1998) The benefits and costs of strong patent protection: a contribution to the current debate. Res Policy 27(3):273–284

Merges RP (1994) Of property rules, Coase, and intellectual property. Columbia Law Rev 94 (8):2655–2673

Merges RP, Nelson RR (1990) On the complex economics of patent scope. Columbia Law Rev 90 (4):839–916

Miller J et al (2019) Sharing of clinical trial data and results reporting practices among large pharmaceutical companies: cross sectional descriptive study and pilot of a tool to improve company practices. BMJ 366:l4217. https://doi.org/10.1136/bmj.l4217

Mueller JM (2001) No 'dilettante affair': rethinking the experimental use exception to patent infringement for biomedical research tools. Washington Law Rev 76:1–66

Nadel L (2003) Encyclopedia of cognitive science, vol 1. Nature Publishing Group, London

Navar AM et al (2016) Use of open access platforms for clinical trial data. JAMA 315 (12):1283–1284. https://doi.org/10.1001/jama.2016.2374

Nevitt SJ et al (2017) Exploring changes over time and characteristics associated with data retrieval across individual participant data meta-analyses: systematic review. BMJ 357:j1390. https://doi.org/10.1136/bmj.j1390

Niehans J (2018) Transaction costs. In: Macmillan Publishers (ed) The new Palgrave dictionary of economics, vol 19, 3rd edn. Palgrave Macmillan, London, pp 13782–13787

OECD (2015) Data-driven innovation: big data for growth and well-being. OECD Publishing, Paris. https://doi.org/10.1787/9789264229358-en

Pammolli F, Magazzini L, Riccaboni M (2011) The productivity crisis in pharmaceutical R&D. Nat Rev Drug Discov 10(6):428–438. https://doi.org/10.1038/nrd3405

Parisi F, Schultz N, Depoorter B (2004) Simultaneous and sequential anticommons. Eur J Econ 17:175–190

Polinsky AM, Shavell S (2008) Law, economic analysis of. In: Durlauf SN, Blume LE (eds) The new Palgrave dictionary of economics, 3rd ed, vol 5. Palgrave Macmillan, London, pp 20–34

Rai AK (1999) Intellectual property rights in biotechnology: addressing new technology. Wake Forest Law Rev 34:827–847

Rathi V et al (2012) Sharing of clinical trial data among trialists: a cross sectional survey. BMJ 345: e7570. https://doi.org/10.1136/bmj.e7570

Reichman JH (2009) Rethinking the role of clinical trial data in international intellectual property law: the case for a public goods approach. Marquette Intellectual Prop Law Rev 13(1):1–68

Reichman JH, Uhlir PF, Dedeurwaerdere T (2016) Governing digitally integrated genetic resources, data, and literature. Global intellectual property strategies for a redesigned microbial research commons. Cambridge University Press, Cambridge

Rockett K (2010) Property rights and invention. In: Hall BH, Rosenberg N (eds) Handbook of the economics of innovation, vol 1. Elsevier, Amsterdam, pp 315–380

Rodwin MA (2012) Independent clinical trials to test drug: the neglected reform. Saint Louis Univ J Health Law Policy 6:113–166
Rose C (1986) The comedy of the commons: custom, commerce, and inherently public property. Univ Chicago Law Rev 53(3):711–781
Scotchmer S (2004) Innovation and incentives. MIT Press, Cambridge
Shapiro SA (1978) Divorcing profit motivation from new drug research: a consideration of proposals to provide the FDA with reliable test data. Duke Law J:155–183
Shavell S (2004) Foundations of economic analysis of law. Harvard University Press, Cambridge
Skovlund E (2009) Statisticians in European regulatory agencies. Pharm Stat 8(4):259–263. https://doi.org/10.1002/pst.367
Smith K (2005) Economic infrastructures and innovation systems. In: Edquist C (ed) Systems of innovation. Technologies, institutions and organizations. Routledge, London pp 86–106
Stewart LA, Tierney JF (2002) To IPD or not to IPD? Advantages and disadvantages of systematic reviews using individual patient data. Eval Health Prof 25(1):76–97. https://doi.org/10.1177/0163278702025001006
Strom BL et al (2016) Data sharing – is the juice worth the squeeze? N Engl J Med 375;17:1608–1609. https://doi.org/10.1056/NEJMp1610336
Sudlow R et al (2016) EFSPI/PSI working group on data sharing: accessing and working with pharmaceutical clinical trial patient level datasets – a primer for academic researchers. BMC Med Res Methodol 16(Suppl 1):73. https://doi.org/10.1186/s12874-016-0171-x
Wallis J, Dollery B (1999) Market failure, government failure, leadership and public policy. Palgrave Macmillan, London
Walsh JP, Arora A, Cohen WM (2003) Effects of research tool patents and licensing on biomedical innovation. In: Cohen WM, Merrill SA (eds) Patents in the knowledge-based economy. National Academies Press, Washington DC, pp 285–340
Yang Y, Adelstein SJ, Kassis AI (2009) Target discovery from data mining approaches. Drug Discov Today 14(3–4):147–154. https://doi.org/10.1016/j.drudis.2008.12.005

Chapter 10
Final Conclusions and the Outlook

Abstract This chapter synthesises the findings of the analysis *de lege lata* and *de lege ferenda* and highlights future research needs.

10.1 Conclusions *de lege lata*

The analysis *de lege lata* examined legal determinants of control over[1] and access to[2] clinical trial data under the applicable framework at EU level. The main findings can be summarised as follows.

While there are no property-type rights in 'raw' patient-level data,[3] trial sponsors can exercise de facto exclusive control over IPD. Such control stems from the obligation under the EU Clinical Trials Regulation—motivated by pharmacovigilance reasons—that trial sponsors have to store and protect all data and information gathered in trials against unauthorised access, in particular, by implementing technical measures of protection. Besides, drug companies' control over IPD can be reinforced through contracts and supported by non-property regimes of protection, such as trade secrets.

When examining access regimes, the analysis focused mainly on situations where access to IPD can be sought for research purposes. The overall tendency towards the broader accessibility of clinical trial data can be observed in the EU, whereby access is currently regulated under the transparency framework and at the level of *summary* results.[4] Provisions governing the accessibility of clinical trial data under the sector legislation are systematically related to the EU Transparency Regulation.[5]

---

[1]Chapter 4 at Sect. 4.2.

[2]Chapter 4 at Sect. 4.3.

[3]As concluded in Chap. 4 at Sect. 4.2.1.2.

[4]Chapter 4 at Sect. 4.3.1.

[5]As summarised in Chap. 4, Table 4.1 'The relationship between the right of access to documents and reservations for CCI under sector regulations'.

D. Kim, *Access to Non-Summary Clinical Trial Data for Research Purposes Under EU Law*, Munich Studies on Innovation and Competition 16,
https://doi.org/10.1007/978-3-030-86778-2_10

Upon examining the interface with personal data protection,[6] no conflict was identified between secondary IPD analysis and data protection obligations. The rules under the GDPR, including the derogations for scientific research, are flexible enough to accommodate the use of IPD for research purposes beyond the original trial protocol. As for IPD held by drug companies, there appears to be no viable legal basis under EU law that could support external researchers' claim for access.

While the importance of promoting transparency in clinical research and regulatory decision-making is undisputable, the analysis found the transparency framework to be limited in several aspects. First, the scope of data accessible through the fundamental right of access to documents is confined to data held by the EMA, which presently excludes IPD. Second, the scope of the exception for the protection of commercial interests remains uncertain. At the same time, the element-by-element assessment of trial data appears to be cumbersome for the EMA, drug sponsors and data requestors. Third, the underlying rationale of the right of access to documents is to improve transparency in policymaking and administrative practices of public authorities within the scope of their mandate and responsibilities. In the case of access to IPD, this implies validating regulatory decisions and procedures concerning drug authorisation and supervision through confirmatory analysis. As the main point of criticism of the current approach, it is submitted that the policy objective of leveraging the research potential of IPD through exploratory analysis goes beyond the transparency framework and is inconsistent with the essence of the right of access to documents.[7]

10.2 Conclusions *de lege ferenda*

The analysis *de lege ferenda* addressed the question of whether policy intervention removing trial sponsors' control over IPD can be justified on the grounds of promoting innovation and, if so, how the rules of access should be designed to protect and balance the multiple interests at stake. This question was examined in light of research on the law and economics of innovation, particularly the theories rationalising how the rights in R&D results should be defined and allocated to promote innovation optimally. The findings can be synthesised as follows.

Clinical trials and clinical trial data are characterised as inherently public goods.[8] Secondary IPD analysis can contribute to both fundamental and applied medical research.[9] Aggregated IPD has considerable research potential to generate knowledge beyond the benefit-risk assessment of investigational products tested in the

[6]Chapter 4 at Sect. 4.2.1.2, subheading 'No Property Rights in IPD as Personal Data'.

[7]As concluded in Chap. 4 at Sect. 4.4.

[8]See Chap. 7 Sect. 7.1.1.1.

[9]Chapter 7 at Sect. 7.1.1.3.

original trials.[10] Access to IPD is an essential prerequisite for realising such potential.

The peculiar characteristics of IPD as the *output* of commercially-sponsored research and a non-rival-in-use research *input* trigger an 'access-innovation' policy dilemma. Two seemingly opposing propositions underlie the crux of the problem: unrestricted access to IPD can promote medical research and innovation and, at the same time, hinder innovation incentives of the research-based drug companies. The analysis found that such dilemmas typically arise with the private provisioning of public goods and stem from the dual implications of knowledge externalities for innovation that prompt a policy trade-off between greater knowledge diffusion and diminished innovation incentives.

The question of whether non-rival-in-use knowledge resources should be treated as an excludable or non-excludable good is highly controversial in the law and economics of innovation.[11] While R&D externalities pose divergent effects on innovation,[12] the choice of a governance regime necessitates a comparative cost-benefit analysis in light of the sector specificities. Accordingly, the normative analysis attempted to qualify and weigh innovation-related costs and benefits of maintaining the status quo where trial sponsors retain exclusive control over IPD vis-à-vis the costs and benefits of removing such control.

Conventionally, where public goods, such as knowledge and innovation, are provided by the private sector, exclusive rights in R&D results are rationalised as a means of internalising R&D benefits to protect and promote innovation incentives of private undertakings.[13] In this regard, the analysis explored the relevance of trial sponsors' de facto exclusive control over IPD for protecting competitive advantage in line with the distinction between two types of competition in the pharmaceutical sector, namely, competition by imitation and competition in innovation, including by drug improvements.[14]

As far as competition by imitation is concerned, the main finding is that control over IPD does not perform the function of protecting drug sponsors' supra-competitive returns on R&D in the presence of policy instruments delaying the generic entry, such as patents, SPCs, test data exclusivity and market exclusivities.[15] Accordingly, as long as third-party access to IPD does not interfere with these forms of protection, it does not 'impede' innovation incentives, as has been argued by drug companies.[16]

---

[10] Chapter 7 at Sect. 7.1.1.2.

[11] As concluded in Chap. 7 at Sect. 7.3.

[12] As summarised in Chap. 7, Table 7.1 'Synthesis of theoretical assumptions about R&D externalities and their implications for the allocation of resources to R&D'.

[13] Chapter 7 at Sect. 7.2.3.3, subheading 'The (Controversial) Role of Patents as a Means to Coordinate Research Efforts'.

[14] See Chap. 8 at Sect. 8.1.4.

[15] As concluded based on the analysis in Chap. 5 at Sects. 5.3; 5.4.1 and 5.4.2, respectively.

[16] See also conclusions in Chap. 5 at Sect. 5.5.

As for competition by product improvement, the important insight is that the pre-market competition in developing ‘follow-on’ drugs is often so intense that secondary analysis of IPD related to the originator drug—if accessed and re-analysed after that drug is authorised for marketing—is unlikely to have a significant impact on the originator’s competitive advantage.[17]

As far as competition by new product is concerned, it appears unclear to what extent trial sponsors’ concerns regarding the diminished competitive advantage due to third-party exploratory IPD analysis can be justified. On the one hand, it cannot be denied that knowledge gained through exploratory IPD analysis can contribute to developing new drugs, including unrelated to those investigational drugs which IPD is re-analysed.[18] On the other hand, such prospects are usually uncertain and distant from the actual commercialisation of the research outcomes.[19] Several potential scenarios were contemplated in this regard. Based on the analysis, protection of data of exploratory nature, such as exploratory endpoints, against premature disclosure was identified as the only instance where a restriction on third-party exploratory IPD analysis could serve as a means of protecting the data holder’s advantage in competition in innovation (hence, innovation incentives).[20]

Given that IPD aggregation is the prerequisite for realising the potential of trial data for generating knowledge, it was examined whether de-identified IPD can be pooled and the usage rights can be allocated effectively on a contractual basis if trial sponsors retain control over data. As shown, effective and efficient data aggregation is unlikely to be achieved through negotiations with individual data holders, particularly due to high transaction costs, uncertainties and imperfect information regarding the prospective outcomes of data analysis.[21] Thus, two distinct innovation-related social costs of trial sponsors’ exclusive control over IPD were qualified as *foregone opportunities* for gaining knowledge through secondary analysis of aggregated data[22] and research *redundancy*, i.e. conducting new studies even though research questions could be answered satisfactorily based on the available data.[23]

The analysis also showed that concerns that IPD non-excludability would cause *wastefully duplicative* research are rather unwarranted. Insights from medical literature suggest that the benefits of multiple analyses of pooled IPD are unlikely to be rivalrous in the sense of generating duplicative results.[24] Rather, robust secondary IPD analysis can promote *diversity* and *multiplicity* of research and research outcomes.

[17] As discussed in Chap. 8 at Sect. 8.1.4.4, subheading ‘The Case of Drug Improvements’.

[18] As shown by the primer on secondary analysis of IPD in Chap. 3 and the specific scenarios in Chap. 8 at Sect. 8.1.4.3.

[19] For the reasons discussed in Chap. 8 at Sect. 8.1.4.4.

[20] Chapter 8 at Sect. 8.1.4.4, subheading ‘Protection of Exploratory Endpoints as Intermediate Research Results’.

[21] As examined in Chap. 9 at Sect. 9.3.

[22] Chapter 8 at Sect. 8.2.3.

[23] Chapter 8 at Sect. 8.3.5.

[24] Chapter 8 at Sect. 8.3.4.

On balance, it is argued that, from an innovation perspective, policy intervention enabling access to IPD can be justified as a means of promoting knowledge diffusion and research cumulativeness and reducing uncertainty in medical research and drug R&D. Upon examining the policy options,[25] it is proposed that, as a default rule, IPD should be transferred to a centralised repository after the trial completion, while competitive concerns of drug sponsors should be hedged by adopting safeguards and reservations.[26] In particular, to protect competitive advantage and, hence, economic incentives, the rules should provide the initial trial sponsor with the priority to conduct secondary exploratory IPD analysis and file patent applications for additional treatment effects observed in a trial.[27] In the EU context, the centralised data repository model can be implemented based on the EU database for clinical trials by amending the relevant provisions under the EU Clinical Trials Regulation as proposed.[28]

10.3 The Outlook

10.3.1 Shifting the Focus from Access to Data Analysis

The EMA transparency policies undeniably marked a milestone in promoting transparency in regulatory decision-making in the EU. However, as long as secondary confirmatory IPD analysis by external researchers occurs in an *ad hoc* manner, data accessibility, in and of itself, does not present a systematic solution for improving research reproducibility. While the EMA has not been conducting IPD analysis routinely, further research is needed to examine how trial findings are validated at the level of national drug authorities and whether additional measures might be necessary at EU level to ensure data robustness.

Furthermore, it appears to be a misconception that, once IPD becomes accessible, it will be broadly utilised for research purposes. Evidence on redundant trials—trials that could have been avoided because the question of interest could have been answered satisfactorily based on the existing data[29]—and evidence on the insufficient consideration of earlier findings in trial design[30] can suggest that the mere availability of data from prior trials does not guarantee that its analysis will be integrated into the subsequent research. In other words, data accessibility is a

[25] As presented in Chap. 9.

[26] Chapter 9 at Sect. 9.6.3.2.

[27] The intersection between IPD accessibility and patent protection is examined in Chap. 5 at Sect. 5.3.

[28] On the proposal for the revision of the relevant provisions under the EU Clinical Trial Regulation, see Chap. 9 at Sect. 9.6.3.

[29] Above (Chap. 8, nn 169–177) and the accompanying text.

[30] Chapter 8 at Sect. 8.3.5.

stepping stone, but, by itself, it does not enable research cumulativeness. While policy discussions, for the most part, focused on access to clinical trial data, further reflection is needed as to whether additional incentives might be necessary to facilitate secondary—both confirmatory and exploratory—IPD analysis.

10.3.2 *Access to IPD as a Case Study on Data-Driven Innovation*

How to maximise the use of data without hindering data generators' innovation incentives has been a central regulatory dilemma in the data-driven economy. The question of how access and usage rights in data as a critical innovation input should be defined and allocated remains subject to debate. The tentative consensus is that, instead of introducing horizontally applicable rules, implementing sector-specific measures directed at the identifiable 'lock-in' constellations presents a more pertinent approach. In this regard, IPD presents a case study on designing access rules while accounting for the nature and characteristics of data and the sector specificities.

At the same time, several factors make the case of access to IPD peculiar. First, clinical trials present perhaps a unique example where scientific (translational) research is, to a large extent, sponsored and carried out by the private sector. Commercial sponsorship causes inherent tension between the appropriability of private returns to innovation and maximising social returns to innovation through knowledge diffusion.[31] Second, exploratory IPD analysis has a considerable potential to facilitate basic medical research and industrial drug R&D,[32] making the delineation and balancing of the intersecting private and public interests especially complex. Third, while multiple regulatory instruments protect returns to drug R&D against imitation in the existing product market, control over IPD can perform the function of protecting the competitive advantage of drug sponsors in competition in R&D.[33] A distinct policy dilemma arises as to how to distinguish between the situations where such control might be justified and where it leads to the lost opportunities in drug research and innovation. In view of these peculiarities, the outcome of the present analysis cannot be generalised to other sectors or types of data.

---

[31] Chapter 7 at Sect. 7.2.3.2, subheading 'A Trade-off Between Knowledge Diffusion and Innovation Incentives'.

[32] Chapter 2.

[33] Chapter 8 at Sect. 8.1.4.4, subheading 'Protection of Exploratory Endpoints as Intermediate Research Results'.

10.3.3 Access to IPD as a Case on R&D Externalities

At a more fundamental level, the case of access to clinical trial data illustrates the central controversy in the law and economics of innovation concerning exclusive control over knowledge, including by IP rights, as a means of promoting innovation. In this regard, the 'access-incentives' dilemma[34] exemplifies general uncertainty in economic literature regarding whether and to what extent R&D externalities can cause a trade-off between knowledge diffusion and the appropriability of returns on R&D (economic incentives).[35] While R&D externalities can impact innovation processes in multiple dynamic ways,[36] assessing their net effect on innovation poses a challenge. At the same time, economists point out that the positive aspects are likely to dominate, especially in settings where innovation is cumulative.[37] This view suggests that society would be better off if IPD, as a source of medical knowledge, is treated as a non-excludable resource and exploratory IPD analysis is not subject to the authorisation of trial sponsors.

IPD analysis can promote the cumulativeness of knowledge, research and drug innovation. The more IPD from earlier trials is available, the more precisely new research hypotheses concerning treatment effects can be formulated, the more efficiently new trials can be designed, and the more aptly clinical uncertainties can be resolved.[38] In this regard, it is curious to observe that the concept of R&D externalities can be applied to justify both exclusive rights in technological knowledge and the non-excludability of research data. While, in the case of technological teachings, the emphasis has been conventionally on internalising 'spillovers' through exclusive (IP) rights, in the case of data, legal rules enabling access can play a more prominent role in 'externalising' social benefits.

---

[34] Chapter 7 at Sect. 7.1.2.

[35] Chapter 7 at Sect. 7.3.1.

[36] See Chap. 7, Table 7.1 'Synthesis of theoretical assumptions about R&D externalities and their implications for the allocation of resources to R&D'.

[37] Above (Chap. 7, nn 145–148) and the accompanying text.

[38] See Chap. 8 at Sects. 8.2.3.2 and 8.2.3.3.

Annex A Statistics on Requests for Access to Documents Held by the EMA (2012–2020)

Affiliation	Number of requests in %								
Not-for-profit organisation	0	0	1	0	0	2	1	3	3
EU Institutions	1	1	1	0	0	0	0	0	0
Regulator outside EU	2	1	0	0	0	0	0	0	0
EU NCA	2	2	1	0	0	0	0	0	0
Patients or Consumer (general public)	8	3	3	3	7	17	12	8	6
Healthcare professional	4	3	2	3	3	6	4	4	4
Academia/Research institute	9	12	8	8	8	8	8	9	9
Legal	14	23	16	11	11	8	9	8	7
Media	17	13	13	7	5	3	4	4	2
Pharmaceutical industry	32	35	47	60	55	44	47	52	55
Consultant	10	4		2	10	11	14	12	14
Other	0	1	7	4	0	0	0	0	0
Total number of requests	281	290	416	701	823	865	822	783	597
Year	2012	2013	2014	2015	2016	2017	2018	2019	2020

Data source: the EMA Annual Reports (EMA. Annual reports and work programmes. https://www.ema.europa.eu/en/about-us/annual-reports-work-programmes. Accessed 22 Jul 2021. The EMA annual reports do not break down the statistics into the types of documents requested from the EMA)

D. Kim, *Access to Non-Summary Clinical Trial Data for Research Purposes Under EU Law*, Munich Studies on Innovation and Competition 16,
https://doi.org/10.1007/978-3-030-86778-2

Annex B Glossary of Terms Related to the Design and Methodology of Randomised Clinical Trials

Arm	A group or subgroup of the trial participants in a clinical trial that receives a specific health intervention (treatment), or no intervention, according to the trial protocol
Case report form	A document containing all data gathered and recorded on individual study subjects according to the variables set out in a trial protocol
Clinical trial	An experimental study designed to evaluate the treatment effect of medical intervention in humans
Confirmatory trial/ 'hypothesis-testing trial'	An interventional study designed to test a predefined research hypothesis and usually conducted in order to generate evidence on the safety and (or) efficacy of an investigational product
Clinical study report	A document prepared according to the standardised structure and submitted to support an application for drug marketing authorisation that contains inter alia the trial protocol, statistical analysis plan, efficacy and safety analyses
Effect size	A quantitative characteristic that correlates with the outcome variable and statistically describes the study hypothesis
Endpoints/outcome measures	Variables that can indicate biologic and pathogenic processes, or pharmacologic responses to a therapeutic intervention
Exploratory (hypothesis-generating) trial	An interventional study that can generate a hypothesis regarding a treatment effect and set the direction for future research
Individual patient data	Data recorded for each individual study subject according to the trial protocol
Meta-analysis	A formal evaluation of the evidence from two or more trials that generated comparable evidence
Null hypothesis	

(continued)

D. Kim, *Access to Non-Summary Clinical Trial Data for Research Purposes Under EU Law*, Munich Studies on Innovation and Competition 16,
https://doi.org/10.1007/978-3-030-86778-2

	An assumption of the absence of the association between the predictor (the indirect measurement of the treatment effect) and the outcome of interest in the target population
Outcome of interest	A measurable effect of a medicinal intervention on the clinical manifestations of a disease (e.g. relief of symptoms, improvement in the quality of life)
Treatment effect	An effect on the state of health that is attributed to an intervention
Power/trial power	Statistical significance of evidence; the probability that the null hypothesis will be rejected if it is not true; the probability that the difference between the study groups will be identified, where such difference, in fact, exists
Protocol/clinical trial protocol	A document setting out objectives, design, methodology, statistical considerations, organisation, and implementation of a clinical trial
Sample size	The number of study subjects enrolled in a trial that can be estimated inter alia based on the calculation of the statistical power of the trial
Source data	All original records of trial results, observations or other evidence gathered in a clinical trial allowing to assess the trial performance vis-à-vis its objectives

The manufacturer's authorised representative in the EU is Springer Nature Customer Service Centre GmbH, Europaplatz 3, 69115 Heidelberg, Germany. If you have any concerns regarding our products, please contact ProductSafety@springernature.com

Printed and bound by CPI Group (UK) Ltd, Croydon, CR0 4YY
30/09/2026
02225290-0013